Newnes Electrical Power Engineer's Handbook

Second Edition

Newnes Electrical Power Engineer's Handbook

Second Edition

D.F. Warne

ELSEVIER

Newnes

AMSTERDAM   BOSTON   HEIDELBERG   LONDON   NEW YORK
OXFORD   PARIS   SAN DIEGO   SAN FRANCISCO
SINGAPORE   SYDNEY   TOKYO

Newnes is an imprint of Elsevier

Newnes is an imprint of Elsevier
Linacre House, Jordan Hill, Oxford OX2 8DP, UK
30 Corporate Drive, Suite 400, Burlington, MA 01803, USA

First Published 2005
Reprinted 2006

British Library Cataloguing in Publication Data
A catalogue record for this book is available from the British Library

Library of Congress Cataloging-in-Publication Data
A catalog record for this book is available from the Library of Congress

ISBN–13: 978-0-7506-6268-0

For information on all Newnes publications
visit our website at www.newnespress.com

Transferred to digital printing in 2009.

Working together to grow
libraries in developing countries

www.elsevier.com | www.bookaid.org | www.sabre.org

ELSEVIER   BOOK AID
International   Sabre Foundation

Contents

Acknowledgements

When I was first asked to update the *Newnes Electrical Engineer's Handbook*, it seemed a relatively straightforward task. But after studying the reviewer's comments which had been assembled by the publisher it became clear that a number of changes in emphasis would be worthwhile. It was also evident that in North America, a clearer distinction has to be made between electrical power engineering and electronics. Since the main target of this book is electrical power equipment, a change of title was agreed and so we now have the *Newnes Electrical Power Engineer's Handbook*.

At the same time, some areas of the technology continue to advance apace, particularly in the structure and operation of power systems, and in emc and power quality. So some chapters needed a substantial overhaul even though the *Newnes Electrical Engineer's Handbook* was published only five years ago.

New contributors have been introduced to handle the various updates, partly because of the retirement of former contributors and partly because of change of emphasis. The co-operation of all the contributors during the preparation of material and through the inevitable differences of pace at which the different sections have been completed, is gratefully acknowledged.

Every work of this type consumes a vast amount of time, with an inevitable sacrifice of personal and social time. Without the patience and understanding of my wife Gill, the completion of the project would have been much more difficult. Her support in this, as in all our ventures, is lovingly acknowledged.

D.F. Warne
October 2004

Chapter 1

Introduction

In many countries the public perception of more traditional aspects of engineering remains at best indifferent and at worst quite negative. Electrical engineering is perhaps seen as mature, unchanging and offering little scope for imagination, with poor prospects for any future career. There is a serious risk in many parts of Europe and North America that substantial areas of knowledge are being lost as large numbers of key experts are retiring without the opportunity to teach and train the specialists of the future. And yet the technology continues to move on, and the understanding of the basic mechanisms of circuits, electromagnetics and dielectrics continues to be as challenging intellectually as it has ever been. There have even been very prominent warnings of the dangers of neglecting the importance of electrical power systems and plant, and of underestimating the value of the skilled engineers necessary to support this infrastructure. The major power failures of the past few years, in the Eastern seaboard of the USA, in Auckland, in Italy, in London and in parts of Scandinavia have highlighted how dependent a modern society is on a reliable source of electrical energy.

So, the need has never been stronger for a basic understanding of principles and a fundamental appreciation of how the major classes of electrical equipment operate. In a handbook, it is not possible to set out a comprehensive treatment but the aim is to provide a balanced overview, and perhaps to engender the interest to pursue areas in more depth. A more complete coverage of all the subjects addressed here can be found in the *Newnes Electrical Engineer's Reference Book*.

The structure of the handbook is, as before, based around three groups of chapters as follows:

* fundamentals and general material
* the design and operation of the main classes of electrical equipment
* special technologies which apply to a range of equipment

The first group covers the fundamentals and principles which run through all aspects of electrical power technology.

The opening chapter deals with the fundamentals of circuit theory and electric and magnetic fields, together with a brief coverage of energy conversion principles.

This is followed by a review of the materials which are crucial to the design and operation of electrical equipment. These are grouped under the headings of magnetic, insulating and conducting materials. In each of these areas, technology continues to move ahead. Further improvement in the performance of permanent magnets is one of the key drivers behind the increasing use of electrical actuators and drives in cars and the miniaturization of whole ranges of domestic and commercial equipment; and the challenges in understanding the behaviour of soft magnetic materials, especially under conditions of distorted supply waveforms, are gradually being overcome. Developments in insulating materials mean that increased reliability can be achieved, and operation at much higher temperatures can be considered. Under the heading of conductors, there are continuing advances in superconductors, which are now able to operate at liquid nitrogen temperatures, and of course semiconductor developments continue to transform the way in which equipment can be controlled.

Finally, in this opening group there is a chapter on measurement and instrumentation. Modern equipment and processes rely increasingly on sensors and instrumentation for control and for condition assessment and diagnostics, so in this chapter there are some changes in coverage, the emphasis now being on sensors and the way in which signals from sensors may be processed.

The next group of eight chapters form the core of the book and they cover the essential groups of electrical equipment found today in commerce and industry.

The opening five chapters here cover generators, transformers, switchgear, fuses and wires and cables. These are the main technologies for the *production and handling* of electrical power, from generation, transmission and distribution at high voltages and high powers down to the voltages found in factories, commercial premises and households. Exciting developments include the advances made in high-voltage switchgear using SF_6 as an insulating and arc-extinguishing medium, the extension of polymer insulation into high-voltage cables and the continuing compaction of miniature and moulded-case circuit breakers. A new section in the wires and cables chapter addresses the growing technology of optical fibre cables. Although the main use for this technology is in telecommunications, which is outside the scope of the book, a chapter on wires and cables would not be complete without it and optical fibres have in any case found a growing number of applications in electrical engineering.

The following four chapters describe different groups of equipment which *use or store* electrical energy. Probably the most important here is electric motors and drives, since these use almost two-thirds of all electrical energy generated. Power electronics is of growing importance not only in the conversion and conditioning of power, most notably in variable-speed motor drives, but also in static power supplies such as emergency standby, and in high-voltage applications in power systems. The range of batteries now available for a variety of applications is extensive and a chapter is set aside for this, including the techniques for battery charging and the emerging and related technology of fuel cells. If fuel cells fulfil their promise and start to play a greater part in the generation of electricity in the future then we can expect to see this area grow, perhaps influencing the generator and power systems chapters in future editions of the handbook.

The final group of four chapters covers subjects which embrace a range of technologies and equipment. There is a chapter on power systems which describes the way in which generators, switchgear, transformers, lines and cables are connected and controlled to transmit and distribute our electrical energy. The privatization of electricity supply in countries across the world continues to bring great changes in the way the power systems are operated, and these are touched upon here, as is the growing impact of distributed generation. The second chapter in this group covers the connected subjects of electromagnetic compatibility and power quality. With the growing number of electronically controlled equipment in use today, it is imperative that precautions are taken to prevent interference and it is also important to understand the issues which are raised by the resulting disturbances in power supply, such as harmonics, unbalance, dips and sags. The next chapter describes the certification and use of equipment for operation in hazardous and explosive environments; this covers a wide range of equipment and several different classes of protection. And finally, but perhaps most importantly, a chapter on health and safety has been added for this edition; this issue rightly pervades most areas of the use of electrical power and this topic is a valuable addition to the handbook.

In most chapters there is a closing section on standards, which influence all aspects of design, specification, procurement and operation of the equipment. At the highest level are the recommendations published by the International Electrotechnical Commission (IEC), which are performance standards, but they are not mandatory unless referred to in a contract. Regional standards in Europe are Euro-Norms (ENs) or Harmonized Documents (HDs) published by the European Committee for Electrotechnical Standardization (CENELEC). CENELEC standards are part of European law and ENs must be transposed into national standards and no national standard may conflict with an HD. Many ENs and HDs are based on IEC recommendations, but some have been specifically prepared to match European legislation requirements such as EU Directives. National standards in the UK are published by the British Standards Institution (BSI). BSI standards are generally identical to IEC or CENELEC standards, but some of them address issues not covered by IEC or CENELEC. In North America, the main regional standards are published by the American National Standards Institute (ANSI) in conjunction with the Institute of Electrical and Electronics Engineers (IEEE). The ANSI/IEEE standards are generally different from IEC recommendations, but the two are becoming closer as a result of international harmonization following GATT treaties on international trade. Coverage of all these groups is attempted in the tables listing the key standards.

Chapter 2

Principles of electrical engineering

Dr D.W. Shimmin
University of Liverpool

2.1 Nomenclature and units

This book uses notation in accordance with the current British and International Standards. Units for engineering quantities are printed in upright roman characters, with a space between the numerical value and the unit, but no space between the decimal prefix and the unit, e.g. 275 kV. Compound units have a space, dot or solidus between the unit elements as appropriate, e.g. 1.5 N m, 9.81 m · s^{-2}, or 300 m/s. Variable symbols are printed in italic typeface, e.g. *V*. For ac quantities, the instantaneous value is printed in lower case italic, peak value in lower case italic with caret (^), and rms value in upper case, e.g. *i*, *î*, *I*. Symbols for the important electrical quantities with their units are given in Table 2.1, and decimal prefix symbols are shown in Table 2.2. Graphical symbols for basic electrical engineering components are shown in Fig. 2.1.

2.2 Electromagnetic fields

2.2.1 Electric fields

Any object can take an *electric* or *electrostatic charge*. When the object is charged positively, it has a deficit of electrons, and when charged negatively it has an excess of electrons. The electron has the smallest known charge, -1.602×10^{-19} C.

Charged objects produce an electric field. The *electric field strength E* (V/m) at a distance *d* (m) from an isolated point charge *Q* (C) in air or a vacuum is given by

$$E = \frac{Q}{4\pi\varepsilon_o d^2} \tag{2.1}$$

where the *permittivity of free space* $\varepsilon_o = 8.854 \times 10^{-12}$ F/m. If the charge is inside an insulating material with *relative permittivity* ε_r, the electric field strength becomes

$$E = \frac{Q}{4\pi\varepsilon_o \varepsilon_r d^2} \tag{2.2}$$

Any charged object or particle experiences a force when inside an electric field. The force *F* (N) experienced by a charge *Q* (C) in an electric field strength *E* (V/m) is given by

$$F = QE \tag{2.3}$$

Table 2.1 Symbols for standard quantities and units

Symbol	Quantity	Unit	Unit symbol
A	Geometric area	square metre	m^2
B	Magnetic flux density	tesla	T
C	Capacitance	farad	F
E	Electric field strength	volt per metre	V/m
F	Mechanical force	newton	N
F_m	Magnetomotive force (mmf)	ampere	A or A·t
G	Conductance	siemens	S
H	Magnetic field strength	ampere per metre	A/m
I	Electric current	ampere	A
J	Electric current density	ampere per square metre	A/m^2
J	Moment of inertia	kilogram metre squared	$kg \cdot m^2$
L	Self-inductance	henry	H
M	Mutual inductance	henry	H
N	Number of turns		
P	Active or real power	watt	W
Q	Electric charge	coulomb	C
Q	Reactive power	volt ampere reactive	VAr
R	Electrical resistance	ohm	Ω
R_m	Reluctance	ampere per weber	A/Wb
S	Apparent power	volt ampere	V·A
T	Mechanical torque	newton metre	N·m
V	Electric potential or voltage	volt	V
W	Energy or work	joule	J
X	Reactance	ohm	Ω
Y	Admittance	siemens	S
Z	Impedance	ohm	Ω
f	Frequency	hertz	Hz
j	Square root of −1		
l	Length	metre	m
m	Mass	kilogram	kg
n	Rotational speed	revolution per minute	rpm
p	Number of machine pole pairs		
t	Time	second	s
v	Linear velocity	metre per second	m/s
ε	Permittivity	farad per metre	F/m
η	Efficiency		
θ	Angle	radian or degree	rad or °
λ	Power factor		
Λ	Permeance	weber per ampere	Wb/A
μ	Permeability	henry per metre	H/m
ρ	Resistivity	ohm metre	$\Omega \cdot m$
σ	Conductivity	siemens per metre	S/m
ϕ	Phase angle	radian	rad
Φ	Magnetic flux	weber	Wb
ψ	Magnetic flux linkage	weber or weber-turn	Wb or Wb·t
ω	Angular velocity or angular frequency	radian per second	rad/s

Table 2.2 Standard decimal prefix symbols

Prefix	Name	Multiple
T	tera	10^{12}
G	giga	10^{9}
M	mega	10^{6}
k	kilo	10^{3}
d	deci	10^{-1}
c	centi	10^{-2}
m	milli	10^{-3}
μ	micro	10^{-6}
n	nano	10^{-9}
p	pico	10^{-12}
f	femto	10^{-15}

Electric field strength is a vector quantity. The direction of the force on one charge due to the electric field of another is repulsive or attractive. Charges with the same polarity repel; charges with opposite polarities attract.

Work must be done to move charges of the same polarity together. The effort required is described by a *voltage* or *electrostatic potential*. The voltage at a point is defined as the work required to move a unit charge from infinity or from earth. (It is normally assumed that the earth is at zero potential.) Positively charged objects have a positive potential relative to the earth.

If a positively charged object is held some distance above the ground, then the voltage at points between the earth and the object rises with distance from the ground, so that there is a *potential gradient* between the earth and the charged object. There is also an electric field pointing away from the object, towards the ground. The electric field strength is equal to the potential gradient, and opposite in direction.

$$E = -\frac{dV}{dx} \tag{2.4}$$

2.2.2 Electric currents

Electric charges are static if they are separated by an insulator. If charges are separated by a conductor, they can move giving an electric current. A current of one ampere flows if one coulomb passes along the conductor every second.

$$I = \frac{Q}{t} \tag{2.5}$$

A given current flowing through a thin wire represents a greater density of current than if it flowed through a thicker wire. The *current density J* (A/m^2) in a wire with cross section area A (m^2) carrying a current I (A) is given by

$$J = \frac{I}{A} \tag{2.6}$$

For wires made from most conducting materials, the current flowing through the wire is directly related to the difference in potential between the ends of the wire.

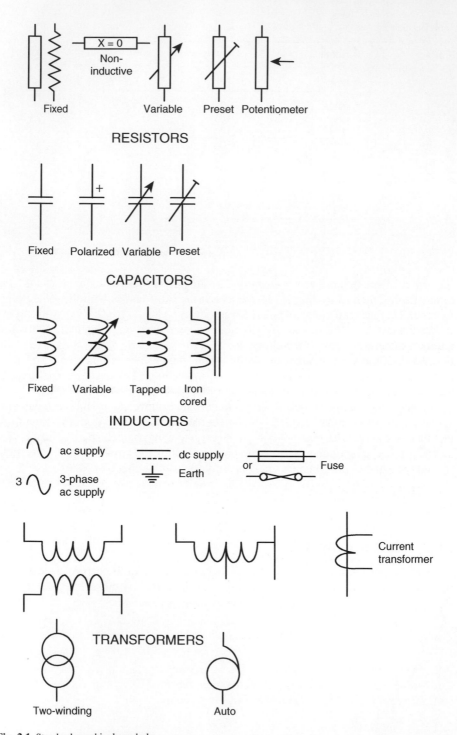

Fig. 2.1 Standard graphical symbols

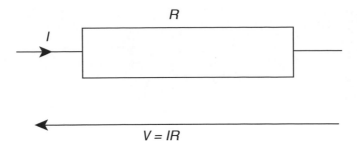

Fig. 2.2 Ohm's law

Ohm's law gives this relationship between the potential difference V (V) and the current I (A) as

$$V = IR \quad \text{or} \quad I = VG \tag{2.7}$$

where R (Ω) is the *resistance*, and G (S) = 1/R is the *conductance* (Fig. 2.2). For a wire of length l and cross section area A, these quantities depend on the *resistivity* ρ ($\Omega \cdot$m) and *conductivity* σ (S/m) of the material

$$R = \rho \frac{l}{A} \quad \text{and} \quad G = \sigma \frac{A}{l} \tag{2.8}$$

For materials normally described as *conductors* ρ is small, while for *insulators* ρ is large. *Semiconductors* have resistivity in between these extremes, and their properties are usually very dependent on purity and temperature.

In metal conductors, the resistivity increases with temperature approximately linearly:

$$R_T = R_{T_0}(1 + \alpha(T - T_0)) \tag{2.9}$$

for a conductor with resistance R_{T_0} at reference temperature T_0. This is explained in more detail in section 3.4.1.

Charges can be stored on conducting objects if the charge is prevented from moving by an insulator. The potential of the charged conductor depends on the *capacitance* C (F) of the metal/insulator object, which is a function of its geometry. The charge is related to the potential by

$$Q = CV \tag{2.10}$$

A simple parallel-plate capacitor, with plate area A, insulator thickness d and relative permittivity ε_r has capacitance

$$C = \frac{\varepsilon_0 \varepsilon_r A}{d} \tag{2.11}$$

2.2.3 Magnetic fields

A flow of current through a wire produces a magnetic field in a circular path around the wire. For a current flowing forwards, the magnetic field follows a clockwise path, as given by the right-hand corkscrew rule (Fig. 2.3). The magnetic field strength

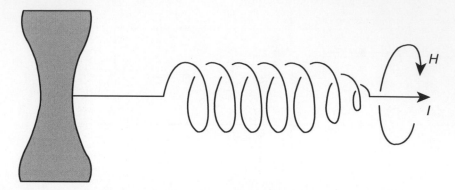

Fig. 2.3 Right-hand corkscrew rule

H (A m^{-1}) is a vector quantity whose magnitude at a distance d from a current I is given by

$$H = \frac{I}{2\pi d} \tag{2.12}$$

For a more complicated geometry, *Ampère's law* relates the number of turns N in a coil, each carrying a current I, to the magnetic field strength H and the distance around the lines of magnetic field l.

$$
\begin{aligned}
Hl &= NI \\
&= F_\mathrm{m}
\end{aligned}
\tag{2.13}
$$

where F_m (ampere-turns) is the *magnetomotive force (mmf)*. This only works for situations where H is uniform along the lines of magnetic field.

The magnetic field produced by a current does not depend on the material surrounding the wire. However, the magnetic force on other conductors is greatly affected by the presence of ferromagnetic materials, such as iron or steel. The magnetic field produces a *magnetic flux density B* (T) in air or vacuum

$$B = \mu_\mathrm{o} H \tag{2.14}$$

where the *permeability of free space* $\mu_\mathrm{o} = 4\pi \times 10^{-7}$ H/m. In a ferromagnetic material with *relative permeability* μ_r

$$B = \mu_\mathrm{o}\mu_\mathrm{r} H \tag{2.15}$$

A second conductor of length l carrying an electric current I will experience a force F in a magnetic flux density B

$$F = BIl \tag{2.16}$$

The force is at right angles to both the wire and the magnetic field. Its direction is given by *Fleming's left-hand rule* (Fig. 2.4). If the magnetic field is not itself perpendicular to the wire, then the force is reduced; only the component of B at right angles to the wire should be used.

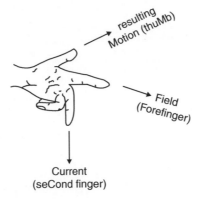

Fig. 2.4 Fleming's left-hand rule

A *magnetic flux* Φ (Wb) corresponding to the flux density in a given cross-section area A is given by

$$\Phi = BA \qquad (2.17)$$

The mmf F_m required to cause a magnetic flux Φ to flow through a region of length l and cross section area A is given by the *reluctance* R_m (A/Wb) or the *permeance* Λ (Wb/A) of the region

$$F_m = \Phi R_m \quad \text{or} \quad \Phi = \Lambda F_m \qquad (2.18)$$

where

$$R_m = \frac{l}{\mu_o \mu_r A} \qquad (2.19)$$

In ideal materials, the flux density B is directly proportional to the magnetic field strength H. In ferromagnetic materials the relation between B and H is non-linear (Fig. 2.5(a)), and also depends on the previous magnetic history of the sample. The *magnetization* or *hysteresis* or *BH* loop of the material is followed as the applied magnetic field is changed (Fig. 2.5(b)). Energy is dissipated as heat in the material as the operating point is forced around the loop, giving *hysteresis loss* in the material. These concepts are developed further in section 3.2.

2.2.4 Electromagnetism

Any change in the magnetic field near a wire generates a voltage in the wire by *electromagnetic induction*. The changing field can be caused by moving the wire in the magnetic field. For a length l of wire moving sideways at speed v (m/s) across a magnetic flux density B, the induced voltage or *electromotive force (emf)* is given by

$$V = Bvl \qquad (2.20)$$

The direction of the induced voltage is given by *Fleming's right-hand rule* (Fig. 2.6). An emf can also be produced by keeping the wire stationary and changing the

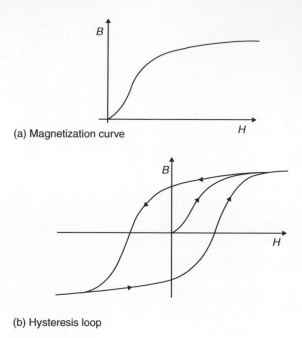

(a) Magnetization curve

(b) Hysteresis loop

Fig. 2.5 Magnetic characteristics

magnetic field. In either case the induced voltage can be found using *Faraday's law*. If a magnetic flux Φ passes through a coil of N turns, the *magnetic flux linkage* ψ (Wb·t) is

$$\psi = N\Phi \tag{2.21}$$

Faraday's law says that the induced emf is given by

$$V = -\frac{d\psi}{dt} \tag{2.22}$$

The direction of the induced emf is given by *Lenz's law*, which says that the induced voltage is in the direction such that, if the voltage caused a current to flow in the wire, the magnetic field produced by this current would oppose the change in ψ. The negative sign indicates the opposing nature of the emf.

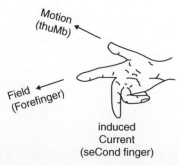

Fig. 2.6 Fleming's right-hand rule

A current flowing in a simple coil produces a magnetic field. Any change in the current will change the magnetic field, which will in turn induce a *back-emf* in the coil. The *self-inductance* or just *inductance* L (H) of the coil relates the induced voltage to the rate of change of current

$$V = L\frac{dI}{dt} \tag{2.23}$$

Two coils placed close together will interact. The magnetic field of one coil will link with the wire of the second. Changing the current in the primary coil will induce a voltage in the secondary coil, given by the *mutual inductance M* (H)

$$V_2 = M\frac{dI_1}{dt} \tag{2.24}$$

Placing the coils very close together, on the same former, gives close coupling of the coils. The magnetic flux linking the primary coil nearly all links the secondary coil. The voltages induced in the primary and secondary coils are each proportional to their number of turns, so that

$$\frac{V_1}{V_2} = \frac{N_1}{N_2} \tag{2.25}$$

and by conservation of energy, approximately

$$\frac{I_1}{I_2} = \frac{N_2}{N_1} \tag{2.26}$$

A two-winding transformer consists of two coils wound on the same ferromagnetic core. An autotransformer has only one coil with tapping points. The voltage across each section is proportional to the number of turns in the section. Transformer action is described in detail in section 6.1.

2.3   Circuits

2.3.1   DC circuits

DC may be supplied by a battery or fuel cell, dc generator or a rectified power supply. The power flowing in a dc circuit is the product of the voltage and current

$$P = VI = I^2R = \frac{V^2}{R} \tag{2.27}$$

Power in a resistor is converted directly into heat.

When two or more resistors are connected in *series*, they carry the same current but their voltages must be added together (Fig. 2.7)

$$V = V_1 + V_2 + V_3 \tag{2.28}$$

As a result, the total resistance is given by

$$R = R_1 + R_2 + R_3 \tag{2.29}$$

When two or more resistors are connected in *parallel*, they have the same voltage but their currents must be added together (Fig. 2.8)

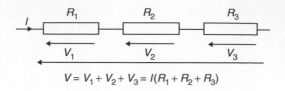

$$V = V_1 + V_2 + V_3 = I(R_1 + R_2 + R_3)$$

Fig. 2.7 Series resistors

$$I = I_1 + I_2 + I_3 \tag{2.30}$$

The total resistance is given by

$$\frac{1}{R} = \frac{1}{R_1} + \frac{1}{R_2} + \frac{1}{R_3} \tag{2.31}$$

A complicated circuit is made of several components of *branches* connected together at *nodes* forming one or more complete circuits, *loops* or *meshes*. At each node, *Kirchhoff's current law* (Fig. 2.9(a)) says that the total current flowing into the node must be balanced by the total current flowing out of the node. In each loop, the sum of all the voltages taken in order around the loop must add to zero, by *Kirchhoff's voltage law* (Fig. 2.9(b)). Neither voltage nor current can be lost in a circuit.

DC circuits are made of resistors and voltage or current sources. A circuit with only two connections to the outside world may be internally complicated. However, to the outside world it will behave as if it contains some resistance and possibly a source of voltage or current. The *Thévenin* equivalent circuit consists of a voltage source and a resistor (Fig. 2.10 (a)), while the *Norton* equivalent circuit consists of a current source and a resistor (Fig. 2.10(b)). The resistor equals the internal resistance of the circuit, the Thévenin voltage source equals the open-circuit voltage, and the Norton current source is equal to the short-circuit current.

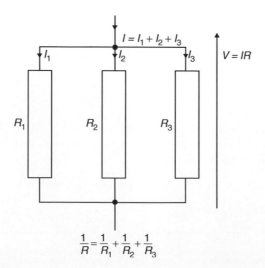

Fig. 2.8 Parallel resistors

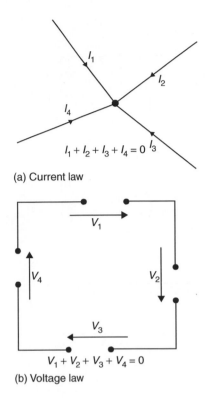

(a) Current law

$I_1 + I_2 + I_3 + I_4 = 0$

$V_1 + V_2 + V_3 + V_4 = 0$

(b) Voltage law

Fig. 2.9 Kirchhoff's laws

Many circuits contain more than one source of voltage or current. The current flowing in each branch, or the voltage at each node, can be found by considering each source separately and adding the results. During this calculation by *superposition*, all sources except the one being studied must be disabled: voltage sources are short-circuited and current sources are open-circuited. In Fig. 2.11, each of the loop currents I_1 and I_2 can be found by considering each voltage source separately and adding the results, so that $I_1 = I_{1a} + I_{1b}$ and $I_2 = I_{2a} + I_{2b}$.

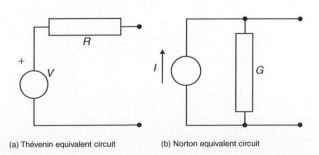

(a) Thévenin equivalent circuit   (b) Norton equivalent circuit

Fig. 2.10 Equivalent circuits

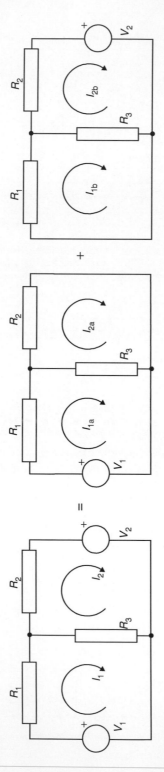

Fig. 2.11 Superposition

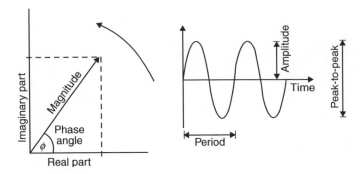

Fig. 2.12 Sinusoidal ac quantities

2.3.2 AC circuits

AC is supplied through a power system from large ac generators or alternators, by a local alternator, or by an electronic synthesis. AC supplies are normally sinusoidal, so that at any instant the voltage is given by

$$v = V_{max}\sin(\omega t - \phi)$$

$$= \hat{v}\sin(\omega t - \phi)$$ (2.32)

V_{max} is the peak voltage or amplitude, ω is the *angular frequency* (rad s⁻¹) and ϕ the *phase angle* (rad). The angular frequency is related to the ordinary *frequency f* (Hz) by

$$\omega = 2\pi f$$ (2.33)

and the *period* is $1/f$. The *peak-to-peak* or *pk–pk* voltage is $2V_{max}$, and the *root mean square* or rms voltage is $V_{max}/\sqrt{2}$. It is conventional for the symbols V and I in ac circuits to refer to the rms values, unless indicated otherwise. AC voltages and currents are shown diagrammatically on a *phasor diagram* (Fig. 2.12).

It is convenient to represent ac voltages using complex numbers. A sinusoidal voltage can be written

$$V = V_{max}\,e^{j\phi}$$ (2.34)

A resistor in an ac circuit behaves the same as in a dc circuit, with the current and voltage in phase and related by the resistance or conductance (Fig. 2.13).

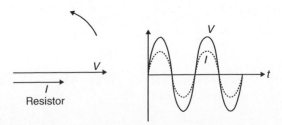

Fig. 2.13 Resistor in an ac circuit

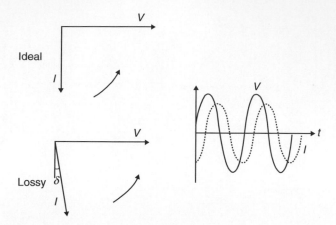

Fig. 2.14 Inductor in an ac circuit

The current in an inductor lags the voltage across it by $90°$ ($\pi/2$ rad) (Fig. 2.14). The ac resistance or *reactance X* of an inductor increases with frequency

$$X_L = \omega L \tag{2.35}$$

The phase shift and reactance are combined in the complex *impedance Z*

$$\frac{V}{I} = Z_L = jX_L = j\omega L \tag{2.36}$$

Inductors in series behave as resistors in series

$$L_s = L_1 + L_2 + L_3 \tag{2.37}$$

and inductors in parallel behave as resistors in parallel

$$\frac{1}{L_p} = \frac{1}{L_1} + \frac{1}{L_2} + \frac{1}{L_3} \tag{2.38}$$

For a capacitor, the current leads the voltage across it by $90°$ ($\pi/2$ rad) (Fig. 2.15). The reactance decreases with increasing frequency

$$X_c = \frac{1}{\omega C} \tag{2.39}$$

In a capacitor, the current leads the voltage, while in an inductor, the voltage leads the current.

The impedance is given by

$$\frac{V}{I} = Z_c = -jX_c = -\frac{j}{\omega C} = \frac{1}{j\omega C} \tag{2.40}$$

Capacitors in series behave as resistors in parallel (eqn 2.41) and capacitors in parallel behave as resistors in series (eqn 2.42)

$$\frac{1}{C_s} = \frac{1}{C_1} + \frac{1}{C_2} + \frac{1}{C_3} \tag{2.41}$$

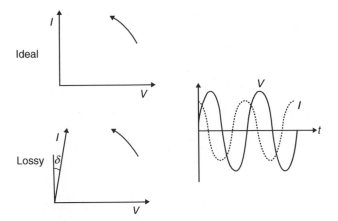

Fig. 2.15 Capacitor in an ac circuit

$$C_p = C_1 + C_2 + C_3 \qquad (2.42)$$

The direction of the phase shift in inductors and capacitors is easily remembered by the mnemonic CIVIL (i.e. C-IV, VI-L). Imperfect inductors and capacitors have some inherent resistance, and the phase lead or lag is less than 90°. The difference between the ideal phase angle and the actual angle is called the *loss angle* δ. For a component of reactance X having a series resistance R

$$\tan(\delta) = \frac{R}{X} \qquad (2.43)$$

The reciprocal of impedance is *admittance*

$$\frac{I}{V} = Y = \frac{1}{Z} \qquad (2.44)$$

Combinations of resistors, capacitors and inductors will have a variation of impedance or admittance with frequency which can be used to select signals at certain frequencies in preference to others. The circuit acts as a *filter*, which can be low-pass, high-pass, band-pass, or band-stop.

An important filter is the *resonant circuit*. A series combination of inductor and capacitor has zero impedance (infinite admittance) at its resonant frequency

$$\omega_o = 2\pi f_o = \frac{1}{\sqrt{LC}} \qquad (2.45)$$

A parallel combination of inductor and capacitor has infinite impedance (zero admittance) at the same frequency.

In practice a circuit will have some resistance (Fig. 2.16), which makes the resonant circuit imperfect. The *quality factor Q* of a series resonant circuit with series resistance R is given by

$$Q = \left(\frac{X_L}{R}\right)_{\omega=\omega_o} = \left(\frac{X_C}{R}\right)_{\omega=\omega_o} \qquad (2.46)$$

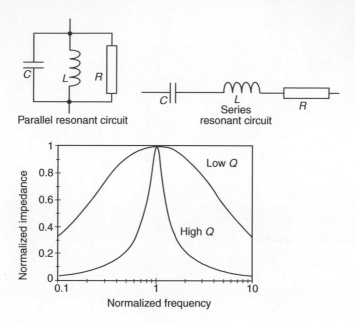

Fig. 2.16 Resonant circuits

and for a parallel circuit with shunt resistance R

$$Q = \left(\frac{R}{X_L}\right)_{\omega=\omega_0} = \left(\frac{R}{X_C}\right)_{\omega=\omega_0} \tag{2.47}$$

A filter with a high Q will have a sharper change of impedance with frequency than one with a low Q, and reduced losses.

AC is generated and distributed using three phases, three equal voltages being generated at $120°$ phase intervals. The *phase voltage* V_P is measured with respect to a common star point or neutral point, and the *line voltage* V_L is measured between the separate phases. In magnitude,

$$V_L = \sqrt{3}\,V_P \tag{2.48}$$

Any three-phase generator, transformer or load can be connected in a *star* or *Y* configuration, or in a *delta* configuration (Fig. 2.17).

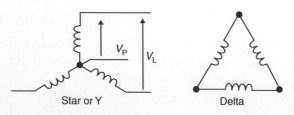

Fig. 2.17 Three-phase connections

2.3.3 Magnetic circuits

A reluctance in a magnetic circuit behaves in the same way as a resistance in a dc electric circuit. Reluctance in series add together

$$R_m = R_{m_1} + R_{m_2} + R_{m_3}$$

or

$$\frac{1}{\Lambda} = \frac{1}{\Lambda_1} + \frac{1}{\Lambda_2} + \frac{1}{\Lambda_3} \qquad (2.49)$$

and for reluctances in parallel

$$\frac{1}{R_m} = \frac{1}{R_{m_1}} + \frac{1}{R_{m_2}} + \frac{1}{R_{m_3}}$$

or

$$\Lambda = \Lambda_1 + \Lambda_2 + \Lambda_3 \qquad (2.50)$$

Transformers and power reactors may contain no air gap in the magnetic circuit. However, motors and generators always have a small air gap between the rotor and stator. Many reactors also have an air gap to reduce saturation of the ferromagnetic parts. The reluctance of the air gap is in series with the reluctance of the steel rotor and stator. The high relative permeability of steel means that the reluctance of even a small air gap can be much larger than the reluctance of the steel parts of the machine. For a total air gap g (m) in a magnetic circuit, the magnetic flux density B in the air gap is given approximately by

$$B \approx \mu_o \frac{F_m}{g} = \mu_o \frac{N_f I_f}{g} \qquad (2.51)$$

where N_f is the number of series turns on the field winding of the machine, and I_f is the current in the field winding (Fig. 2.18).

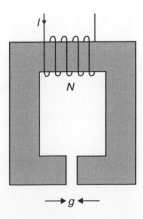

Fig. 2.18 Air-gap magnetic circuit

2.4   Energy and power

2.4.1   Mechanical energy

According to Newton's third law of motion, mechanical force causes movement in a straight line, such that the force F (N) accelerates a mass m (kg) with acceleration a (m/s^2)

$$F = ma \tag{2.52}$$

Rotational movement depends in the same way upon torque T (N m) accelerating a moment of inertia J (kg m^2) with angular acceleration α (rad/s^2)

$$T = J\alpha \tag{2.53}$$

Movement is often opposed by friction. Friction forces and torques always work against the movement. Friction between dry surfaces has a maximum value, depending on the contact force. Once the driving force exceeds the limiting friction force, the system will move and the friction force says constant. Viscous damping gives a restraining force which increases with the speed. Friction between lubricated surfaces is mainly a viscous effect.

Objects store potential energy when they are lifted up. The stored energy W (J) of a mass m is proportional to the height h (m) above ground level, when the acceleration due to gravity $g = 9.81$ m/s

$$W = mgh \tag{2.54}$$

Moving objects have kinetic energy depending on their linear speed v (m/s) or angular speed ω (rad/s)

$$W = \frac{1}{2}mv^2$$
$$= \frac{1}{2}J\omega^2 \tag{2.55}$$

Mechanical work is done whenever an object is moved a distance x (m) against an opposing force, or through an angle θ (rad) against an opposing torque

$$W = Fx$$
$$= T\theta \tag{2.56}$$

Mechanical power P (W) is the rate of doing work

$$P = Fv$$
$$= T\omega \tag{2.57}$$

2.4.2   Electrical energy

In electrical circuits, electrical potential energy is stored in a capacitance C charged to a voltage V

$$W = \frac{1}{2}CV^2 \tag{2.58}$$

while kinetic energy is stored in an inductance L carrying a current I

$$W = \frac{1}{2}LI^2 \tag{2.59}$$

Capacitors and inductors store electrical energy. Resistors dissipate energy and convert it into heat. The power dissipated and lost to the electrical system in a resistor R has already been shown in eqn 2.27.

Electrical and mechanical systems can convert and store energy, but overall the total energy in a system is conserved. The overall energy balance in an electromechanical system can be written as

electrical energy in + mechanical energy in
    = electrical energy lost in resistance
        + mechanical energy lost in friction
        + magnetic energy lost in steel core   (2.60)
        + increase in stored mechanical energy
        + increase in stored electrical energy

The energy balance is sometimes illustrated in a power flow diagram (Fig. 2.19).

The overall efficiency of a system is the ratio of the useful output power to the total input power, in whatever form

$$\eta = \frac{P_{out}}{P_{in}} \tag{2.61}$$

In an ac circuit, the instantaneous power depends on the instantaneous product of voltage and current. For sinusoidal voltage and current waveforms, the *apparent power* S (VA) is the product of the rms voltage and the rms current.

$$S = VI \tag{2.62}$$

When the voltage waveform leads the current waveform by an angle ϕ, the average, *active* or *real power* P (W) is

$$P = VI\cos(\phi) \tag{2.63}$$

The factor relating the real power to the apparent power is the *power factor* λ

$$\lambda = \cos(\phi) \tag{2.64}$$

The real power is converted mainly into heat or mechanical power. In addition there is an oscillating flow of instantaneous power, which is stored and then released each

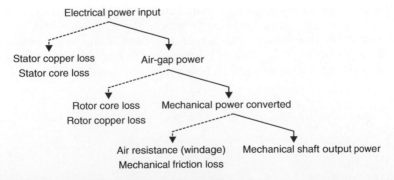

Fig. 2.19 Power flow diagram (example of a typical motor)

cycle by the capacitance and inductance in the circuit. This imaginary or *reactive power Q* (VAr) is given by

$$Q = VI\sin(\phi) \tag{2.65}$$

By convention an inductive circuit (where the current lags the voltage) absorbs VAr, while a capacitive circuit (where the current leads the voltage) acts as a source of VAr. The relationship between S, P and Q is

$$S^2 = P^2 + Q^2 \tag{2.66}$$

In a three-phase circuit, the total real power is the sum of the power flowing into each phase. For a balanced three-phase circuit with phase-neutral voltage V_P and phase current I_P the total power is the sum of the powers in each phase

$$P = 3V_P I_P \cos(\phi) \tag{2.67}$$

In terms of the line voltage V_L, the real power is

$$P = \sqrt{3}\, V_L I_P \cos(\phi) \tag{2.68}$$

Similar relations hold for the reactive and apparent power.

The power in a three-phase circuit can be measured using three wattmeters, one per phase, and the measurements added together (Fig. 2.20(a)). If it is known that the load is balanced, with equal current and power in each phase, the measurement can be made with just two wattmeters (Fig. 2.20(b)). The total power is then

$$W = W_1 + W_2 \tag{2.69}$$

The two-wattmeter arrangement also yields the power factor angle

$$\tan(\phi) = \frac{\sqrt{3}(W_1 - W_2)}{(W_1 + W_2)} \tag{2.70}$$

2.4.3   Per-unit notation

Power systems often involve transformers which step the voltage up or down. Transformers are very efficient, so that the output power from the transformer is only slightly less than the input power. Analysis and design of the circuit is made easier if the circuit values are normalized, such that the normalized values are the same on both sides of the transformers. This is accomplished using the *per-unit* system. A given section of the power system operates at a certain *base voltage* V_B. Across transformers the voltage steps up or down according to the turns ratio. The base voltage will be different on each side of the transformer. A section of the system is allocated a base apparent power rating or *base VA* $V \cdot A_B$. This will be the same on both sides of the transformer. Combining these base quantities gives a base impedance Z_B and base current I_B.

$$Z_B = \frac{V_B^2}{V \cdot A_B}$$

$$I_B = \frac{V \cdot A_B}{V_B} \tag{2.71}$$

The base impedance and base current change across a transformer.

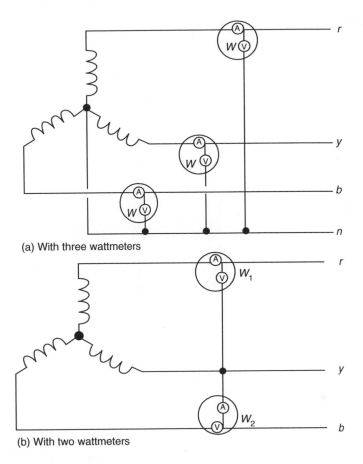

(a) With three wattmeters

(b) With two wattmeters

Fig. 2.20 Three-phase power measurement

All voltages and impedances in the system are normalized using the appropriate base value. The resulting normalized quantities are per-unit values. Once the circuit has been converted to per-unit values, the transformers have no effect on nominal tap position, and can be replaced by their own equivalent per-unit impedance.

The per-unit values are sometimes quoted as *per cent* values, by multiplying by 100 per cent.

A particular advantage of the per-unit and per cent notation is that the resulting impedances are very similar for equipment of very different sizes.

2.4.4 Energy transformation effects

Most electrical energy is generated by electromagnetic induction. However, electricity can be produced by other means. Batteries use electrochemistry to produce low voltages. An electrolyte is a solution of chemicals in water such that the chemical separates into positively and negatively charged ions when dissolved. The charged ions react with the conducting electrodes and release energy, as well as give up their charge (Fig. 2.21). A fixed *electrode potential* is associated with the reaction at each electrode; the difference between the two electrode potentials drives a current around an external circuit. The electrolyte must be sealed into a safe container to make a suitable battery.

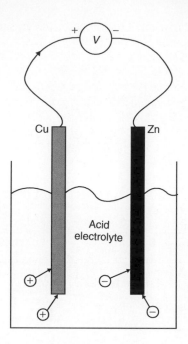

Fig. 2.21 Simple cell

'Dry' cells use an electrolyte in the form of a gel or thick paste. A *primary cell* releases electricity as the chemicals react, and the cell is discarded once all the active chemicals have been used up, or the electrodes have become contaminated. A *secondary cell* uses a reversible chemical reaction, so that it can be recharged to regenerate the active chemicals. The *fuel cell* is a primary cell which is constructed so that the active chemicals (fuel) pass through the cell, and the cell can be used for long periods by replenishing the chemicals. Large batteries consist of cells connected in series or parallel to increase the output voltage or current. The main types of primary, secondary and fuel cell are described in sections 12.2 to 12.5.

Electricity can be generated directly from heat. When two different materials are used in an electrical circuit, a small electrochemical voltage (contact potential) is generated at the junction. In most circuits these contact potentials cancel out around the circuit and no current flows. However, the junction potential varies with temperature, so that if one junction is at a different temperature from the others, the contact potentials will not cancel out and the net circuit voltage causes current to flow (*Seebeck effect*). The available voltage is very small, but can be made more useful by connecting many pairs of hot and cold junctions in series. The *thermocouple* is used mostly for measurement of temperature by this effect, rather than for the generation of electrical power (see section 4.4.1). The efficiency of energy conversion is greater with semiconductor junctions, but metal junctions have a more consistent coefficient and are preferred for accurate measurements. The effect can be reversed with suitable materials, so that passing an electric current around the circuit makes one junction hotter and the other colder (*Peltier effect*). Such miniature heat pumps are used for cooling small components.

Certain crystalline chemicals are made from charged ions of different sizes. When a voltage is applied across the crystal, the charged ions move slightly towards the side of opposite polarity, causing a small distortion of the crystal. Conversely, applying a force so as to distort the crystal moves the charged ions and generates a voltage. This *piezoelectric effect* is used to generate high voltages from a small mechanical force, but very little current is available. The use of piezoelectric sensors for the measurement of mechanical pressure and force is described in section 4.2.4. Ferromagnetic materials also distort slightly in a magnetic field. The *magnetostrictive effect* produces low frequency vibration (hum) in ac machines and transformers.

Electricity can be produced directly from light. The *photovoltaic effect* occurs when light falls on suitable materials, releasing electrons from the material and generating electricity. The magnitude of the effect is greater with short wavelength light (blue) than long wavelength light (red), and stops altogether beyond a wavelength threshold. Light falling on small photovoltaic cells is used for light measurement, communications and for proximity sensors. On a larger scale, semiconductor solar cells are being made with usable efficiency for power generation.

Light is produced from electricity in incandescent filament bulbs, by heating a wire to a sufficiently high temperature that it glows. Fluorescent lights produce an electrical discharge through a low-pressure gas. The discharge emits ultraviolet radiation, which causes a fluorescent coating on the inside of the tube to glow.

References

2A.   Professional brief edited by Burns, R.W., Dellow, F. and Forbes, R.G., *Symbols and Abbreviations for use in Electrical and Electronic Engineering*. The Institution of Electrical Engineers, 1992.

2B.   BS 3939:1985, *Graphical symbols for electrical power, telecommunications and electronics diagrams*, BSI, 1985.

2C.   BS 5555:1993 (ISO 1000:1992), *SI units and recommendations for the use of their multiples and certain other units*, BSI, 1993.

2D.   BS 5775:1993 (ISO 32:1992), *Quantities, units and symbols*, Part 5: *electricity and magnetism*, BSI, 1993.

2E.   Smith, R.J. and Dorf, R.C., *Circuits, Devices and Systems* (5th edn), Wiley, 1992, ISBN 0-471-55221-6.

2F.   Hughes, E. (revised Smith, I.M.), *Hughes Electrical Technology*, Longman Scientific and Technical, 1995, ISBN 0-582-22696-1.

2G.   Bird, J.O., *Higher Electrical Technology*, Butterworth-Heinemann, 1994, ISBN 0-7506-01019.

2H.   Breithaupt, J., *Understanding Physics for Advanced Level*, S. Thornes, 1990, ISBN 0-7487-0510-4.

2I.   Del Toro, V., *Electrical Engineering Fundamentals*, Prentice-Hall, 1986, ISBN 0-13-247131-0.

Chapter 3

Materials for electrical engineering

Professor A.G. Clegg
Magnet Centre, University of Sunderland

A.G. Whitman
With amendments by
J.E. Neal
Krempel Group

3.1 Introduction

Most types of electrical equipment rely, for their safe and efficient performance, on an electrical circuit and the means to keep this circuit isolated from the surrounding materials and environment. Many types of equipment also have a magnetic circuit, which is linked to the electrical circuit by the laws outlined in Chapter 2.

The main material characteristics of relevance to electrical engineering are therefore those associated with conductors for the electrical circuit, with the insulation system necessary to isolate this circuit and with the specialized steels and permanent magnets used for magnetic circuit.

Other properties, such as mechanical, thermal and chemical properties are also relevant, but these are often important in specialized cases and their coverage is best left to other books which address these areas more broadly. The scope of this chapter is restricted to the main types and characteristics of conductors, insulation systems and magnetic materials which are used generally in electrical plant and equipment.

3.2 Magnetic materials

All materials have magnetic properties. These characteristic properties may be divided into five groups as follows:

- diamagnetic
- paramagnetic
- ferromagnetic
- antiferromagnetic
- ferrimagnetic

Only ferromagnetic and ferrimagnetic materials have properties which are useful in practical applications.

Ferromagnetic properties are confined almost entirely to iron, nickel and cobalt and their alloys. The only exceptions are some alloys of manganese and some of the rare earth elements.

Ferrimagnetism is the magnetism of the mixed oxides of the ferromagnetic elements. These are variously called ferrites and garnets. The basic ferrite is magnetite, or Fe_3O_4, which can be written as $FeO \cdot Fe_2O_3$. By substituting the FeO with other divalent oxides, a wide range of compounds with useful properties can be produced. The main advantage of these materials is that they have high electrical resistivity which minimizes eddy currents when they are used at high frequencies.

The important parameters in magnetic materials can be defined as follows:

- *permeability* – this is the flux density B per unit of magnetic field H, as defined in eqns 2.14 and 2.15. It is usual and more convenient to quote the value of relative permeability μ_r, which is $B/\mu_o H$. A curve showing the variation of permeability with magnetic field for a ferromagnetic material is given in Fig. 3.1. This is derived from the initial magnetization curve and it indicates that the permeability is a variable which is dependent on the magnetic field. The two important values are the *initial permeability*, which is the slope of the magnetization curve at $H = 0$, and the *maximum permeability*, corresponding to the knee of the magnetization curve.
- *saturation* – when sufficient field is applied to a magnetic material it becomes saturated. Any further increase in the field will not increase the magnetization and any increase in the flux density will be due to the added field. The *saturation magnetization* is M_s in amperes per metre and J_s or B_s in tesla.
- *remanence, B_r and coercivity, H_c* – these are the points on the hysteresis loop shown in Fig. 3.2 at which the field H is zero and the flux density B is zero, respectively. It is assumed that in passing round this loop, the material has been saturated. If this is not the case, an inner loop is traversed with lower values of remanence and coercivity.

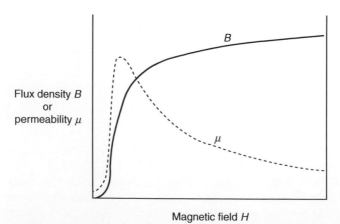

Fig. 3.1 Magnetization and permeability curves

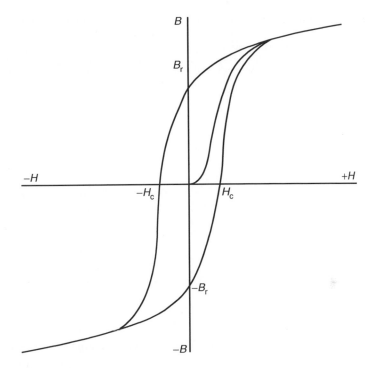

Fig. 3.2 Magnetization and hysteresis curves

Ferromagnetic and ferrimagnetic materials have moderate to high permeabilities, as seen in Table 3.1. The permeability varies with the applied magnetic field, rising to a maximum at the knee of the $B-H$ curve and reducing to a low value at very high fields. These materials also exhibit magnetic hysteresis, where the intensity of magnetization of the material varies according to whether the field is being increased in a positive sense or decreased in a negative sense, as shown in Fig. 3.2. When the magnetization is cycled continuously around a hysteresis loop, as for example when the applied field arises from an alternating current, there is an energy loss proportional to the area of the included loop. This is the *hysteresis loss*, and it is measured in joules per cubic metre. High hysteresis loss is associated with permanent magnetic characteristics exhibited by materials commonly termed *hard* magnetic materials, as these often have hard mechanical properties. Those materials with low hysteresis loss are termed *soft* and are difficult to magnetize permanently.

Ferromagnetic or ferrimagnetic properties disappear reversibly if the material is heated above the Curie temperature, at which point it becomes *paramagnetic*, that is effectively non-magnetic.

3.2.1   Soft (high permeability) materials

There is a wide variety of soft magnetic materials for applications from constant dc field through 50/60 Hz up to microwave frequencies. For the bulk of applications iron, steel or cast iron are used. They have the advantage of low cost and strength, but they should only be used for dc applications since they have low electrical resistivity which would result in eddy currents if used in alternating fields.

Table 3.1 Properties of soft magnetic materials

	Maximum permeability μ_{max} ($\times 10^{-3}$)	Saturation magnetization J_s (T)	Coercivity H_c (A/m)	Curie temperature (°C)	Resistivity (Ω m $\times 10^8$)
Sheet steels					
3% Si grain-oriented	90	2.0	6–7	745	48
2.5% Si non-oriented	8	2.0	40	745	44
<0.5% Si non-oriented	8	2.1	50–100	770	12
Low-carbon iron	3–10	2.1	50–120	770	12
Nickel iron					
78% Ni	250–400	0.8	1.0	350	40
50% Ni	100	1.5–1.6	10	530	60
Amorphous					
Iron-based	35–600	1.3–1.8	1.0–1.6	310–415	120–140
Ferrite					(Ω m)
Mn Zn	1.8–3.0	0.45–0.50	15–27	110–220	0.05–10
Ni Zn	0.12–1.5	0.20–0.35	30–300	250–400	>10^4
Garnet	–	0.3–1.7	–	11–280	>10^6

In choosing a high-permeability material for a particular application there may be special considerations. If the frequency of the applied voltage is 10 kHz or above, then ferrites or garnets will normally be used. For constant-field applications mild steel is used, and low-carbon iron is used where the highest permeability or lowest coercivity is needed. For 50/60 Hz power transformers, grain-oriented silicon steel is used, but there is now serious competition from amorphous strip and although this is more expensive the core loss is significantly lower than that of silicon steel. For the highest permeability and lowest coercivity in specialist applications up to about 10 kHz nickel iron would be the preferred choice although its cost may be prohibitive.

3.2.1.1   Sheet steels

There are a number of grades of sheet steel and these comprise by far the greatest part of the soft, high-permeability material used. These metallic materials have a comparatively low electrical resistivity and they are used in sheet (or lamination) form because this limits the flow of eddy currents and the losses that result. The range of sheet thickness used for 50/60 Hz applications is 0.35–0.65 mm for non-oriented materials and 0.13–0.35 mm for grain-oriented silicon steels. For higher frequencies of 400–1000 Hz thicknesses of 0.05–0.20 mm are used.

In grain-oriented steel an increasing level of silicon content reduces the losses but also reduces the permeability; for many applications a 3 per cent silicon content represents a good balance. The effect of silicon is to increase the electrical resistivity of steel; not only does this reduce eddy current losses but it also improves the stability of the steel and aids the production of grain orientation. The sheet is subject to cold rolling and a complex annealing treatment to produce the grain orientation in the rolling direction and this also gives improved magnetic permeability in that direction. The properties are further improved at the annealing stage with a glass film on the surface that holds the steel in a state of tension and provides electrical insulation between laminations in a core. A phosphate coating is applied to complete the tensioning and insulation. The grain size in the resulting sheets is comparatively large and the domain boundaries are quite widely spaced. Artificial grain boundaries can be produced by laying down lines of ablated spots on the steel surface; the resulting stress and atomic disruption pattern pins domain walls and leads to a smaller wall spacing. Various methods have been used to produce this ablating, including spark and laser techniques. All these processes are applied within the steel manufacturer's works and the resulting steel is often referred to as *fully processed*. The main application of this grain-oriented material is in power transformers where low power loss is important, since the transformer is always connected even when its loading is at a minimum.

For rotating machinery, especially motors rated more than about 100 kW, non-oriented steels with a lower silicon content are used. Whilst efficiency remains important in this application, high permeability is now also important in order to minimize magnetizing current and to maximize torque. This material is also used in smaller transformers, chokes for fluorescent tubes, metres and magnetic shielding.

For smaller motors silicon-free non-oriented steels are often used. These are produced by the steel manufacturer and have a relatively high carbon content. The carbon makes the sheet sufficiently hard for punching into laminations and after punching, the material is de-carburized and annealed to increase the grain size. Because of the need for this secondary processing, the materials are often known as *semi-processed*. These materials are generally cheaper than silicon steels but in their finished form they have a much higher permeability; in small motors particularly this can be more important than efficiency. Other applications include relays and magnetic clutches.

Silicon steels are also produced in the form of bars, rods or wires for relays, stepping motors and gyroscope housings. The tensile strength of the silicon steels may be improved by the addition of alloying elements such as manganese; this type of material is used in highly stressed parts of the magnetic circuit in high-speed motors or generators.

3.2.1.2 Amorphous alloys

This class of alloys, often called *metallic glasses*, is produced by rapid solidification of the alloy at cooling rates of about a million degrees centigrade per second. The alloys solidify with a glass-like atomic structure which is a non-crystalline frozen liquid. The rapid cooling is achieved by causing the molten alloy to flow through an orifice onto a rapidly rotating water-cooled drum. This can produce sheets as thin as 10 μm and a metre or more wide.

There are two main groups of amorphous alloys. The first is the iron-rich group of magnetic alloys, which have the highest saturation magnetization among the amorphous alloys and are based on inexpensive raw materials. Iron-rich alloys are currently being used in long-term tests in power transformers in the USA. The second is the cobalt-based group, which have very low or zero magnetostriction, leading to the highest permeability and the lowest core loss. Cobalt-based alloys are used for a variety of high-frequency applications including pulsing devices and tape recorder heads, where their mechanical hardness provides excellent wear resistance.

All of these alloys have a resistivity which is higher than that of conventional crystalline electrical steels. Because of this, eddy current losses are minimized both at 50/60 Hz and at higher frequencies. The alloys have other advantages including flexibility without loss of hardness, high tensile strength and good corrosion resistance.

3.2.1.3 Nickel iron alloys

The very high magnetic permeability and low coercivity of nickel iron alloys is due to two fundamental properties, which are *magnetostriction* and *magnetic anisotropy*. At a nickel content of about 78 per cent both of these parameters are zero. Magnetostriction has been briefly referred to in section 2.4.4; it is the change of dimensions in a material due to magnetization, and when this is zero there are no internal stresses induced during magnetizing. Magnetic anisotropy is the difference between magnetic behaviour in different directions; when this is zero the magnetization increases steeply under the influence of a magnetic field, independent of crystal direction. Alloys with about 78 per cent nickel content are variously called *Mumetal* or *Permalloy*.

Commercial alloys in this class have additions of chromium, copper and molybdenum in order to increase the resistivity and improve the magnetic properties. Applications include special transformers, circuit breakers, magnetic recording heads and magnetic shielding.

Fifty per cent nickel iron alloys have the highest saturation magnetization of this class of materials, and consequently in the best flux-carrying capacity. They have a higher permeability and better corrosion resistance than silicon iron materials, but are more expensive. A wide range of properties can be produced by various processing techniques. Severe cold reduction produces a cube texture and a square hysteresis loop in annealed strip, and the properties can be tailored by annealing the material below the Curie temperature in a magnetic field. Applications for this material include chokes, relays and small motors.

3.2.1.4 Ferrites and garnets

Ferrites are iron oxide compounds containing one or more other metal oxides. The important high magnetic permeability materials are manganese zinc ferrite and nickel zinc ferrite. They are prepared from the constituent oxides in powder form, preferably of the same particle size intimately mixed. The mixture is fired at about 1000°C and this is followed by crushing, milling and then pressing of the powder in a die or extrusion to the required shape. The resulting compact is a black brittle ceramic and any subsequent machining must be by grinding. The materials may be prepared with high permeability, and because their high electric resistivity limits eddy currents to a negligible level, they can be used at frequencies up to 20 MHz as the solid core of inductors or transformers.

A combination of hysteresis, eddy current and residual losses occur, and these components may be separately controlled by composition and processing conditions, taking into account the required permeability and the working frequency.

The saturation flux density of ferrites is relatively low, making them unsuitable for power applications. Their use is therefore almost entirely in the electronics and telecommunications industry where they have now largely replaced laminated alloy and powder cores.

Garnets are used for frequencies of 100 MHz and above. They have resistivities in excess of 10^8 Ω m compared with ferrite resistivity which is up to 10^3 Ω m. Since eddy current losses are limited by resistivity, they are greatly reduced in garnets. The basic Yttrium Iron Garnet (YIG) composition is $3Y_2O_3 \cdot 5Fe_2O_3$. This is modified to obtain improved properties such as very low loss and greater temperature stability by the addition of other elements including Al and Gd. The materials are prepared by heating of the mixed oxides under pressure at over 1300°C for up to 10 hours. Garnets are used in microwave circuits in filters, isolators, circulators and mixers.

3.2.1.5   Soft magnetic composites

Soft magnetic composites (SMC) consist of iron or iron alloy powder mixed with a binder and a small amount of lubricant. The composite is pressed into the final shape that can be quite complex. The lubricant reduces the friction during pressing and aids the ejection of the part from the die. After pressing, the parts are either cured at 150 to 275°C or heat treated at 500°C.

The isolation of the particles within the binder minimizes the eddy currents and makes the components very suitable for medium frequency applications up to 1 kHz or more. They have the added advantages of being isotropic and can be used for components with quite complex shape.

3.2.2   Hard (permanent magnet) materials

The key properties of a permanent magnet material are given by the demagnetization curve, which is the section of the hysteresis curve in the second quadrant between B_r and $-H_c$. It can be shown that when a piece of permanent magnet material is a part of a magnetic circuit, the magnetic field generated in a gap in the circuit is proportional to $B \times H \times V$, where B and H represent a point on the demagnetization curve and V is the volume of permanent magnet. To obtain a given field with a minimum volume of magnet the product $B \times H$ must therefore be a maximum, and the $(BH)_{max}$ value is useful in comparing material characteristics.

The original permanent magnet materials were steels, but these have now been superceded by better and more stable materials including *Alnico*, *ferrites* and *rare earth* alloys. The magnetic properties of all the permanent magnet materials are summarized in Table 3.2.

3.2.2.1   Alnico alloys

A wide range of alloys with magnetically useful properties is based on the Al–Ni–Co–Fe system. These alloys are characterized by high remanence, high available energy and moderately high coercivity. They have a low and reversible temperature coefficient of about $-0.02\%/°C$ and the widest useful temperature range (up to over 500°C) of any permanent magnet material.

The alloys are produced either by melting or sintering together the constituent elements. Anisotropy is achieved by heating to a high temperature and allowing the material to cool at a controlled rate in a magnetic field in the direction in which the magnets are to be magnetized. The properties are much improved in this direction at the expense of properties in the other directions. This is followed by a tempering treatment in the range 650–550°C. A range of coercivities can be produced by varying the cobalt content. The properties in the preferred direction may be further improved by producing an alloy with columnar crystals.

3.2.2.2   Ferrites

The permanent magnet ferrites are also called ceramics and they are mixtures of ferric oxide and an oxide of a divalent heavy metal, usually barium or strontium. These ferrites are made by mixing together barium or strontium carbonate with iron oxide. The mixture is fired and the resulting material is milled to a particle size of about 1 μm. The powder is then pressed to the required shape in a die and anisotropic magnets are produced by applying a magnetic field in the pressing direction. The resulting compact is then fired.

Table 3.2 Characteristics of permanent magnet materials

	Remanence B_r (T)	Energy product $(BH)_{max}$ (kJ/m^3)	Coercivity H_c (kA/m)	Relative recoil permeability μ_{rec}	Curie temperature (°C)	Resistivity (Ω m $\times 10^8$)
Alnico						
Normal anisotropic	1.1–1.3	36–43	46–60	2.6–4.4	800–850	50
High coercivity	0.8–0.9	32–46	95–150	2.0–2.8	800–850	50
Columnar	1.35	60	60–130	1.8	800–850	50
Ferrites (ceramics)						
Barium isotropic	0.22	8	130–155 (a)	1.2	450	10^{12}
Barium anisotropic	0.39	28.5	150 (a)	1.05	450	10^{12}
Strontium anisotropic	0.36–0.43	24–34	240–300 (a)	1.05	450	10^{12}
La, Co, Sr anisotropic	0.42–0.45	33–38	260–320 (a)	1.05	450	10^{12}
Bonded ferrite						
Isotropic	0.14	3.2	90 (a)	1.1	450	*
Anisotropic	0.23–0.27	10–14	180 (a)	1.05	450	*
Rare earth						
SmCo$_5$ sintered	0.9	160	600–660 (b)	1.05	>700	50–60
SmCo$_5$ bonded	0.5–0.6	56–64	400–460 (b)	1.1	>700	*
Sm$_2$Co$_{17}$ sintered	1.1	150–240	600–700 (c)	1.05	>700	75–85
NdFeB sintered	0.9–1.5	160–460	750–920 (d)	1.05	310	140–160
NdFeB bonded	0.25–0.8	10–95	180–460 (e)	1.05	310	*

Intrinsic coercivities: (a) 160–380 kA/m
                                        (b) 800–1500 kA/m
                                        (c) 600–1300 kA/m
                                        (d) 950–3000 kA/m
                                        (e) 460–1300 kA/m

*Dependent on the resistivity of the polymer bond

3.2.2.3 Rare earth alloys

The $(BH)_{max}$ values that can be achieved with rare earth alloys are 4–6 times greater than those for Alnico or ferrite.

The three main permanent magnet rare earth alloys are samarium cobalt (SmCo$_5$ and Sm$_2$Co$_{17}$) and neodymium iron boron (NdFeB). These materials may be produced by alloying the constituent elements, or more usually by reducing a mixture of the oxides together in a hydrogen atmosphere using calcium as the reducing agent. The alloy is then milled to a particle size of about 10 μm, pressed in a magnetic field and sintered in vacuum.

The first alloy to be available was SmCo$_5$, but this has gradually been replaced by Sm$_2$Co$_{17}$ because of its lower cost and better temperature stability. The more recently developed NdFeB magnets have the advantage of higher remanence B_r and higher $(BH)_{max}$, and they are lower in cost because the raw materials are cheaper. The disadvantage of NdFeB materials is that they are subject to corrosion and they suffer from a rapid change of magnetic properties (particularly coercivity) with temperature.

Corrosion can be prevented by coating the magnets and the properties at elevated temperature may be improved by small additions of other elements.

3.2.2.4 Bonded magnets

Ferrites and rare earth magnets are also produced in bonded forms. The magnet powder particles are mixed with the bond and the resulting compact can be rolled, pressed or injection-moulded. For a flexible magnet the bonds may be rubber, or for a rigid magnet they may be nylon, polypropylene, resin or other polymers. Although the magnetic properties are reduced by the bond, they can be easily cut or sliced and in contrast to the sintered magnets they are not subject to cracking or chipping. Rolled or pressed magnets give the best properties as some anisotropy may be induced, but injection moulding is sometimes preferred to produce complex shapes which might even incorporate other components. Injection moulding also produces a precise shape with no waste of material.

3.2.2.5 Applications

Permanent magnets have a very wide range of applications and virtually every part of industry and commerce uses them to some extent. At one time Alnico was the only available high-energy material, but it has gradually been replaced by ferrites and rare earth alloys, except in high-stability applications. Ferrites are much cheaper than Alnico but because of their lower flux density and energy product a larger magnet is often required. However, about 70 per cent of magnets used are ferrite. They find bulk applications in loudspeakers, small motors and generators and a wide range of electronic applications. The rare earth alloys are more expensive but despite this and because of their much greater strength they are being used in increasing quantities. They give the opportunity for miniaturization, and Figs 3.3 and 3.4 show the range of applications for permanent magnets in the home and in a car.

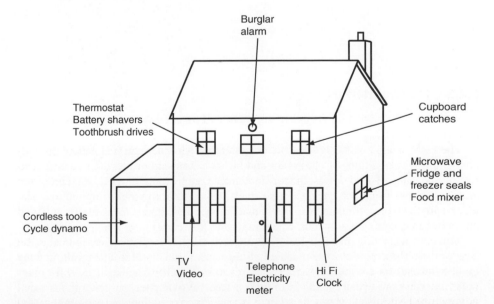

Fig. 3.3 Magnets in the home

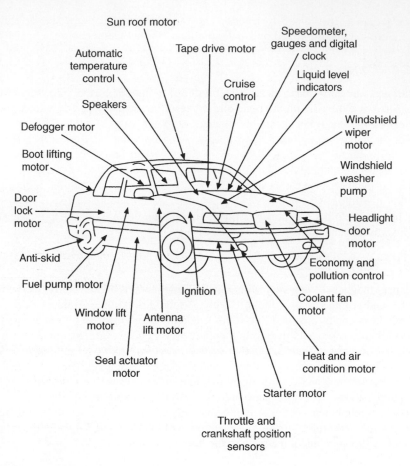

Fig. 3.4 Magnets in the car

3.2.3   Other materials

In addition to the two main groups of soft and hard magnetic materials there are other materials that meet special needs.

The feebly magnetic steels are austenitic, and their virtually non-magnetic properties are achieved by additions of chromium and nickel to low-carbon steel. To attain a relative permeability of 1.05 or less, the recommended composition is 18 per cent chromium and 10 per cent nickel, or greater. These steels, which have minimum strength requirements, are used for non-magnetic parts of machinery, for magnetic measuring equipment and for minesweeping equipment, where magnetic flux can actuate magnetic mines.

Magnetic recording makes use of fine magnetic particles which are embedded in the tape or disc. These particles are made up of metal or of iron or chromium oxide, and the choice depends on a compromise between price and quality. The heads used for magnetic recording are usually made from high-permeability ferrites, but amorphous metal is now also being used. Magnetic storage is a rapidly expanding area with higher and higher information densities being achieved.

3.2.4   Standards

The main national and international standard for magnetic materials is IEC 60404, to which BS 6404 is equivalent. This standard has many parts with specifications for the properties of silicon steels, nickel irons and permanent magnets. Also included are measurement standards for these materials. The BS standards are currently being renumbered with BSEN numbers. These standards have recently been reviewed by Stanbury (reference 3D).

3.3   Insulating materials

The reason for using insulating materials is to separate electrically the conducting parts of equipment from each other and from earthed components. Earthed components may include the mechanical casing or structure that is necessary to enable the equipment to be handled and to operate. Whereas the 'active' parts of the equipment play a useful role in its operation, the insulation is in many ways a necessary evil. For example in an electric motor the copper of the winding and the steel core making up the magnetic circuit are the active components and both contribute to the power output of the motor; the insulation which keeps these two components apart contributes nothing, in fact it takes up valuable space and it may be considered by the designer as not much more than a nuisance.

For these reasons, insulating materials have become a design focus in many types of electrical equipment, with many companies employing specialists in this field and carrying out sophisticated life testing of insulation systems. Such is the importance attached to this field that major international conferences on the subject are held regularly, for instance by the IEEE in USA, IEE and Electrical Insulation Association (EIA) in UK and the European Electrical Insulation Association (EEIM) in Europe, all of which publish the papers presented. Conferences are also held in Canada, India and South Africa.

The simplest way to define an insulating material is to state what it is not. It is not a good conductor of electricity and it has a high electrical resistance that decreases with rising temperature, unlike conductors. The following are the most important properties of insulating materials:

- *volume resistivity*, which is also known as specific resistance
- *relative permittivity* (or dielectric constant), which is defined as the ratio of the electric flux density produced in the material to that produced in vacuum by the same electric field strength. The definitions have been set down in eqns 2.1 and 2.2. Relative permittivity can be expressed as the ratio of the capacitance of a capacitor made of that material to that of the same capacitor using vacuum as its dielectric (see eqn 2.11).
- *dielectric loss* (or electrical dissipation factor), which is defined as the ratio of the power loss in a dielectric material to the total power transmitted through it. It is given by the tangent of the loss angle and is commonly known as *tan delta*. Tan delta has been defined in eqn 2.43.

The volume resistivity, relative permittivity and tan delta values for a range of insulating materials are shown in Table 3.3.

The most important characteristic of an insulating material is its ability to withstand electric stress without breaking down. This ability is sometimes known as its

Table 3.3 Representative properties of typical insulating materials

	Volume resistivity (Ω m)	Relative permittivity	Tan delta (at 50 Hz)
Vacuum	Infinity	1.0	0
Air	Infinity	1.0006	0
Mineral insulating oil	10^{11}–10^{13}	2.0–2.5	0.0002
Pressboard	10^{8}	3.1	0.013
Dry paper	10^{10}	1.9–2.9	0.005
Oiled paper	–	2.8–4.0	0.005
Porcelain	10^{10}–10^{12}	5.0–7.0	–
E-glass	10^{16}	6.1–6.7	0.002–0.005
Polyester resin	10^{14}–10^{16}	2.8–4.1	0.008–0.041
Epoxy resin	10^{12}–10^{15}	3.5–4.5	0.01
Mica	10^{11}–10^{15}	4.5–7.0	0.0003
Micapaper	10^{13}–10^{17}	5.0–8.7	0.0003
PETP film	10^{18}	3.3	0.0025
Aramid paper	10^{16}	2.5–3.5	0.005–0.020
Epoxy glass laminate	–	4.5–4.7	0.008
Silicone glass laminate	–	4.5–6.0	0.003
Polystyrene	10^{15}	2.6	0.0002
Polyethylene	10^{15}	2.3	0.0001
Methyl methacrylate	10^{13}	2.8	0.06
Polyvinyl chloride	10^{11}	5.0–7.0	0.1
Fused quartz	10^{16}	3.9	–

dielectric strength, and is usually quoted in kilovolts per millimetre (kV/mm). Typical values may range from 5 to 100 kV/mm, but it is dependent on a number of other factors which include the speed of application of the electric field, the length of time for which it is applied, temperature and whether ac or dc voltage is used.

Another significant aspect of all insulating materials that dominates the way in which they are categorized is the maximum temperature at which they will perform satisfactorily. Generally speaking, insulating materials deteriorate over time more quickly at higher temperatures and the deterioration can reach a point at which the insulation ceases to perform its required function. This characteristic is known as ageing, and for each material it has been usual to assign a maximum temperature beyond which it is unwise to operate if a reasonable life is to be achieved. The main gradings or classes of insulation as defined in IEC 60085:1984 and its UK equivalent BS 2757:1986(1994) are listed in Table 3.4. Where a thermal class is used to describe

Table 3.4 Thermal classes for insulation

Thermal class	Operating temperature (°C)
Y	90
A	105
E	120
B	130
F	155
H	180
200	200
220	220
250	250

an item of electrical equipment, it normally represents the maximum temperature found within that product under rated load and other conditions. However, not all the insulation is necessarily located at the point of maximum temperature, and insulation with a lower thermal classification may be used in other parts of the equipment.

The ageing of insulation depends not only on the physical and chemical properties of the material and the thermal stress to which it is exposed, but also on the presence and degree of influence of mechanical, electrical and environmental stresses. The processing of the material during manufacture and the way in which it is used in the complete equipment may also significantly affect the ageing process. The definition of a useful lifetime will also vary according to the type and usage of equipment; for instance the running hours of a domestic appliance and a power station generator will be very different over a 25-year period. All of these factors should therefore influence the choice of insulating material for a particular application.

There is therefore a general movement in the development of standards and methods of testing for insulating materials towards the consideration of combinations of materials or *insulating systems*, rather than focusing on individual materials. It is not uncommon to consider life testing in which more than one form of stress is introduced; this is known as *multifunctional* or *multifactor testing*.

Primary insulation is often taken to mean the main insulation, as in the PVC coating on a live conductor or wire. *Secondary* insulation refers to a second 'line of defence' which ensures that even if the primary insulation is damaged, the exposed live component does not cause an outer metal casing to become live. Sleeving is frequently used as a secondary insulation.

Insulating materials may be divided into basic groups which are *solid dielectrics*, *liquid dielectrics*, *gas* and *vacuum*. Each is covered separately in the following sections.

3.3.1 Solid dielectrics

Solid dielectric insulating materials have in the past (for instance in BS 5691-2:1995 and IEC 60216-2:1990) been subdivided into three general groups as follows:

- solid insulation of all forms not undergoing a transformation during application
- solid sheet insulation for winding or stacking, obtained by bonding superimposed layers
- insulation which is solid in its final state, but is applied in the form of a liquid or paste for filling, varnishing or bonding

A more convenient and up-to-date way to subdivide this very large group of materials is used by the IEC Technical Committee 15: Insulating Materials, and is as follows:

- inorganic (ceramic and glass) materials
- plastic films
- flexible insulating sleeving
- rigid fibrous reinforced laminates
- resins and varnishes
- pressure-sensitive adhesive tapes
- cellulosic materials
- combined flexible materials
- mica products

This subdivision is organized on the basis of application and is therefore more help-ful to the practising engineer. A brief description of each of these classes of material is given in the following sections.

3.3.1.1   Inorganic (ceramic and glass) materials

A major application for materials in this category is in high-voltage overhead lines as suspension insulators (see Figs 13.2 and 13.8), or as bushings on high-voltage transformers and switchgear (see Fig. 6.14). In either case the material is formed into a series of flanged discs to increase the creepage distance along the surface of the complete insulator. Ceramic materials are used for a number of reasons, which include:

- ease of production of a wide range of shapes
- good electrical breakdown strength
- retention of insulating characteristics in the event of surface damage

3.3.1.2   Plastic films

Materials such as polyethylene terephthalate (PETP, more commonly referred to as polyester), polycarbonate, polyimide and polyethylene naphthalate (PEN) have been used as films, depending on the operating temperature requirement, in a variety of applications. These include the insulation between foils in capacitors, slot insulation in rotating electrical machines (either itself or as a composite with other sheet materials) and more recently as a backing for mica-based products used in the insulation of high-voltage equipment. Plastic films are used in applications requiring dimensional stability, high dielectric strength, moisture resistance and physical toughness.

3.3.1.3   Flexible insulating sleeving

This fulfils a number of requirements including the provision of primary or secondary electrical insulation of component wiring, the protection of cables and components from the deleterious effects of mechanical and thermal damage, and as a rapid and low-cost method of bunching and containing cables. Sleevings may find application in electrical machines, transformers, domestic and heating appliances, light fittings, cable connections, switchgear and as wiring harnesseses in domestic appliances and in vehicles. They are used because of their ease of application, flexibility and high dielec-tric strength and they lend themselves to an extremely wide range of formats. These include shrink sleeving, expandable constructions and textile-reinforced grades incor-porating glass, polyester, aramid or keramid yarns, or sometimes a combination together with a coating of acrylic, silicone or polyurethane resins, for low and high voltage and temperatures across the range $-70°C$ to $+450°C$.

3.3.1.4   Rigid fibrous reinforced laminates

In the manufacture of most electrical equipment, there is a need for items machined out of solid board or in the form of tubes and rods. These items can take the form of densified wood or laminates of paper, woven cotton, glass or polyester, or glass or polyester random mats laminated together with a thermosetting resin which might be phenolic, epoxy, polyester, melamine, silicone or polyimide, depending upon the prop-erties required. Rigid boards are used because they are capable of being machined to size, and they retain their shape and properties during their service life, unlike unseasoned timber which was used in early equipment.

3.3.1.5   Resins and varnishes

In addition to their use in the laminates outlined in section 3.3.1.4, a wide range of varnishes and resins are used by themselves in the impregnation and coating of electrical equipment in order to improve its resistance to working conditions, to enhance its electrical characteristics and to increase its working life. At first many resins and varnishes were based on naturally occurring materials such as bitumen, shellac and vegetable oils, but now they are synthetically produced in a comprehensive range of thermoplastic, thermosetting and elastomeric forms. The more common types are phenolic, polyester, epoxy, silicone and polyimide, and these can be formulated to provide the most suitable processing and the required final characteristics. Varnishes and resins are used because of their ability to impregnate, coat and bond basic insulating materials; this assists in the application of the insulating materials and it significantly improves their service life and their ability to withstand dirt and moisture.

3.3.1.6   Pressure-sensitive adhesive tapes

Certain types of pressure-sensitive adhesive (PSA) tapes have become so much a part of modern life that the trade names have been absorbed into everyday language. The usefulness of PSA in short lengths for holding down, sealing or locating applies equally in the field of insulation and a range of tapes has been developed which is based on paper, films (as in section 3.3.1.2) or woven glass cloth, coated with suitable adhesive such as rubber, silicone or acrylic.

3.3.1.7   Cellulosic materials

Materials in the form of papers, pressboards and presspapers continue to play a vital role in oil-filled power transformers. Included in this area are also other materials which are produced by paper-making techniques, but which use aramid fibres; these materials find wide application in high-temperature and dry-type transformers as well as in other types of electrical equipment. Cellulose materials are mainly used in conjunction with oil, and it is their porous nature which lends itself to successful use in transformers and cables.

3.3.1.8   Combined flexible materials

In order to produce suitable materials with the required properties such as tear strength, electric strength and thermal resistance at an acceptable price, a range of laminated or combined flexible sheet products has been developed. These employ cellulosic, polyester and aramid papers or fleeces, also glass and polyester fabrics as well as other materials, in combination with many of the plastic films already referred to, in a range of forms to suit the application. These products are used in large quantities in low-voltage electric motors and generators.

3.3.1.9   Mica products

Materials based on mica, a naturally occurring mineral, play a central part in the design and manufacture of high-voltage rotating machines. Originally the material was in the form of mica splittings, but at present the industry uses predominantly micapaper, which is produced by breaking down the mica into small platelets by chemical or mechanical means, producing a slurry and then feeding this through a traditional paper-making machine. The resulting micapaper, when suitably supported by a woven glass or film backing (see section 3.3.1.2) and impregnated with epoxy or a similar resin, is used to insulate the copper bars which make up the stator winding of the machine. Micapaper is used in the ground insulation of the winding which is shown as part of the stator slot section illustrated in Fig. 3.5.

Micapaper products are available in a resin-rich form, in which all the necessary resin for consolidation of the insulation around the winding is included within the material. This consolidation is usually carried out in large steam or electrically heated presses into which an insulated coil side or bar can be placed; heat and pressure are then applied as necessary to cure fully the resin-rich micapaper insulation. Alternatively, micapaper products are available for use with Vacuum Pressure Impregnation (VPI), in which most of the resin is introduced after winding the machine. The use of VPI eliminates the need to consolidate the insulation in a press; consolidation is achieved either by the use of hydrostatic forces or by ensuring full impregnation of the bars or coils already placed or wound into slots. A large electrical machine stator is shown being lowered into a VPI tank in Fig. 3.6.

Mica-based products dominate high-voltage insulation systems because of their unique combination of properties, which include the following:

- high dielectric strength
- low dielectric loss at high frequency
- high surface and volume resistivity
- excellent resistance to corona discharge and electric arc erosion
- temperature capability from −273°C to 1000°C
- flame resistance
- excellent chemical resistance
- high resistance to compressive forces

3.3.1.10   Textile insulation

Although the use of fully varnished fabric is becoming less common, products using glass and polyester-based yarn, and to a lesser extent cotton and rayon, are still in use.

A much larger range of unvarnished narrow-fabric products, more commonly called woven tapes, exists and these products use various combinations of glass, polyester and aramid yarns to meet specific applications. Primarily these tapes are used as finishing on top of other insulation such as micapaper, in order to provide a tough outer surface which can readily be coated with a final varnish or paint finish. When manufactured on modern needle looms they are an economic proposition.

Woven tapes are used because of their ease of application, good conformity and bedding down. An example of the complex shapes that they can be used to cover is shown in Fig. 3.7, which illustrates the endwinding of a high-voltage electrical machine.

3.3.1.11   Elastomers and thermoplastics

There is a very wide range of polymeric and rubber-like insulation materials. These have traditionally been dealt with in connection with electric cables and IEC and BSI reflect this by dealing with them separately in Technical Committee 20: Electrical Cables.

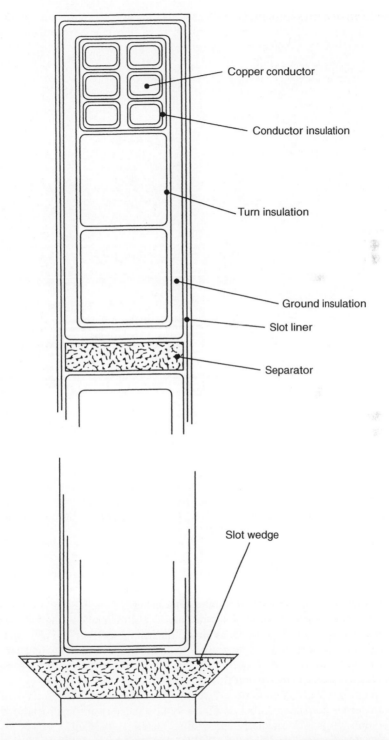

Fig. 3.5 Stator coil section of a high-voltage electrical machine (courtesy of ALSTOM Electrical Machines Ltd)

Fig. 3.6 A large vaccum and pressure impregnation facility capable of treating stators up to 100 MVA (courtesy of ALSTOM Electrical Machines Ltd)

Some elastomers such as silicone have found application in sleeving, traction systems and increasingly as overhead line insulators, but the bulk of their application continues to be related to cables. The leading materials such as PVC, MDPE, XLPE and EPR are therefore referred to in Chapter 9.

3.3.2 Liquid dielectrics

A liquid dielectric remains in the liquid state throughout its working life, unlike resins and varnishes which become solid after processing.

The principal uses of liquid dielectrics are as a filling and cooling medium for transformers, capacitors and rheostats, as an arc-quenching medium in switchgear and as an impregnant of absorbent insulation used mainly in transformers, switchgear, capacitors and cables. The important properties of dielectric liquids are therefore electric strength, viscosity, chemical stability and flashpoint.

Typical materials include highly refined hydrocarbon mineral oils obtained from selected crude petroleum, silicone fluids, synthetic esters and hydrocarbons with high molecular weight. A specially interesting material for cables has been the waxy *Mineral Insulating Non-Draining (MIND)* compound which has been used in paper-insulated cables; this is described in section 9.2.3. A group of polychlorinated biphenyls (PCBs) has been used in transformers, but these materials are now being

Fig. 3.7 Stator endwinding bracing system for a 4-pole, 11 kV, 8 MW induction motor (courtesy of ALSTOM Electrical Machines Ltd)

replaced and carefully disposed of because of their toxic nature and their resistance to biological and chemical degradation.

3.3.3 Gas insulation

Two gases already in common use for insulation are nitrogen and sulphur hexafluoride (SF_6). Nitrogen is used as an insulating medium in some sealed transformers, while SF_6 is finding increasing use in transmission and distribution switchgear because of its insulating properties and its arc-extinguishing capabilities; this is described further in sections 7.4.1 and 7.5.3.2.

3.3.4   Vacuum insulation

Vacuum insulation is now used in a range of medium-voltage switchgear. Like SF_6, it has both insulating and arc-extinguishing properties. The action of vacuum in a circuit breaker is explained in section 7.4.1.

3.3.5   Standards

A selection of national and international standards covering the insulation field is given in Table 3.5. The majority of the IEC standards with their EN and BSEN equivalents consist of three parts as follows:

- Part 1: Definitions, classsification and general requirements
- Part 2: Methods of test
- Part 3: Specifications for individual materials

Table 3.5 National and international standards relating to insulating materials

IEC	EN/HD	BS	Subject	North American
60085	HD 566S1	2757	Thermal evaluation and classification of electrical insulation	
60216	HD 611.1S1	5691	Determination of thermal endurance properties of electrical insulating materials	
60243	EN 60243	EN 60243	Methods of test for electric strength of solid insulating materials	
60371	EN 60371	EN 60371	Specifications for insulating materials based on mica	
60394		5689	Varnish fabrics for electrical purposes	
60454	EN 60454	EN 60454	Specifications for pressure-sensitive adhesive tapes for electrical purposes	
60455	EN 60455	EN 60455 5664	Specifications for solventless polymerizable resinous compounds used for electrical insulation	
60464	EN 60464	5629	Specifications for insulating varnishes containing solvent	
60626	EN 60626	EN60626	Combined flexible materials for electrical insulation	
60641	EN 60641	EN 60641	Specification for pressboard and presspaper for electrical purposes	
60672	EN 60672	EN 60672	Specifications for ceramic and glass insulating materials	
60674	EN 60674	EN 60674	Specifications for plastic films for electrical purposes	
60684	EN 60684	EN 60684	Specifications for flexible insulating sleevings	NEMA VS-1
60819	EN 60819	EN 60819-1 7824	Non-cellulosic papers for electrical purposes	
60893	EN 60893	EN 60893	Specifications for industrial rigid laminated sheets based on thermosetting resins for electrical purposes	NEMA LI-1
61067	EN 61067	EN 61067	Specifications for glass and glass polyester fibre woven tapes	
61068	EN 61068	EN 61068	Polyester fibre woven tapes	
61212	EN 61212	EN 61212 6128	Industrial rigid laminated tubes and rods based on thermosetting resins for electrical purposes	

3.4   Conducting materials

Strictly, conducting materials fall into three groups, which are conductors, semiconductors and imperfect insulators. Insulators have been discussed in section 3.3, so the focus here is on conductors and semiconductors.

3.4.1   Conductors

In general, metals and alloys are conductors of electricity. The conductivity in metals such as copper and aluminium is due to electrons that are attracted to the positive terminal when a voltage is applied. The freedom with which the electrons can move determines the conductivity and resistivity. The restraints on electron movements are impurities, stresses and thermal lattice vibrations; therefore, to obtain the highest conductivity the metal must be very pure and in the annealed state. With increasing temperature the thermal lattice vibrations increase and conductivity is therefore reduced.

The principal material for commercial applications as conductors are the pure metals aluminium and copper, although very widely used are the alloys of these metals, with small additions of other elements to improve their properties for particular applications. Table 3.6 shows for comparison typical values of the key parameters for the two metals.

3.4.1.1   Copper and its alloys

Copper has the highest electrical and thermal conductivity of the common industrial metals. It has good mechanical properties, is easy to solder, is readily available and has high scrap value. It is widely used in wire form, and Table 3.7 gives information for the commonly used wire sizes.

The electrical resistance of copper, as of all other pure metals, varies with temperature. The variation is sufficient to reduce the conductivity of pure copper at 100°C to about 76 per cent of its value at 20°C.

The resistance R_{t1} at temperature t_1 is given by the relationship

$$R_{t1} = R_t [1 + \alpha_t(t_1 - t)] \tag{3.1}$$

where α_t is the constant-mass temperature coefficient of resistance of copper at the reference temperature t (°C). Although resistance may be regarded for practical

Table 3.6 Comparison of the typical properties of aluminium and copper

	Aluminium	Copper	Units
Electrical conductivity at 20°C	37.2×10^6	58.1×10^6	S/m
Electrical resistivity at 20°C	2.69	1.72	$\mu\Omega$ cm
Density at 20°C	2.69	8.93	kg/dm³
Temperature coefficient of resistance (0–100°C)	4.2×10^{-3}	3.93×10^{-3}	per °C
Thermal conductivity (0–100°C)	238	599	W/m °C
Mean specific heat (0–100°C)	0.909	0.388	J/g °C
Coefficient of linear expansion (0–100°C)	23.5×10^{-6}	16.8×10^{-6}	per °C
Melting point	660	1085	°C

Table 3.7 IEC and BSI recommended sizes of annealed copper wire

Nominal diameter (mm)	Cross-sectional area (mm^2)	Weight (kg/km)	Resistance (Ω/km)	Current rating at 1000A/in.^{2} (A)
5.000	19.64	174.56	0.8703	30.43
4.750	17.72	157.54	0.9646	27.47
4.500	15.90	141.39	1.0750	24.65
4.000	12.57	111.71	1.3602	19.48
3.750	11.045	98.19	1.5476	17.12
3.550	9.898	87.99	1.7269	15.43
3.350	8.814	78.38	1.9393	13.66
3.150	7.793	69.28	2.193	12.08
3.000	7.069	62.84	2.418	10.96
2.800	6.158	54.74	2.776	9.544
2.650	5.515	49.03	3.099	8.549
2.500	4.909	43.64	3.482	7.609
2.360	4.374	38.89	3.908	6.780
2.240	3.941	35.03	4.338	6.108
2.120	3.530	31.38	4.843	5.471
2.000	3.142	27.93	5.441	4.869
1.900	2.835	25.21	6.029	4.375
1.800	2.545	22.62	6.718	3.944
1.700	2.270	20.18	7.531	3.518
1.600	2.011	17.874	8.502	3.116
1.500	1.767	15.710	9.673	2.739
1.400	1.539	13.685	11.10	2.386
1.320	1.368	12.166	12.49	2.121
1.250	1.227	10.910	13.93	1.902
1.180	1.094	9.722	15.63	1.695
1.120	0.9852	8.758	17.35	1.527
1.060	0.8825	7.845	19.37	1.368
1.000	0.7854	6.982	21.76	1.217
0.950	0.7088	6.301	24.12	1.099
0.900	0.6362	5.656	26.87	0.9861
0.850	0.5675	5.045	30.12	0.8796
0.800	0.5027	4.469	34.01	0.7791
0.750	0.4418	3.927	38.69	0.6848
0.710	0.3959	3.520	43.18	0.6137
0.630	0.3117	2.771	54.84	0.4832
0.560	0.2463	2.190	69.40	0.3818
0.500	0.1964	1.746	87.06	0.3043
0.450	0.1590	1.414	107.5	0.2465
0.400	0.1257	1.117	136.0	0.1948
0.355	0.0990	0.880	172.7	0.1534
0.315	0.0779	0.693	219.3	0.1208
0.280	0.0616	0.547	277.6	0.0954
0.250	0.0491	0.436	348.2	0.0761
0.224	0.0394	0.350	433.8	0.0611
0.200	0.0314	0.279	544.1	0.0487

Table 3.7 (*contd*)

Nominal diameter (mm)	Cross-sectional area (mm²)	Weight (kg/km)	Resistance (Ω/km)	Current rating at 1000A/in.² (A)
0.180	0.0255	0.227	671.8	0.0396
0.160	0.0201	0.179	850.2	0.0312
0.140	0.0154	0.137	1110	0.0239
0.125	0.0123	0.109	1393	0.0190
0.112	0.00985	0.0876	1735	0.0153
0.100	0.00785	0.0698	2176	0.0122
0.090	0.00636	0.0566	2687	0.0099
0.080	0.00503	0.0447	3401	0.0078
0.071	0.00396	0.0352	4318	0.0061
0.063	0.00312	0.0277	5484	0.0048

purposes as a linear function of temperature, the value of the temperature coefficient is not constant, but depends upon the reference temperature as given in eqn 3.2.

$$\alpha_t = \frac{1}{1+1/\alpha_o} = \frac{1}{1+234.45} \tag{3.2}$$

At 20°C, the value of α_{20} which is given by eqn 3.2 is 0.00393/°C, which is the value adopted by IEC. Multiplier constants and their reciprocals correlating the resistance of copper at a standard temperature with the resistance at other temperatures may be obtained from tables which are included in BS 125, BS 1432–1434 and BS 4109.

Cadmium copper, chromium copper, silver copper, tellurium copper and sulphur copper find wide application in the electrical industry where high conductivity is required. The key physical properties of these alloys are shown in Table 3.8. It can be seen that some of the alloys are deoxidized and some are 'tough pitch' (oxygen containing) or deoxidized. Tough pitch coppers and alloys become embrittled at elevated temperatures in a reducing atmosphere, and where such conditions are likely to be met, oxygen-free or deoxidized materials should be used.

Cadmium copper has greater strength than ordinary copper under both static and alternating stresses and it has better resistance to wear. It is particularly suitable for the contact wires in electric railways, tramways, trolley buses, gantry cranes and similar equipment, and it is also used in overhead telecommunication lines and transmission lines of long span. It retains its hardness and strength in temperatures at which high-conductivity materials would soften, and is used in electrode holders for spot and seam welding of steel, and it has also been used in commutator bars for certain types of motor. Because it has a comparatively high elastic limit in the work-hardened condition, it is also used in small springs required to carry current, and it is used as thin hard-rolled strip for reinforcing the lead sheaths of cables which operate under internal pressure. Castings of cadmium copper have some application in switchgear components and in the secondaries of transformers for welding machines. Cadmium copper can be soft soldered, silver soldered and brazed in the same way as ordinary copper, although special fluxes are required under certain conditions, and these should contain fluorides. Since it is a deoxidized material there is no risk of embrittlement by reducing gases during such processes.

Table 3.8 Selected physical properties of copper alloys

Property	Cadmium copper	Chromium copper	Silver copper	Tellurium copper	Sulphur copper
Content	0.7–1.0% Cadmium	0.4–0.8% Chromium	0.03–0.1% Silver	0.3–0.7% Tellurium	0.3–0.6% Sulphur
Tough pitch (oxygen-containing) or deoxidized	Deoxidized	Deoxidized	Tough pitch or deoxidized	Tough pitch or deoxidized	Deoxidized
Modulus of elasticity ($10^9\,\mathrm{N\,m^{-2}}$)	132	108	118	118*	118
Resistivity at 20°C ($10^{-8}\,\Omega\,\mathrm{m}$)					
Annealed	2.2–1.9		1.74–1.71	1.76**	1.81
Solution heat treated	–	4.9	–	–	–
Precipitation hardened		2.3–2.0	–	–	–
Cold worked	2.3–2.0	–	1.78	1.80	1.85

*Tough pitch
**Solution heat treated or annealed

Chromium copper is particularly suitable for high-strength applications such as spot and seam types of welding electrodes. Strip and, to a lesser extent, wire are used for light springs which carry current. In its heat-treated state, the material can be used at temperatures up to 350°C without risk of deterioration of properties, and it is used for commutator segments in rotating machines where the temperatures are higher than normal. In the solution heat-treated condition, chromium copper is soft and can be machined; in the hardened state it is not difficult to cut but it is not free-machining like leaded brass or tellurium copper. Joining methods similar to cadmium copper are applicable, and chromium copper can be welded using gas-shielded arcs.

Silver copper has the same electrical conductivity as ordinary high-conductivity copper, but its softening temperature, after hardening by cold work, is much higher and its resistance to creep at moderately elevated temperatures is enhanced. Since its outstanding properties are in the work-hardened state, it is rarely required in the annealed condition. Its principal uses are in electrical machines which operate at high temperatures or are exposed to high temperatures in manufacture. Examples of the latter are soft soldering or stoving of insulating materials. Silver copper is available in hard-drawn or rolled rods and sections, especially those designed for commutator segments, rotor bars and similar applications. Silver copper can be soft soldered, silver soldered, brazed or welded without difficulty but the temperatures involved in all these processes are sufficient to anneal the material, if in the cold-worked condition. Because the tough pitch material contains oxygen as dispersed particles of cuprous oxide, it is also important to avoid heating it to brazing and welding temperatures in a reducing atmosphere. In the work-hardened state, silver copper is not free-cutting, but it is not difficult to machine.

Tellurium copper offers free-machining, high electrical conductivity, with retention of work hardening at moderately elevated temperatures and good corrosion resistance. It is unsuitable for most types of welding, but gas-shielded arc welding and resistance welding can be done with care. A typical application is magnetron bodies, which are often machined from solid. Tellurium copper can be soft soldered, silver soldered and brazed without difficulty. For tough pitch, brazing should be done in an inert or slightly oxidizing atmosphere since reducing atmospheres are conducive to embrittlement. Deoxidized tellurium copper is not subject to embrittlement.

Sulphur copper is free-machining and does not have the tendency of tellurium copper to form coarse stringers in the structure which can affect accuracy and finish. It has greater resistance to softening than high-conductivity copper at moderately high temperatures and gives good corrosion resistance. Sulphur copper has applications in all machined parts requiring high electrical conductivity, such as contacts and connectors; its joining characteristics are similar to those of tellurium copper. It is deoxidized with a controlled amount of phosphorus and therefore does not suffer from hydrogen embrittlement in normal torch brazing, but long exposure to a reducing atmosphere can result in loss of sulphur and consequent embrittlement.

3.4.1.2   Aluminium and its alloys

For many years aluminium has been used as a conductor in most branches of electrical engineering. Several aluminium alloys are also good conductors, combining strength with acceptable conductivity. Aluminium is less dense and cheaper than copper, and its price is not subject to the same wide fluctuations as copper. World production of aluminium has steadily increased over recent years to overtake that of copper, which it has replaced in many electrical applications.

There are two specifications for aluminium, one for pure metal grade 1E and the other for a heat-treatable alloy 91E. Grade 1E is available in a number of forms which are extruded tube (E1E), solid conductor (C1E), wire (G1E) and rolled strip (D1E). The heat-treatable alloy, which has moderate strength and a conductivity approaching that of aluminium, is available in tubes and sections (E91E).
The main application areas are as follows.

Busbars.   Although aluminium has been used as busbars for many years, only recently has it been accepted generally. The electricity supply industry has now adopted aluminium busbars as standard in 400 kV substations, and they are also used widely in switchgear, plating shops, rising mains and in UK aluminium smelting plants. Sometimes busbars are tin-plated in applications where joints have to be opened and re-made frequently.

Cable.   The use of aluminium in wires and cables is described at length in Chapter 9. Aluminium is used extensively in cables rated up to 11 kV and house wiring cable above 2.5 mm^2 is also available with aluminium conductor.

Overhead lines.   The Aluminium Conductor Steel Reinforced (ACSR) conductor referred to in section 13.3 and Fig. 13.3 is the standard adopted throughout the world, although in USA Aluminium Conductor Aluminium alloy wire Reinforced (ACAR) is rapidly gaining acceptance; it offers freedom from bimetallic corrosion and improved conductance for a given cross section.

Motors.   Aluminium is cast into the cage rotors of induction motors forming the rotor bars and end rings. Motor frames are often die-cast or extruded from aluminium, and shaft-driven cooling fans are sometimes of cast aluminium.

Foil windings. These are suitable for transformers, reactors and solenoids. They offer better space factor than a wire-wound copper coil, the aluminium conductor occupying about 90 per cent of the space, compared to 60 per cent occupied by copper. Heat transfer is aided by the improved space factor and the reduced insulation that is needed in foil windings, and efficient radial heat transfer ensures an even temperature gradient. Windings of transformers are described in greater depth in section 6.2.2.

Heating elements. These have been developed in aluminium but they are not widely used at present. Applications include foil film wallpaper, curing concrete and possibly soil warming.

Heat sinks. They are an ideal application for aluminium because of its high thermal conductivity and the ease of extrusion or casting into solid or hollow shapes with integral fins. They are used in a variety of applications such as semiconductor devices and transformer tanks. The low weight of aluminium heat sinks make them ideal for pole-mounted transformers and there is the added advantage that the material does not react with transformer oil to form a sludge.

3.4.1.3   Resistance alloys

There are many alloys with high resistivity, the two main applications being resistors and heating elements.

Alloys for standard resistors are required to have a low temperature coefficient of resistivity in the region of room temperature. The traditionally used alloy is Manganin, but this has increasingly been replaced by Ni–Cr–Al alloys with the trade names Karma and Evanohm. The resistivity of these alloys is about 1.3 $\mu\Omega$ m and the temperature coefficient is $\pm 0.5 \times 10^{-5}/°C$. For lower precision applications copper–nickel alloys are used, but these have a lower resistivity and a relatively high thermo emf against copper.

For heating elements in electric fires, storage heaters and industrial and laboratory furnaces there is a considerable range of alloys available. A considerable resistivity is required from the alloy in order to limit the bulk of wire required, and the temperature coefficient of resistivity must be small so that the current remains reasonably constant with a constant applied voltage. The Ni–Cr alloys are used for temperatures up to 1100°C, and Cr–Fe–Al alloys are used up to 1400°C. Ceramic rods are used for higher temperatures and silicon carbide may be used up to 1600°C. For even higher temperatures, the cermets $MoSi_2$ and Zircothal are used. The maximum temperature at which the materials may be used depends on the type of atmosphere.

3.4.2   Semiconductors

A semiconductor is able to conduct electricity at room temperature more readily than an insulator but less readily than a conductor. At low temperatures, pure semiconductors behave like insulators. When the temperature of a semiconductor is increased, or when it is illuminated, electrons are energized and these become conduction electrons. Deficiencies or 'holes' are left behind; these are said to be carriers of positive electricity. The resulting conduction is called intrinsic conduction.

The common semiconductors include elements such as silicon, germanium and selenium and compounds such as indium arsenide and gallium antimonide. Germanium was originally used for the manufacture of semiconductor devices, but because of

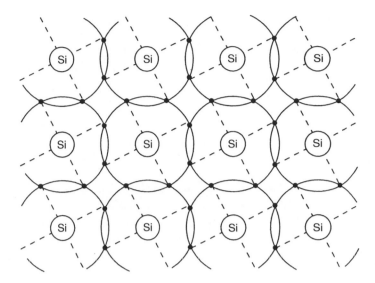

Fig. 3.8 Atoms in a silicon crystal

resistivity problems and difficulty of supply it was replaced by silicon which is now the dominant material for production of all active devices such as diodes, bipolar transistors, MOSFETs, thyristors and IGBTs. The principles of operation of the MOSFET and IGBT are described briefly in section 11.3.

Both silicon and germanium are group IV elements of the periodic table, having four electrons in their outer orbit. This results in a diamond-type crystal giving a tight bond of the electrons. Figure 3.8 shows the atoms in a silicon crystal. Each atom is surrounded by eight electrons, four of which are its own and four are from neighbouring atoms; this is the maximum number in an orbit and it results in a strong equilibrium. It is for this reason that pure crystals of silicon and germanium are not good conductors at low temperature.

3.4.2.1 *Impurity effects and doping*

The conductivity of group IV semiconductors like silicon can be greatly increased by the addition of small amounts of elements from group V (such as phosphorus, arsenic or tin) of group III (such as boron, aluminium, gallium or indium).

Phosphorus has five electrons in its outer shell and when an atom of phosphorus replaces an atom of silicon it generates a free electron, as shown in Fig. 3.9. This is called doping. The extra electrons are very mobile; when a voltage is applied they move very easily and a current passes. If 10^{16} phosphorus atoms/cm^3 are added to a pure crystal, the electron concentration is greatly increased and the conductivity is increased by a factor of about a million. The impurities are called donor atoms and the material is an impurity semiconductor. This is called an n-type semiconductor, and n represents the excess of free electron carriers.

If the material is doped with group III atoms such as indium, then a similar effect occurs. This is shown in Fig. 3.10. The missing electron forms a 'hole' in the structure which acts as a positive carrier. This structure is known as a p-type semiconductor and p represents the excess of positive carriers. The impurities are called acceptor atoms.

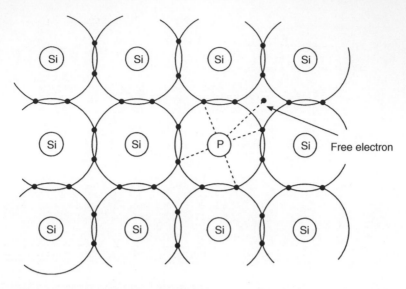

Fig. 3.9 Representation of an n-type semiconductor

A single crystal containing both n-type and p-type regions can be prepared by introducing the donor and acceptor impurities into molten silicon at different stages of the crystal formation. The resultant crystal has two distinct regions of p-type and n-type material, and the boundary joining the two areas is known as a p–n junction. Such a junction may also be produced by placing a piece of donor impurity material against the surface of a p-type crystal or a piece of acceptor impurity material against an n-type crystal, and applying heat to diffuse the impurity atoms through the outer layer.

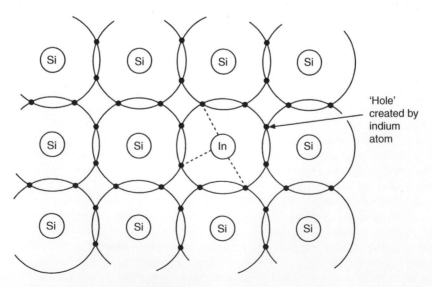

Fig. 3.10 Representation of an p-type semiconductor

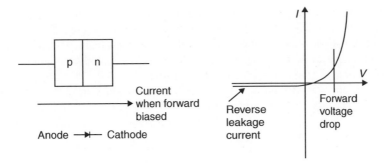

Fig. 3.11 Construction, symbol and characteristic of a semiconductor diode

When an external voltage is applied, the n–p junction acts as a rectifier, permitting current to flow in only one direction. If the p-type region is made positive and the n-type region negative, a current flows through the material across the junction, but when the potential difference is reversed, no current flows. This characteristic is shown in Fig. 3.11.

3.4.2.2   The transistor

Many types of device can be built with quite elaborate combinations and constructions based around the n–p and p–n junction. Further information on these devices may be found in reference 3G.

Possibly the most important single device is the transistor, in which a combination of two or more junctions may be used to achieve amplification. One type, known as the n–p–n junction transistor, consists of a very thin layer of p-type material between two sections of n-type material, arranged in a circuit shown in Fig. 3.12. The n-type material at the left of the diagram is the emitter element of the transistor, constituting the electron source. To permit the flow of current across the n–p junction, the emitter has a small negative voltage with respect to the p-type layer (or base component) that controls the electron flow. The n-type material in the output circuit serves as the collector element, which has a large positive voltage with respect to the base in order to prevent reverse current flow. Electrons moving from the emitter enter the base, are

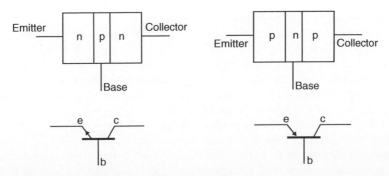

Fig. 3.12 Construction and symbols for n–p–n and p–n–p transistors

attracted to the positively charged collector, and flow through the output circuit. The input impedance, between the collector and the base is low, whereas the output impedance between the collector and the base is high. Therefore, small changes in the voltage of the base cause large changes in the voltage drop across the collector resistance, making this type of transistor an effective amplifier.

Similar in operation to the n–p–n type is the p–n–p junction transistor also shown in Fig. 3.12. This also has two junctions and is equivalent to a triode vacuum tube.

Other types, such as the n–p–n–p junction transistor, provide greater amplification than these two-junction transistors.

3.4.2.3   Printed circuits and integrated circuits

A printed circuit is an electrical circuit made by printing and bonding conducting material as a network of fine threads on a thin ceramic or polymer insulating sheet. This replaces the wiring used in conventional circuits. Other elements such as transistors, resistors and capacitors can be deposited onto the same base as the printed circuit.

An integrated circuit is effectively a combination of many printed circuits. It is formed as a single unit by diffusing impurities into single-crystal silicon, which then serves as a semiconductor material, or by etching the silicon by means of electron beams. Several hundred integrated circuits (ICs) are made at a time on a thin wafer several centimetres in diameter, and the wafer is subsequently sliced into individual ICs called chips.

In large scale integration (LSI), several thousand circuit elements such as resistors and transistors are combined in a 5 mm square area of silicon no more than 0.5 mm thick. Over 200 such circuits can be arrayed on a silicon wafer 100 mm in diameter.

In very large scale integration (VLSI), hundreds of thousands of circuit elements fit onto a single silicon chip. Individual circuit elements on a chip are interconnected by thin metal or semiconductor films which are insulated from the rest of the circuit by thin dielectric layers. This is achieved by the formation of a silicon dioxide layer on the silicon wafer surface, silicon dioxide being an excellent dielectric. Metal Oxide Semiconductor Field Effect Transistors (MOSFETs) are made using this technique. These transistors are used for high-frequency switching applications and for random access memories in computers. They have very high speed and low power consumption.

3.4.2.4   The microprocessor

The microprocessor is a single chip of silicon which has the ability to control processes. It can form the central processing unit (CPU) of a small computer and it can be used in a wide range of other applications. A microprocessor may incorporate from a thousand up to several hundred thousand elements. It typically contains a read-only memory (ROM), that is a memory that can be read repeatedly but cannot be changed, but it may also have some random access memory (RAM) for holding transient data. Also present in a microprocessor are registers for holding computing instructions, for holding the 'address' of each instruction in turn and for holding data, and a logic unit. Interfaces for connecting with external memories and other systems are included as required.

The microprocessors used in personal computers have been the subject of intensive development during the 1990s. The speed of operation is usually defined as a frequency and chips with frequencies of 3 GHz or higher are now available; this corresponds to an individual operation time of 0.33 nanoseconds. Personal computers with hard disk capacities of 80 Gb and laptop computers of 60 Gb can now be obtained. The amount of information that can be transferred in parallel and held in registers is known as a bit, and 64-bit processors are now available.

3.4.3 Superconductors

The ideal superconducting state is characterized by two fundamental properties, which are the disappearance of resistance when the temperature is reduced to a critical value, and the expulsion of any magnetic flux in the material when the critical temperature (T_c) is reached. Superconductivity was first discovered in the element mercury, in 1911. Other elements have subsequently been found to exhibit superconductivity and theories have been developed to explain the phenomenon. The critical temperatures for these materials were typically about 10 K ($-263°C$), which meant that they had to be cooled with liquid helium at 4 K. In general these materials have been of academic interest only because they could only support a low current density in a low magnetic field without losing their superconducting properties.

In the 1950s a new class of materials was discovered. These are the metallic alloys, the most important among them being niobium titanium and niobium tin. The highest critical temperature achieved by these materials is 23.2 K and they can be used to produce magnetic flux densities of over 15 T. The main commercial application for these low-T_c superconductors is for magnets in medical imaging equipment which require the high fields to excite magnetic resonance in nuclei of hydrogen and other elements. The magnet or solenoid of the magnetic resonance imaging (MRI) unit has an internal diameter of about 1.2 m and the patient to be examined is put into this aperture. The image from the resonance test shows unexpected concentrations of fluids or tissue and enables a diagnosis. Superconducting magnets producing high magnetic fields are also used in magnetic research and in high-energy physics research; other applications such as dc motors and generators, levitated trains, cables and ac switches have been explored but the complexity and high cost of providing the liquid helium environment prevented commercial development in most cases.

In late 1986 a ceramic material LaBaCuO was discovered to be superconducting at 35 K and in 1987 the material YBaCuO was found to have a critical temperature of 92 K. Since that time the critical temperatures of these new high temperature super-conducting (HTS) materials has progressively increased to over 130 K. Examples of these are BiSrCaCuO (with a T_c of 106 K), ThBaCaCuO (T_c of 125 K) and HgBaCaCuO (T_c of 133 K). The enormous significance of these discoveries is that these materials will be superconducting in liquid nitrogen, which has a boiling point of 77 K and is much easier and cheaper to provide than helium.

Much work has been directed towards finding materials with higher T_c values but this has remained at 133 K for some time. However, considerable effort with resulting success has been directed to the production of suitable HTS conductors. The HTS material is very brittle and it is deposited using laser deposition onto a suitable substrate tape. The tape is 3 mm wide and cables of up to 600 m in length have been produced.

There are many trials being made of the application of the HTS cables throughout the world including USA, Europe and Japan. There are prototypes of power trans-formers, underground power cables, large motors and generators, and fault current limiters in active development and in use. The electricity supply of the City of Geneva in Switzerland is completely provided by power transformers wound with HTS conductors. Detroit is being re-equipped with HTS power cable for its transmission system and copper cables weighing over 7 tons are being replaced with HTS cables of less than 0.12 tons. These and other developments will help to establish the long-term

Table 3.9 Standards for conducting materials

	BS	BS EN	IEC	Subject of Standard
Copper	–	12166		Copper and copper alloys. Wire for general purposes.
		13601	60356	Specifications for copper for electrical purposes.
	1434			Copper sections for commutators.
	4109	13602		Specifications for copper for electrical purposes. Wire for general electrical purposes and for insulated cables and flexible cords.
	7884			Specifications for copper and copper–cadmium stranded conductors for overhead electric traction and power transmission systems.
Aluminium	215 pts 1&2		60207	Aluminium conductors
	2627			Wrought aluminium for electrical purposes – wire
	2897			Wrought aluminium for electrical purposes – strip with drawn or rolled edges.
	2898			Wrought aluminium for electrical purposes – bars, tubes and sections.
	–	59183	60208	Aluminium alloy stranded conductors for overhead power transmission.
	3988		60121	Wrought aluminium for electrical purposes – solid conductors for insulated cables.
	6360		60228 60228A	Specifications for conductors in insulated cables and cords.
Semiconductor	4727pt 1 group 05		60050 (521)	Semiconductor terminology.
	6493	60747	60747	This has many parts with specifications for diodes, transistors, thyristors and integrated circuits.

feasibility of the HTS material. It is expected that there will be definite power saving from the use of HTS.

Small-scale applications which use HTS material include SQUIDS (Superconducting QUantum Interference DeviceS) which measure very low magnetic fields. They are applied in measurements in biomagnetism (investigations of electrical activity in the heart, brain and muscles) and in geophysics for the study of rock magnetism and anomalies in the earth's surface.

3.4.4 Standards

Each country has in the past had its own standards for materials. Over the past twenty years or so there has been a movement towards international standards which for electrical materials are produced by IEC. When an IEC standard is produced the member countries copy this standard and issue it under their own covers (Table 3.9).

Acknowledgement

My thanks are due to the staff of the University of Sunderland and of the City of Sunderland College for help with this chapter.

References

3A.   Laughton, M.A. and Warne, D.F. *Electrical Engineer's Reference Book* (16th edn), Newnes, 2003.

3B.   Brandes, E.A. and Brook, G.R. (eds), *Smithells Metals Reference Book* (7th edn), Butterworth-Heinemann, 1992.

3C.   McCaig, M. and Clegg, A.G. *Permanent Magnets in Theory and Practice* (2nd edn), Pentech and Wiley, 1987.

3D.   Stanbury, H. *The Development of International Standards for Magnetic Alloys and Steels,* UK Magnetics Society, 2004.

3E.   Reeves, E.A. *Newnes Electronics Engineer's Pocket Book,* BSI, 2004.

3F.   Narlikar, A.V. *High Temperature Superconductivity,* Springer Verlag, 2004.

3G.   Warnes, L.A.A. *Electronic and Electrical Engineering,* Macmillan, 1998 (good chapters on semiconductors).

Chapter 4

Measurements and instrumentation

E.A. Parr
Thamesteel

4.1 Introduction

4.1.1 Definition of terms

Measured variable and *process variable* arc both terms for the physical quantity that is to be measured. The *measured value* is the actual value which is measured or recorded, in engineering units (e.g. the level is 1252 mm).

A *primary element* or *sensor* is the device which converts the measured value into a form suitable for further conversion into an instrumentation signal. An orifice plate is a typical sensor. A *transducer* is a device which converts a signal from one quantity to another (e.g. a Pt100 temperature transducer converts a temperature to a resistance). A *transmitter* is a transducer which gives a standard instrumentation signal (e.g. 4–20 mA) as an output signal.

4.1.2 Range, accuracy and error

Measuring span, *measuring interval* and *range* are terms which describe the difference between the lower and upper limits that can be measured. *Turndown* is the ratio between the upper limit and the lower limits where the specified accuracy can be obtained.

Error is a measurement of the difference between the measured value and the true value and accuracy is the maximum error which can occur between the process variable and the measured value when the transducer is operating under specified conditions. Error can be expressed in many ways. The commonest are absolute value, as a percentage of the actual value, or as a percentage of full scale.

Many devices have an inherent coarseness in their measuring capabilities. A wire wound potentiometer, for example, can only change its resistance in small steps. *Resolution* is used to define the smallest step in which a reading can be made.

In many applications, the accuracy of a measurement is less important than its consistency. *Repeatability* is defined as the difference in readings obtained when the same measuring point is approached several times from the same direction.

Hysteresis occurs when the measured value depends on the direction of approach. Mechanical backlash and stiction are common causes of hysteresis.

The accuracy of a transducer will be adversely affected by environmental changes, particularly temperature cycling, and will degrade with time. Both of these effects will be seen as a *zero shift* or a change of sensitivity (known as a *span error*).

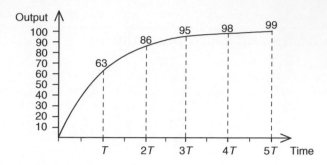

Fig. 4.1 Step response of first-order lag

4.1.3 Dynamic effects

A sensor cannot respond instantly to changes in the measured process variable. Commonly the sensor will change as a first-order lag with time constant T, as shown in Fig. 4.1.

For a step change in input, the output reaches 63 per cent of the final value in time T, and it follows that a significant delay may occur for a dynamically changing input signal.

A second order response occurs when the transducer is analogous to a mechanical spring/viscous damper. The step response depends on both the damping factor b and the natural frequency ω_n. The former determines the overshoot and the latter the speed of response. For values of $b < 1$ damped oscillations occur. The case where $b = 1$ is called *critical damping*. For $b > 1$, the system behaves as two first-order lags in series.

Intuitively, $b = 1$ is the ideal value but this may not always be true. If an overshoot to a step input signal can be tolerated a lower value of b will give a faster response and settling time within a specified error band. The signal enters the error band then overshoots to a peak which is just within the error band as shown in Fig. 4.2. Many instruments have a damping factor of 0.7 which gives the fastest response time to enter and stay within a 5 per cent settling band.

4.1.4 Signals and standards

The signals from most primary sensors are very small and in a form which is inconvenient for direct processing. Commercial transmitters are designed to give a standard output signal.

The most common electrical standard is the 4–20 mA loop, which uses a variable current with 4 mA representing one end of the signal range and 20 mA the other.

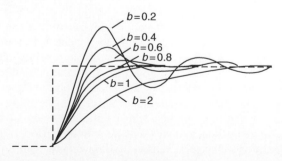

Fig. 4.2 Step response of second-order lag for various values of b

This use of a current gives excellent noise reduction and errors from different earth potentials around the plant are avoided. Because the signal is transmitted as a current, rather than a voltage, line resistance has no effect.

Several display or control devices can be connected in series provided the total loop resistance does not rise above some value specified for the transducer (typically 1 kΩ). Transducers using 4–20 mA can be current sourcing (with their own power supply) as shown in Fig. 4.3(a) or they may be designed for two-wire operation with the signal line carrying both the supply and the signal as shown in Fig. 4.3(b). The use of the offset zero of 4 mA allows many sensor or cable faults to be detected. A cable break or sensor fault will cause the loop current to fall to zero and these fault conditions can then be detected by the display or control device.

4.1.5 P&ID symbols

Instruments and controllers are usually part of a large control system. The system is generally represented by a *Piping & Instrumentation Drawing* (P&ID) which shows the devices, their locations and the method of interconnection. The basic symbols and a typical example are shown in Fig. 4.4.

The devices are represented by circles called balloons containing a unique tag which has two parts. The first part is two or more letters describing the function. The second part is a number which uniquely identifies the device, for example FE127 is flow sensor number 127. It is a good practice for devices to have their P&ID tag physically attached to them to aid maintenance.

4.2 Pressure

There are four types of pressure measurement. *Differential pressure* is the difference between two pressures applied to the transducer. *Gauge pressure* is made with respect to atmospheric pressure. It can be considered as a differential pressure measurement with the low pressure leg left open, and is usually denoted by the suffix 'g' (e.g. 37 psig). Most pressure transducers in hydraulic and pneumatic systems indicate gauge pressure. *Absolute pressure* is made with respect to a vacuum, so

$$\text{Absolute pressure} = \text{gauge pressure} + \text{atmospheric pressure} \qquad (4.1)$$

Atmospheric pressure is approximately 1 bar, 100 kPa or 14.7 psi.

Head pressure is used in liquid level measurement and refers to pressure in terms of the height of a liquid (e.g. inches water gauge). It is effectively a gauge pressure measurement, as in a vented vessel any changes in atmospheric pressure will affect both legs of the transducer equally giving a reading which is directly related to the liquid height. The head pressure is given by:

$$P = \rho g h \qquad (4.2)$$

where P is the pressure in pascals, ρ is the density (kg m^{-2}), g is the acceleration due to gravity (9.8 ms^{-2}) and h is the column height in metres.

4.2.1 The manometer

Manometers give a useful insight into the principle of pressure measurement. If a U-tube is part filled with liquid, and differing pressures are applied to both legs (Fig. 4.5), the liquid will fall on the high pressure side and rise on the low pressure side until the

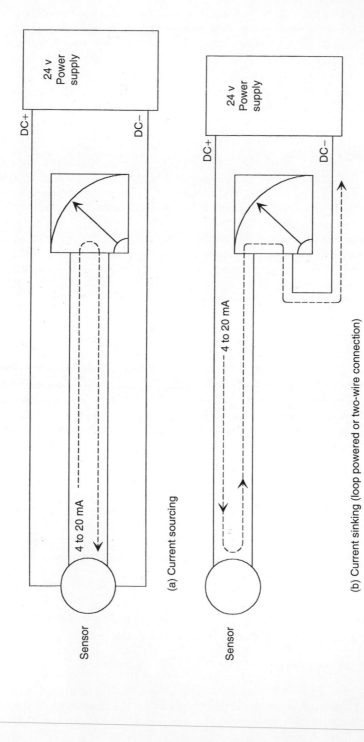

Fig. 4.3 Current sourcing and current sinking 4–20 mA loops

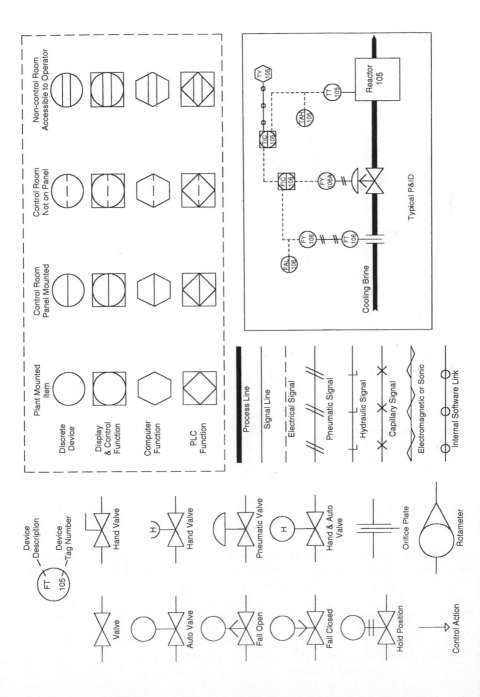

Fig. 4.4 Piping & Instrumentation Drawing (P&ID) symbols

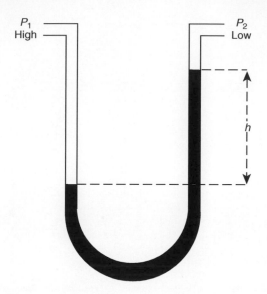

Fig. 4.5 U-tube manometer

head pressure of the liquid matches the pressure difference. If the two levels are separated by a height h then

$$h = (P_1 - P_2)/\rho g \tag{4.3}$$

where P_1 and P_2 are the pressures (in pascals) and ρ and g are as in eqn 4.2.

4.2.2   Elastic sensing elements

The Bourdon tube is the most common pressure indicating device. The tube is made by flattening a circular cross section tube to the section shown in Fig. 4.6 and bending it into a C-shape. One end is fixed and connected to the pressure to be measured. The other end is closed and left free. If pressure is applied to the tube it will tend to straighten, causing the free end to move up and to the right. This motion is converted to a circular motion for a pointer with a quadrant and pinion linkage. A Bourdon tube inherently measures gauge pressure.

Bourdon tubes are usable up to about 50 MPa. Where an electrical output is required, the tube can be coupled to a potentiometer or Linear Variable Differential Transformer (LVDT).

Diaphragms can also be used to convert a pressure differential into a mechanical displacement. Various arrangements are shown in Fig. 4.7. The displacement can be measured by LVDTs (section 4.5.3) or strain gauges (section 4.7.2).

4.2.3   Piezo elements

The *piezoelectric* effect occurs in quartz crystals. An electrical charge appears on the faces when a force is applied. The charge, q, is directly proportional to the applied force, which can be related to pressure with suitable diaphragms. The resulting charge is converted to a voltage by the circuit shown in Fig. 4.8 which is called a *charge amplifier*. The output voltage is given by:

$$V = -q/C \tag{4.4}$$

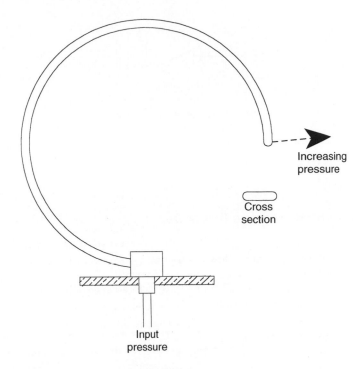

Fig. 4.6 Pressure measurement using a Bourdon tube

As q is proportional to the force, V is proportional to the force applied to the crystal. In practice the charge leaks away, even with FET amplifiers and low leakage capacitors and piezoelectric transducers are thus unsuitable for measuring static pressures, but they have a fast response and are ideal for measuring fast dynamic pressure changes.

A related effect is the piezoresistive effect which results in a change in resistance with applied force. Piezoresistive devices are connected into a Wheatstone bridge.

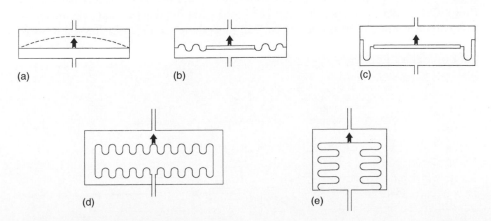

Fig. 4.7 Various types of diaphragm

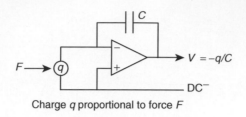

Charge q proportional to force F

Fig. 4.8 The charge amplifier

4.3 Flow

4.3.1 Differential pressure flowmeters

If a constriction is placed in a pipe (Fig. 4.9) the speed must be higher through the restriction to maintain equal mass flow at all points. The energy in a unit mass of fluid has three components which are kinetic energy, potential energy from the height of the fluid and flow energy caused by the fluid pressure. Flow energy is given by P/ρ where P is the pressure and ρ is the density.

In Fig. 4.9 the pipe is horizontal, so the potential energy is the same at all points. As the flow velocity increases through the restriction, the kinetic energy will increase and, for conservation of energy, the flow energy (i.e. the pressure) must fall. Hence eqn 4.5, which is the basis of all differential flowmeters.

$$V_1^2 + P_1/\rho = V_2^2 + P_2/\rho \tag{4.5}$$

Calculation of the actual pressure drop is complex, especially for compressible gases, but is generally of the form:

$$Q = \sqrt[K]{\Delta P} \tag{4.6}$$

where K is a constant for the restriction and ΔP is the differential pressure.

The most common differential pressure flowmeter is the orifice plate shown in Fig. 4.10. This is inserted into the pipe with upstream tapping point at D and downstream tapping point at $D/2$ where D is the pipe diameter. The plate should be drilled with a small hole to release bubbles (for liquid flow) or drain condensate (for gas flow). Orifice plates suffer from a loss of pressure on the downstream side; this *head loss* can be as high as 50 per cent. The Dall tube, shown in Fig. 4.11 has a lower loss of around 5 per cent but is bulky and more expensive. Another low-loss device is the pitot tube (Fig. 4.12).

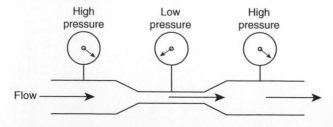

Fig. 4.9 The relationship between flow and pressure

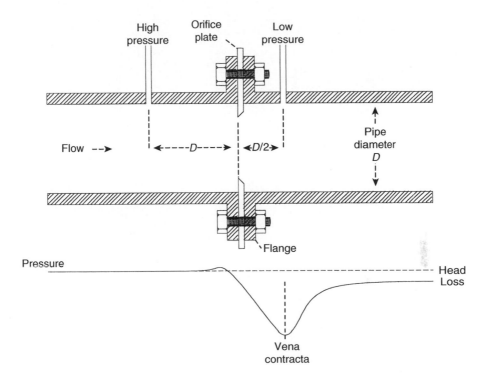

Fig. 4.10 The orifice plate

The transmitter should be mounted with a manifold block as shown in Fig. 4.13. Valves B and C are isolation valves. Valve A is an equalizing valve. In normal operation, A is closed and B and C are open. Valves D and E are vent valves used to relieve locked-in pressure prior to the removal of the transmitter.

Conversion of the pressure to an electrical signal requires a differential pressure transmitter and a linearizing square root unit. This square root extraction imposes a

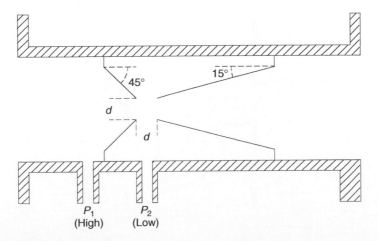

Fig. 4.11 The Dall tube

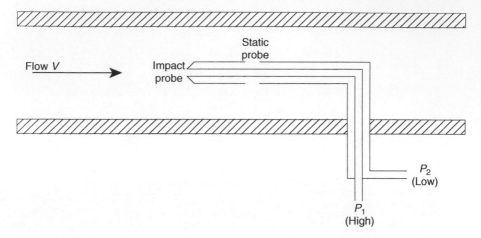

Fig. 4.12 The Pitot tube

limit on the turndown as zeroing errors are magnified. Although the accuracy and turndown of differential flowmeters is poor (typically 4 per cent and 4:1) they remain the most common type because of their robustness, low cost and ease of installation.

4.3.2 Turbine flowmeters

The arrangement of a turbine flowmeter is shown in Fig. 4.14. Within a specified flow range, (usually with about a 10:1 turndown for liquids, 20:1 for gases,) the rotational speed is directly proportional to the flow velocity. The blades are of ferromagnetic material and pass below a variable reluctance transducer producing an output approximating to a sine wave of the form shown in eqn 4.7.

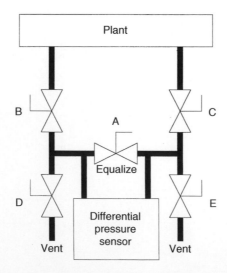

Fig. 4.13 Piping and hand-operated maintenance valves for a differential pressure transducer

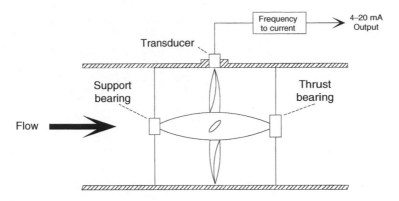

Fig. 4.14 The turbine flowmeter

$$E = A \,\omega \sin(N\,\omega\,t) \tag{4.7}$$

where A is a constant, ω is the angular velocity and N is the number of blades. Both the output amplitude and the frequency are proportional to the flow, although the frequency is normally used.

Turbine flowmeters are relatively expensive. They are less robust than other flowmeters and are particularly vulnerable to damage from suspended solids. Their main advantages are a linear output and a good turndown ratio, and the pulse output can also be used directly for flow totalization.

4.3.3 Vortex shedding flowmeters

If a bluff body is placed in a flow, vortices detach themselves at regular intervals from the alternate downstream edges as shown in Fig. 4.15. At Reynolds numbers in excess of 1000, the volumetric flow rate is directly proportional to the observed frequency of vortex shedding.

The vortices manifest themselves as sinusoidal pressure changes which can be detected by a sensitive diaphragm on the bluff body or by a downstream modulated ultrasonic beam. The vortex shedding flowmeter can work at low Reynolds numbers, has an excellent turndown (typically 15:1), no moving parts and minimal head loss.

Fig. 4.15 Vortex shedding flowmeter

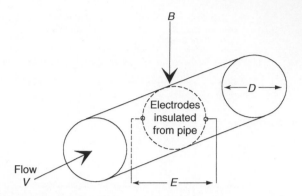

Fig. 4.16 Electromagnetic flowmeter

4.3.4   Electromagnetic flowmeters

In section 2.2.4 (eqn 2.20) it is seen that when a conductor of length l moves with velocity v perpendicular to a magnetic field of flux density B, the induced voltage is given by Bvl. This principle is used in the electromagnetic flowmeter. In Fig. 4.16 a conductive fluid passes down a pipe with mean velocity v through an insulated pipe section. A magnetic field B is applied perpendicular to the flow. Two electrodes are placed into the pipe sides to form, via the fluid, a moving conductor of length D relative to the field where D is the pipe diameter. A voltage is generated across the electrodes which is proportional to the mean flow velocity across the pipe.

In practice an AC field is used to minimize electrolysis and to reduce errors from DC thermoelectric and electrochemical voltages which are of the same order of magnitude as the induced voltage.

Electromagnetic flowmeters are linear and have an excellent turndown of about 15:1. There is no practical size limit and no head loss. An insulated pipe section is required, with earth bonding either side of the metre to avoid damage from any welding which may occur in normal service. They can only be used on fluids with a conductivity in excess of $1\,\text{mS m}^{-1}$, which permits use with many common liquids but excludes their use with gases.

4.3.5   Ultrasonic flowmeters

The Doppler effect occurs when there is relative motion between a sound transmitter and a receiver as shown in Fig. 4.17(a). If the transmitted frequency is f_t, V_s is the speed of sound and v is the relative speed, the observed received frequency, f_r is given by eqn 4.8.

$$f_r = f_t (v + V_s)/V_s \qquad (4.8)$$

A Doppler flowmeter injects an ultrasonic sound wave (typically a few hundred kilohertz) at an angle θ into a fluid moving in a pipe as shown in Fig. 4.17(b). A small part of this beam is reflected back from small bubbles and solid matter and is picked up by a receiver mounted alongside the transmitter. The frequency is subject to changes as it moves upstream against the flow and as it moves back with the flow. The received frequency is thus dependent on the flow velocity and the injection angle. As v is much smaller than V_s, the frequency shift is directly proportional to the flow velocity.

The Doppler flowmeter measures mean flow velocity, is linear, and can be installed or removed without the need to break into the pipe. The turndown of about 100:1 is

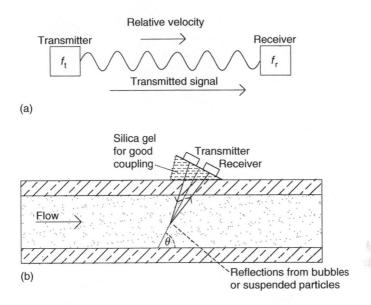

(a)

(b)

Silica gel for good coupling

Transmitter

Receiver

Flow

Reflections from bubbles or suspended particles

Fig. 4.17 Ultrasonic flowmeter

the best of all flowmeters. Assuming the measurement of mean flow velocity is acceptable it can be used at all Reynolds numbers.

4.3.6 Mass flowmeters

Many modern mass flowmeters are based on the Coriolis effect. In Fig. 4.18 an object of mass m is required to move with linear velocity v from point A to point B on a surface which is rotating with angular velocity ω. If the object moves in a straight line as viewed by a static observer, it will appear to veer to the right when viewed by an

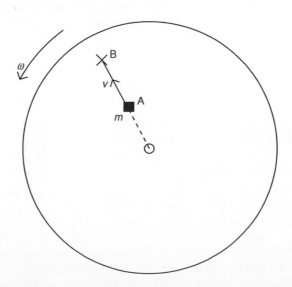

Fig. 4.18 The Coriolis force

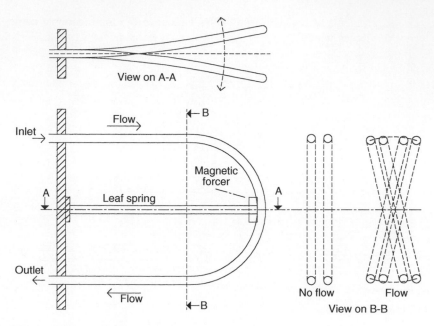

Fig. 4.19 Coriolis force mass flowmeter

observer on the disc. If the object is to move in a straight line as seen by an observer on the disc a force must be applied to the object as it moves out. This force is known as the *Coriolis force* and is given by

$$F = 2m\omega v \qquad (4.9)$$

where m is the mass, ω is the angular velocity and v is the linear velocity.

Coriolis force also occurs for sinusoidal motion. This effect is used as the basis of a Coriolis flowmeter shown in Fig. 4.19. The flow passes through a 'C' shaped tube which is attached to a leaf spring and vibrated sinusoidally by a magnetic forcer. The Coriolis force arises from the angular motion induced into the two horizontal pipe sections with respect to the fixed base. Any flow in the top pipe is in the opposite direction to the flow in the bottom pipe and the Coriolis force causes a rolling motion to be induced as shown. The resultant angular deflection is proportional to the mass flow rate.

4.4 Temperature

4.4.1 The thermocouple

The Seebeck effect and its use in thermocouples has been outlined in section 2.4. The voltage, typically a few millivolts, is a function of the temperature difference between the measuring and the meter junction.

In practice, the meter will be remote from the measuring point. If normal copper cables were used to link the meter and the thermocouple, the temperature of the joints would not be known and further voltages, and hence errors, would be introduced. The thermocouple cables must therefore be taken back to the meter. Two forms of cable are used; *extension cables* which are essentially identical to the thermocouple cable or *compensating cables* which are cheaper and match the thermocouple characteristics over a limited temperature range.

Because the indication is a function of the temperature at both ends of the cable, correction must be made for the local meter temperature. A common method, called *cold junction compensation*, measures the local temperature by some other method such as a resistance thermometer and applies a correction.

Thermocouples are non-linear devices and linearizing circuits must be provided if readings are to be taken over an extended range.

Although a thermocouple can be made from any dissimilar metals, common combinations have evolved with well-documented characteristics. These are identified by single letter codes. Table 4.1 gives the details of common thermocouple types.

4.4.2   The resistance thermometer

The variation of resistance with temperature in metals has been described in section 3.4.1.1. This change in resistance can be used to measure temperature. Platinum is widely used because the relationship between resistance and temperature is fairly linear. A standard device is made from a coil of wire with a resistance of 100 Ω at 0°C, giving rise to the common name of a Pt100 sensor. These can be used over the temperature range –200°C to 800°C. At 100°C a Pt100 sensor has a resistance of 138.5 Ω, and the 38.5 Ω change from its resistance at 0°C is called the *fundamental interval*. The current through the sensor must be limited in order to avoid heating effects. Errors from the cabling resistance can be overcome by the use of the three-wire connection of Fig. 4.20.

A *thermistor* is a more sensitive device comprising a semiconductor crystal whose resistance changes dramatically with temperature. Devices are obtainable which decrease or increase resistance for increasing temperature, the former being more common. The relationship is very non-linear, a typical device giving 300 kΩ at 0°C and 5 kΩ at 100°C.

Although very non-linear, thermistors can be used for measurement over a limited range, but their high sensitivity and low cost make them specially useful where a signal is required if the temperature rises above or falls below a preset value. Electrical machines often have thermistors embedded in the windings to give warning of overheating.

4.4.3   The pyrometer

The basis of the pyrometer is shown in Fig. 4.21. The object whose surface temperature is to be measured is viewed through a fixed aperture and part of the radiation emitted by the object falls on the temperature sensor causing its temperature to rise. The temperature of the object is then inferred from the rise in temperature seen by the sensor.

The sensor size must be very small, typically of the order of 1 mm diameter. Often a circular ring of thermocouples connected in series (a *thermopile*) is used. Alternatively a small resistance thermometer (a *bolometer*) may be used. Some pyrometers measure the radiation directly using photoelectric detectors.

The temperature measurement is independent of distance from the object, provided the field of view is full. Figure 4.22 shows that the source is radiating energy uniformly in all directions, so the energy received from a point is proportional to the solid angle subtended by the sensor. This decreases as the square of the distance, but as the sensor moves away the scanned area also increases as the square of the distance. These two effects cancel to give a reading which is independent of the distance.

Table 4.1 Common thermocouple types

Type	Positive material	Negative material	µV/°C (i)	Usable range	Comments
E	Chromel 90% Nickel, 10% Chromium	Constantan 57% Copper, 43% Nickel	68.00	0–800°C	Highest output thermocouple
T	Copper	Constantan	46.00	−185 to 300°C	Used for cryogenics and mildly oxidizing or reducing atmospheres. Often used for boiler flues
K	Chromel	Alumel (ii)	42.00	0 to 1100°C	General purpose. Widely used
J	Iron	Constantan	46.00	20 to 700°C	Used with reducing atmospheres. Tends to rust
R	Platinum with 13% Rhodium	Platinum	8.00	0 to 1600°C	High temperatures (e.g. iron foundries and steel making). Used in UK in preference to type S
S	Platinum with 10% Rhodium	Platinum	8.00	0 to 1600°C	As type R. Used outside UK
V	Copper	Copper/Nickel	–		Compensating cable for type K to 80°C
U	Copper	Copper/Nickel	–		Compensating cable for types R and S to 50°C

Notes: (i) µV/°C is typical over range.
(ii) Alumel is an alloy comprising 94% Nickel, 3% Manganese, 2% Aluminium and 1% Silicon.

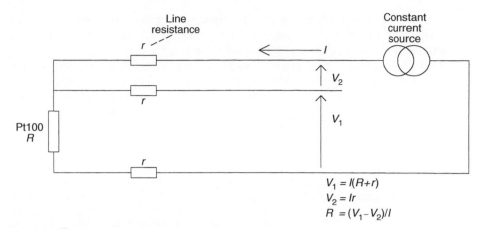

$$V_1 = I(R+r)$$
$$V_2 = Ir$$
$$R = (V_1 - V_2)/I$$

Fig. 4.20 Three-wire Pt100 temperature transducer

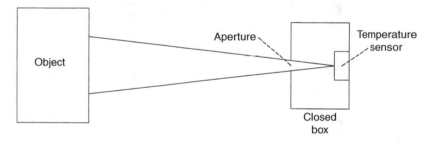

Fig. 4.21 The principle of the pyrometer

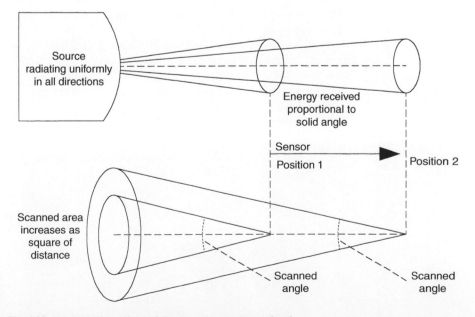

Fig. 4.22 The cancelling effects of distance on a pyrometer signal

There are various ways in which the temperature can be deduced from the received radiation. The simplest measures the total energy received from the object, which is proportional to T^4 where T is the temperature in kelvin.

A lack of knowledge of the emissivity of the surface can introduce error, but this maybe reduced by filters which restrict the measuring range to frequencies where the emissivity approaches unity. An alternative method is to take measurements at two different frequencies (that is, two different colours) and compare the relative intensities to give an indication of temperature.

4.5 Position

4.5.1 The potentiometer

The potentiometer can directly measure angular or linear displacements and is the simplest form of position transducer.

Potentiometers have finite resolution which is determined by the wire size or grain size of the track. Since they are mechanical devices, they can also suffer from stiction, backlash and hysteresis. Their failure mode also needs consideration; a track break can cause the output signal to be fully high above the failure point and fully low below it. In closed-loop position control systems this manifests itself as a high-speed dither around the break point.

4.5.2 The synchro and resolver

Figure 4.23 shows a transformer whose secondary can be rotated with respect to the primary. At angle θ the output voltage will be given by eqn 4.10,

$$V_o = K V_i \cos \theta \tag{4.10}$$

where K is a constant. The output amplitude is dependent on the angle, and the signal can be in phase (from $\theta = 270°$ to $90°$) or anti-phase (from $\theta = 90°$ to $270°$). This principle is the basis of synchros and resolvers.

A synchro link consists of a transmitter and a control transformer connected as shown in Fig. 4.24. Although this appears to be a three-phase circuit, it is fed from a single-phase supply, often at 400 Hz. The voltage applied to the transmitter induces in-phase or anti-phase voltages in the windings as described earlier and causes currents

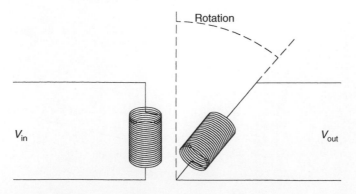

Fig. 4.23 The principle of servos and resolvers

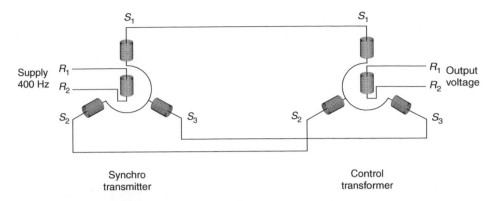

Fig. 4.24 Synchro transmitter and a control transformer

to flow through the stator windings at the control transformer. An ac voltage will be induced in the rotor of the control transformer, if it is not perpendicular to the field. The magnitude of this voltage is related to the angle error and the phase to the direction. The ac output signal must be converted to dc by a phase sensitive rectifier. The *Cowan rectifier* of Fig. 4.25(a) is commonly used. Another common circuit is the positive/negative amplifier of Fig. 4.25(b), the electronic polarity switch being driven by the reference supply.

A resolver has two stator coils at right angles and a rotor coil as shown in Fig. 4.26. The voltages induced in the two stator coils are simply

$$V_1 = KV_i \cos \theta \qquad (4.11)$$

$$\text{and } V_2 = KV_i \sin \theta \qquad (4.12)$$

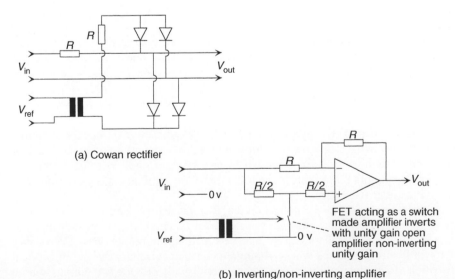

(a) Cowan rectifier

FET acting as a switch
made amplifier inverts
with unity gain open
amplifier non-inverting
unity gain

(b) Inverting/non-inverting amplifier

Fig. 4.25 Phase sensitive rectifiers

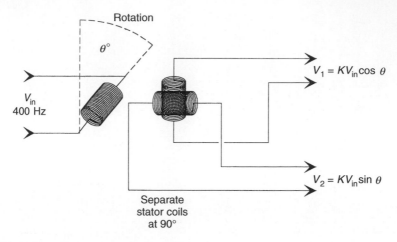

Fig. 4.26 Angular measurement with a resolver

4.5.3 The linear variable differential transformer (LVDT)

An LVDT is used to measure small displacements, typically less than a few millimetres. It consists of a transformer with two secondary windings and a movable core as shown in Fig. 4.27. At the centre position, the voltages induced in the secondary windings are equal but of opposite phase giving zero output signal. As the core moves away from the centre one induced voltage becomes larger, giving a signal whose amplitude is proportional to the displacement and whose phase shows the direction. A dc output signal can be obtained with a phase sensitive rectifier.

4.5.4 The shaft encoder

Shaft encoders give a digital representation of an angular position, and exist in two forms.

An *absolute encoder* gives a parallel output signal, typically 12 bits (one part in 4096) per revolution. Binary Coded Decimal (BCD) outputs are also available. A simplified four-bit encoder would therefore operate as shown in Fig. 4.28(a). Most absolute encoders use a coded wheel similar to Fig. 4.28(b) moving in front of a set of photocells. A simple binary coded shaft encoder can give anomalous readings as the outputs change state. There are two solutions to this problem. The first uses an additional track, called an anti-ambiguity track, which is used by the encoder's internal logic to inhibit changes around transition points. The second solution uses a *unit distance code*, such as the *Gray code* which only changes one bit for each step and hence has no ambiguity. The conversion logic from Gray to binary is usually contained within the encoder itself.

Incremental encoders give a pulsed output, each pulse representing a fixed angular distance. These pulses are counted by an external counter to give an indication of the position. Simple encoders can be made by reading the output from a proximity detector in front of a toothed wheel. A single-pulse output train carries no directional information and commercial incremental encoders usually provide two outputs shifted 90° in phase as shown in Fig. 4.29(a). For clockwise rotation, A will lead B as shown in Fig. 4.29(b) and the positive edge of A will occur when B is low. For anti-clockwise

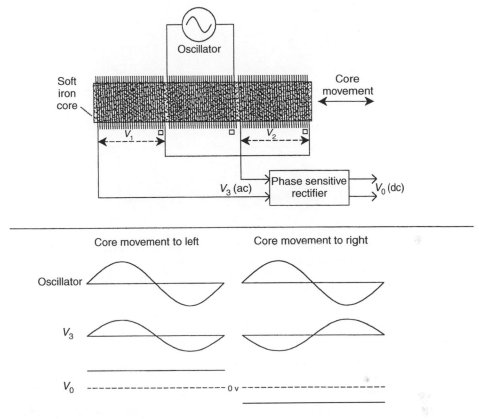

Fig. 4.27 Measurement of linear displacement with an LVDT

rotation B will lead A. The two-output incremental encoder can thus be used to follow reversals without cumulative error, but can still lose its datum position after a power failure.

4.5.5 The variable capacitance transducer

Very small deflections obtained in weighing systems or pressure transducers can be converted into an electrical signal by using a varying capacitance between two plates.

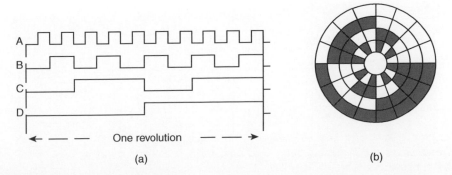

(a)   (b)

Fig. 4.28 An absolute position encoder

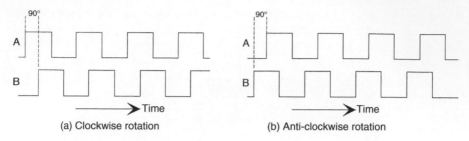

Fig. 4.29 Incremental encoder signals

The capacitance C of a parallel plate capacitor is given by eqn 4.13,

$$C = \varepsilon A d \tag{4.13}$$

where ε is the permittivity of the material between the plates, A the area and d the separation. A variation in capacitance can be achieved by sliding in a dielectric to vary ε, or by moving the plates horizontally to change A or by moving the plates apart to change d. Although the change is linear for ε and A, the effect is small.

4.5.6 Laser distance measurement

There are two broad classes of laser distance measurement.

In the *triangulation laser* shown in Fig. 4.30, a laser beam is used to produce a very bright spot on the target object. This is viewed by an imaging device which can be considered as a linear array of tiny photocells. The position of the spot on the image will vary according to the distance as shown, and the distance is found by simple triangulation trigonometry.

The second type of measurement is used at longer distances. For very long distances, in surveying for example, the time taken for a pulse to reach the target and return is measured. At shorter distances a continuous amplitude modulated laser beam

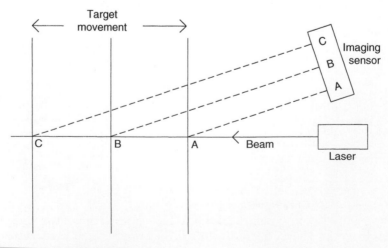

Fig. 4.30 Laser position measurement

is sent and the phase shift between transmitted and received signals is used to calculate the distance. The latter method can give ambiguous results, so the two techniques are often combined to give a coarse measurement from time of flight which is fine-tuned by the phase shift.

4.5.7 Proximity switches and photocells

Limit switches are often used to indicate positional state. In mechanical form these devices are bulky, expensive and prone to failure where the environment is hostile.

Proximity switches can be considered to be solid-state limit switches. The most common type is constructed around a coil whose inductance changes in the presence of a metal surface. A simple two-wire AC device acts like a switch, having low impedance and capable of passing several hundred milliamperes when covered by metal, and a high impedance with a typical leakage of 1 mA when uncovered. Sensing distances up to 20 mm are feasible, but 5 mm to 10 mm is more common. DC-powered switches use three wires, two for the supply and one for the output. DC sensors use an internal high frequency oscillator to detect the inductance change which gives a much faster response. Operation at several kilohertz is easily achieved.

Inductive detectors only work with metal targets. Capacitive proximity detectors work on the change in capacitance caused by the target and accordingly work with wood, plastic, paper and other common materials. Their one disadvantage is that they need adjustable sensitivity to cope with different applications.

Ultrasonic sensors can also be used, operating on the same principle as the level sensor described in section 4.6.3. Ultrasonic proximity detectors often have adjustable near and far limits so that a sensing window can be set.

Photocells (or PECs) are another possible solution. The presence of an object can be detected either by the breaking of a light beam between a light source and a PEC or by reflecting a light beam from the object which is detected by the PEC. In the second approach (called a *retro-reflective PEC*) light is seen for an object present. It is usual to use a modulated light source to minimize the effects of change in ambient light and to prevent interaction between adjacent PECs.

4.6 Level

4.6.1 The pressure-based system

The absolute pressure at the bottom of a tank has two components and is given in eqn 4.14,

$$P = \rho g h + \text{atmospheric pressure} \tag{4.14}$$

where ρ is the density, g is the gravitational acceleration and h, the depth of the liquid.

4.6.2 Electrical probe

Capacitive probes can be used to measure the depth of liquids and solids. A rod, usually coated with PVC or PTFE, is inserted into the tank as shown in Fig. 4.31 and the capacitance to the tank wall is measured. This capacitance has two components, C_1 above the surface and C_2 below the surface. As the level rises C_1 will decrease and C_2 will increase. The two capacitors are in parallel, but as liquids and solids have a

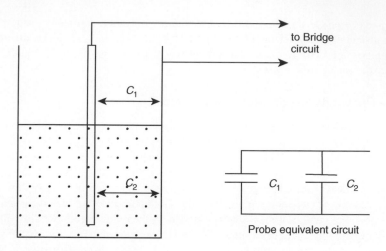

Fig. 4.31 Capacitive level measurement

higher dielectric constant than vapour, the total capacitance will rise for increasing level.

If the resistance of a liquid is reasonably constant, the level can be inferred by measuring the resistance between the two submerged metal probes. A bridge measuring circuit is again used with an ac supply to avoid electrolysis and plating effects. Corrosion can be a problem with many liquids and stainless steel is often used, but the technique works well with water. Cheap probes using this principle are available for stop–start level control applications.

4.6.3 Ultrasonic transducers

Ultrasonic methods use high frequency sound produced by the application of a suitable AC voltage to a piezoelectric crystal. Frequencies in the range 50 kHz to 1 MHz can be used, the lower frequencies being more common in industry. The principle is shown in Fig. 4.32. An ultrasonic pulse is emitted by a transmitter, is reflected from a surface and detected by a receiver. The time of flight is given by eqn 4.15,

$$T = 2d/v \qquad (4.15)$$

where v is the velocity of sound in the medium above the surface. The velocity of sound in air is about 3000 ms^{-1}, so for a tank whose depth d can vary from 1 m to 10 m the delay will vary from about 7 ms (full) to 70 ms (empty).

Two methods are used to measure the delay. The simplest and most commonly used in industry is a narrow pulse. The receiver will see several pulses, one almost immediately through the air, the required surface reflection and spurious reflections from sides, the bottom and 'rogue' objects above the surface such as steps and platforms. The measuring electronics normally provides adjustable filters and upper and lower limits to reject unwanted readings.

Pulse-driven systems lose accuracy when the time of flight is small. For distance below a few millimetres a swept frequency is used, and a peak in the response will be observed when the path difference is a multiple of the wavelength,

$$d = v/2f \qquad (4.16)$$

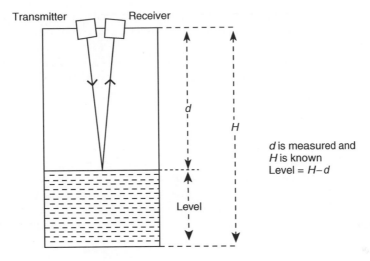

Fig. 4.32 Ultrasonic level measurement

where v is the velocity of propagation and f is the frequency at which the peak occurs. This can be ambiguous, since peaks will also be observed at integer multiples of the wavelength.

4.6.4 Nucleonic methods

Radioactive isotopes spontaneously emit gamma or beta radiation. As this radiation passes through material it is attenuated and this allows a level measurement system to be constructed in one of the ways shown in Fig. 4.33. In each case the intensity of the received radiation is dependent on the level, being maximum when the level is low. This is particularly attractive for aggressive materials or extreme temperatures and pressures since the measuring equipment can be placed outside the vessel.

The biological effects of radiation are complex and there is no definitive safe level; all exposure must be considered harmful and legislation is based around the concept of '*as low as reasonably practical*' (ALARP).

Nucleonic level detection also requires detectors. Two types are commonly used, the *Geiger–Muller* tube and the *scintillation counter*. Both produce a semi-random pulse stream, the number of pulse received in a given time being dependent on the

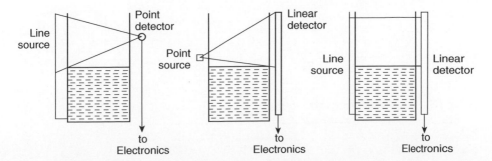

Fig. 4.33 Various methods of level measurement using radioactive sources

strength of the radiation. This pulse chain must be converted to a dc voltage by suitable filtering to give a signal which is dependent on the level of material between the source and the detector.

Conceptually nucleonic level measurement is simple and reliable, but public mistrust and the complex legislation can make its application troublesome.

4.7   Weighing and force

There are two basic techniques in use. In the *force balance system*, the weight of the object is opposed by some known force which will equal the weight. Industrial systems use hydraulic or pneumatic pressure to balance the load, the pressure required being directly proportional to the weight. The more common method is *strain weighing*, which uses the gravitational force from the load to cause a measurable change in the structure. Its simplest form is the spring balance where the deflection is proportional to the load.

4.7.1   Stress and strain

The strain shown by a material increases with stress as shown in Fig. 4.34. Over region AB the material is elastic; the relationship is linear and the material returns to its original dimension when the stress is removed. Beyond point B, plastic deformation occurs and the change is not reversible; the relationship is now non-linear, and with increasing stress the object fractures at point C. Region AB is used for strain measurement and point B is the *elastic limit*. The inverse slope of the line AB is the *modulus of elasticity*, more commonly known as *Young's modulus,* which is defined in eqn 4.17.

$$\text{Young's modulus} = \frac{\text{Stress}}{\text{Strain}} = \frac{F/A}{\Delta L/L} \qquad (4.17)$$

When an object experiences strain, it displays a change in length and a change in cross section. This is defined by *Poisson's ratio, v.* If an object has a length L and width W

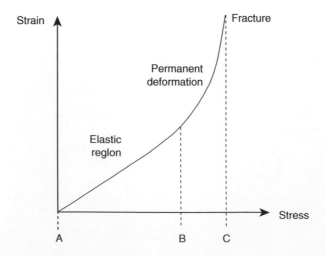

Fig. 4.34 Relationship between stress and strain

in its unstrained state and experiences changes ΔL and ΔW when strained, Poisson's ratio is defined in eqn 4.18.

$$v = \frac{\Delta W / W}{\Delta L / L} \qquad (4.18)$$

Typically v is between 0.2 and 0.4 and this can be used to calculate the change in the cross-sectional area.

4.7.2 The strain gauge

Devices based on the change of resistance with load are called *strain gauges*.

The electrical resistance of a conductor is given by $\rho L/A$ as explained in section 2.2.2 and eqn 2.8. When a conductor suffers stress, the change in its length and area results in a change in resistance. Under tensile stress, the length increases and the cross-sectional area decreases, both contributing to increased resistance. Ignoring the second-order effects, the change in resistance is given by eqn 4.19,

$$\Delta R / R = G\,\Delta L / L \qquad (4.19)$$

where G is a constant, the *gauge factor*. Here, $\Delta L/L$ is the strain E, giving

$$\Delta R = GRE \qquad (4.20)$$

Practical strain gauges consist of a thin small as shown in Fig. 4.35, with a pattern to increase the conductor length and hence the gauge factor. The gauge is attached to the stressed member with epoxy resin and experiences the same strain as the member. Modern gauges are photo-etched from metallized film deposited onto a polyester or plastic backing. A typical strain gauge will have a gauge factor of 2, a resistance of 120 Ω and will operate up to 1000 μ strain, which will result in a resistance change of 0.24 Ω.

Strain gauges must reject strains in unwanted directions. A gauge has two axes, an active axis along which the strain is applied and a passive axis (usually at 90°) along which the gauge is least sensitive. The relationship between these is defined by the *cross-sensitivity*, where

$$\text{cross-sensitivity} = \frac{\text{sensitivity on passive axis}}{\text{sensitivity on active axis}} \qquad (4.21)$$

Cross-sensitivity is typically about 0.002.

4.7.3 Bridge circuits

In Fig. 4.36 the strain gauge R_g is connected into a classical Wheatstone bridge and the small change in resistance is superimposed on the large unstrained resistance.

Fig. 4.35 A typical strain gauge

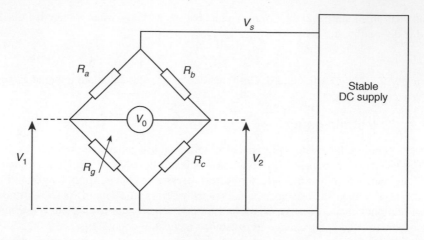

Fig. 4.36 Connection of a strain gauge into a Wheatstone bridge

The actual resistance of a strain gauge is of little value, but the change caused by the applied load is important. If R_b and R_c are made equal, and R_a is made equal to the unloaded resistance of the strain gauge, voltages V_1 and V_2 will both be half the supply voltage and V_0 will be zero. If a load is applied to the strain gauge such that its resistance changes by fractional change x (i.e. $x = \Delta R/R$) it can be shown that

$$V_0 = V_s x/2(2 + x) \tag{4.22}$$

In a normal circuit x will be very small when compared with 2. For the earlier example x has the value $0.24/120 = 0.002$, so eqn 4.22 can be simplified to:

$$V_0 = V_s \Delta R \times x/4R \tag{4.23}$$

But from eqn 4.20, ΔR is EGR, giving

$$V_0 = EGV_s/4 \tag{4.24}$$

For small values of x, the output voltage is thus linearly related to the strain.

For a 24 V supply, 1000 μ strain and a gauge factor of 2, 12 mV is obtained. This output voltage must be amplified before it can be used and care must be taken to avoid common mode noise, so the differential amplifier circuit of Fig. 4.37 is commonly used.

Resistance changes from temperature variation (section 4.4) are of a similar magnitude to resistance changes from strain. The simplest way of eliminating unwanted temperature effects is to use two gauges arranged as shown in Fig. 4.38(a). One gauge has its active axis and the other gauge the passive axis aligned with the load. If these are connected into a bridge circuit as shown, both the gauges will exhibit the same resistance change from temperature and these will cancel, leaving the output voltage purely dependent on the strain.

In the arrangement (Fig. 4.38(b)) four gauges are used, two of which experience compressive strain and two tensile strain. Here all the four gauges are active, giving

$$V_0 = E \times G \times V_s \tag{4.25}$$

Again temperature compensation occurs because all gauges are at the same temperature.

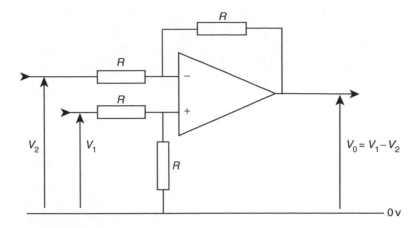

Fig. 4.37 A differential amplifier

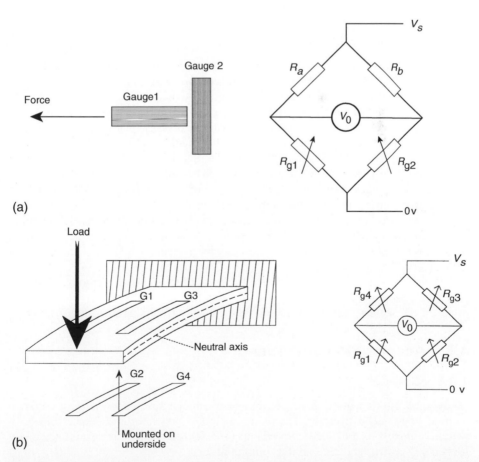

Fig. 4.38 (a) Two gauge bridge, used to give temperature compensation (b) A shear beam load cell using four strain gauges

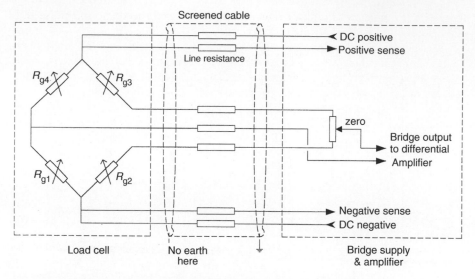

Fig. 4.39 Full cabling diagram for a load cell with compensation for cable resistance

4.7.4 The load cell

A load cell converts a force into a strain which can then be converted to an electrical signal by the strain gauges. A load cell will typically have four gauges, two in compression and two in tension. These will typically be connected as shown in Fig. 4.39. The supply at the bridge is measured to compensate for the cable resistance.

Mechanical coupling of the load requires care. A pressure plate applies the load to the members carrying the strain gauges through a knuckle and avoids error from slight misalignment. A flexible diaphragm provides a seal against dust and weather and a small gap ensures that shock and overloads will reach the load cell bottom without damage to the gauges. Maintenance must ensure that this gap is not closed, causing the strain gauge member to carry only part of the load.

Multi-cell weighing systems can be used, with the readings from each cell being summed electronically. Three-cell systems inherently spread the load across all the cells. In four-cell systems the support structure must ensure that all cells are in contact with the load at all times.

Usually the load cells are the only route to the ground from a weighing platform, so it is advisable to provide a flexible earth strap for electrical safety and to provide a route for any welding current which might arise from repairs or later modification.

4.8 Acceleration and speed

4.8.1 Speed

Many applications require the measurement of angular speed and the most common method is the tachogenerator. This is a simple DC generator, the output voltage of which is proportional to speed. Speeds up to 10 000 rpm can be measured, the upper limit being centrifugal force on the commutator.

The pulse tachometer is also becoming common. This is the same in form as the incremental encoder described in section 4.5.4. It produces a constant amplitude pulse train whose frequency is directly related to the rotational speed and this pulse train can be converted into a voltage in three ways.

The first, basically analogue, method fires a fixed width monostable which gives a mark-space ratio which is speed dependent and the monostable output is taken through a low-pass filter to give an output proportional to the speed. The maximum achievable speed is determined by the monostable pulse width.

The second, digital, method counts the number of pulses in a given time. This effectively averages the speed over the count period, which is chosen to give a reasonable balance between resolution and speed of response. Counting over a fixed time is not suitable for slow speeds as adequate resolution can only be obtained with a long duration sample time. At slow speeds, therefore, the period of the pulses is often directly timed giving an average speed per pulse, the time being inversely proportional to speed.

The third, digital, speed control systems, often use both of the last two methods and switch between them according to the speed.

Linear velocity can be measured using *Doppler shift*. This operates in a similar way to the ultrasonic flowmeter described in section 4.3.5 and allows remote measurement of velocity. The transmitted and received frequencies are mixed to give a beat frequency equal to the frequency shift and hence proportional to the speed of the target object. The beat frequency can be measured by counting the number of cycles in a given time.

4.8.2 Accelerometers and vibration transducers

Accelerometers and vibration transducers consist of a seismic mass linked to a spring as shown in Fig. 4.40(a). The movement is damped by the dashpot giving a second-order response defined by the natural frequency and the damping factor, typically chosen to be 0.7.

If the frame of the transducer is moved sinusoidally, the relationship between the mass displacement and the frequency will be as shown in Fig. 4.40(b). In region A, the device acts as an accelerometer and the mass displacement is proportional to the acceleration.

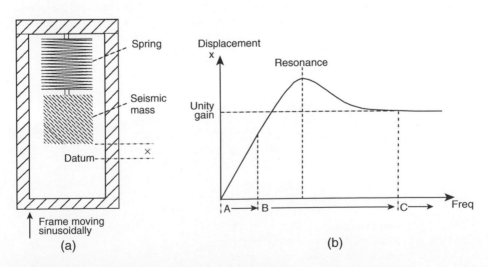

Fig. 4.40 Principle of vibration transducers and accelerometers

In region C, the mass does not move, and the displacement with respect to the frame is solely dependent on the frame movement, but is shifted by 180°. In this region the device acts as a vibration transducer. Typically, region A extends to about one-third of the natural frequency and region C starts at about three times the natural frequency. In both cases the displacement is converted to an electrical signal by LVDTs or strain gauges.

4.9 Current

The most common method of measuring current is the current transformer (CT) shown in Fig. 4.41. An ac current passes through the primary of a transformer and the secondary, which has a large number of turns, is short-circuited. Often there is just one primary turn and the primary current is then reduced by the number of secondary turns. For example a 200:5 A current transformer would have 40 turns. A CT must always have an impedance connected across the secondary terminals since otherwise a very high voltage (in the ratio N_s to N_p) will occur.

The current output is usually passed through a low-value resistor and the resultant ac voltage is amplified and rectified to give a dc voltage proportional to the current. The rectification and filtering required can introduce a small delay of a few milliseconds which may be undesirable, particularly in dc drives where a fast response is needed.

The simplest dc current measurement method is a shunt. The current flow through the shunt will produce a small potential, which is of a few millivolts, which can be amplified by an isolated dc amplifier to give a usable current signal with good dynamic response. If a shunt cannot be mounted, the DC-CT can be used. A single cable (or bus bar) with the dc current to be measured passes through a toroidal transformer core as in Fig. 4.42. The transformer primary is energized from a fixed high frequency oscillator, and induces voltages in the secondary winding which can be rectified. Increasing dc current will induce a dc flux into the core which reduces the coupling between the primary and secondary. Although rectification and filtering is still required, the higher frequency of the oscillator significantly reduces the delay and improves the response.

A further method uses *Hall effect* transducers. The Hall effect occurs in some materials when a magnetic field is applied to a conductor carrying a dc current. This causes a voltage to appear across the faces of the material at 90° to both the field and the current as shown in Fig. 4.43. In a Hall effect current sensor, a constant current is

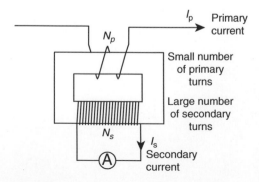

Fig. 4.41 The current transformer

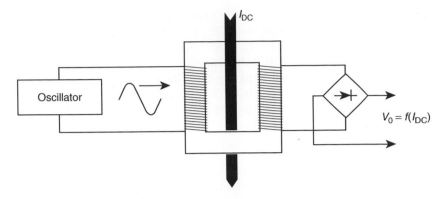

Fig. 4.42 The dc current transformer

passed through the Hall effect material and the magnetic field is derived from the current to be measured. The direction of the current is provided directly by the shunt and the Hall effect transducer. In both cases, the voltage changes polarity if the current reverses direction. With the CT and the DC-CT the direction has to be inferred by other means.

4.10   HART and Fieldbus

A Fieldbus is a way of interconnecting several devices (e.g. transducers, controllers, actuators etc.) via a simple and cheap serial cable. Devices on the network are identified by addresses which allow messages to be passed between them.

Of particular interest in instrumentation is the *HART* (Highway Addressable Remote Transducer) protocol. HART was originally introduced by Rosemount but is

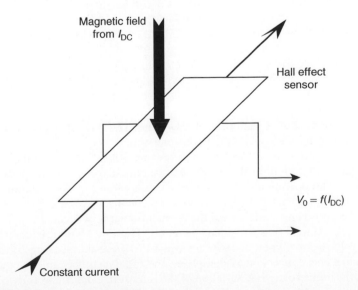

Fig. 4.43 Hall effect current measurement

now an open protocol. It is a very simple master–slave system; devices only speak when requested and the operation is always master requests, slave replies. Up to fifteen slaves can be connected to each master.

HART can work in two ways. In its simplest, and probably the most common form of point-to-point, it superimposes the serial communication data on to a standard 4–20 mA loop signal. *Frequency shift keying* is used with frequencies of 1200 Hz for a '1' and 2200 Hz for a '0'. These frequencies are far too high to affect the analogue instrumentation so the analogue signal is still used in the normal manner. The system operates with one master (usually a computer, Programmable Logic Controller (PLC) or hand-held programming terminal) and one slave (a transducer or an actuator). This approach allows HART to be retro-fitted onto the existing cabling and instrumentation schemes, the serial data then allows much more information to be conveyed. HART devices can be remotely configured and monitored allowing very simple diagnostics and quick replacement after a failure. In addition much more plant data can be passed from the transducer. With a HART programming terminal connected to the line, the transducer can also be made to send fixed currents to aid loop checks and fault diagnosis. The second method of using HART is in a normal *Fieldbus multidrop system*. Here each device has an identifying address and it sends data to the master on request. Usually devices (and their relevant parameters) are polled on a regular cyclic scan. The multidrop brings all the cost savings from simple cabling and makes the system easy to modify and expand.

An advantage of a HART system is that the communicator can be connected in any of the ways shown in Fig. 4.44. A resistor of at least $250\,\Omega$ must be connected somewhere in the loop, and the HART communicator connects across the lines. If only one line is accessible this can be opened and a temporary $250\,\Omega$ resistor fitted as shown, but suitable precautions against the opening of the loop must be taken.

4.11   Data acquisition

Transducers often use digital signal processing. This makes them less prone to drift and simplifies interfacing to computers and *Programmable Logic Controllers* (PLCs).

A binary number can represent an analogue signal. An 8 bit number, for example, can represent a decimal number from 0 to 255 (or -128 to $+127$ if bipolar operation is used). A 12-bit number similarly has a range of 0 to 4095.

A device which converts a digital number to an analogue voltage is called a *Digital to Analogue Converter*, or DAC. This is also the heart of the *Analogue to Digital Converter*, or ADC, which performs the conversion of an analogue signal to a digital number.

Two common DAC circuits are shown in Fig. 4.45. In each case the output voltage is determined by the binary pattern on the switches. The R–2R ladder circuit is particularly well-suited to integrated circuit construction.

The two most common ADC circuits are the *Ramp ADC* and the *Successive Approximation ADC*. Both of these compare the output voltage from a DAC with the input voltage. The operation of the Ramp ADC, shown in Fig. 4.46, commences with a start command which sets FF1 and resets the counter to zero. FF1 gates pulses to the counter which then counts up and the counter output is connected to a DAC whose output voltage ramps up as the count increases. The DAC output is compared with the input voltage, and when the two are equal FF1 is reset, blocking further pulses and indicating the conversion is complete. The binary number in the counter now represents the input voltage. A variation of the Ramp ADC known as a *Tracking ADC* uses an up–down counter that continuously follows the input voltage. The Ramp ADC

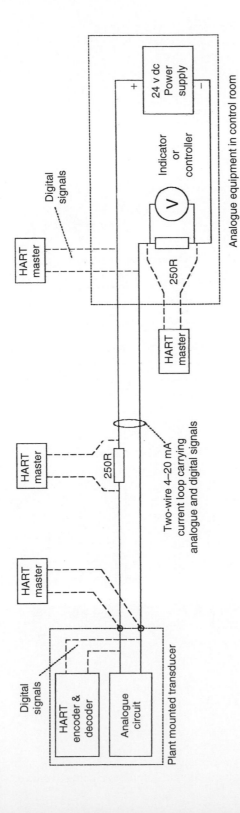

Fig. 4.44 HART and the 4–20 mA loop

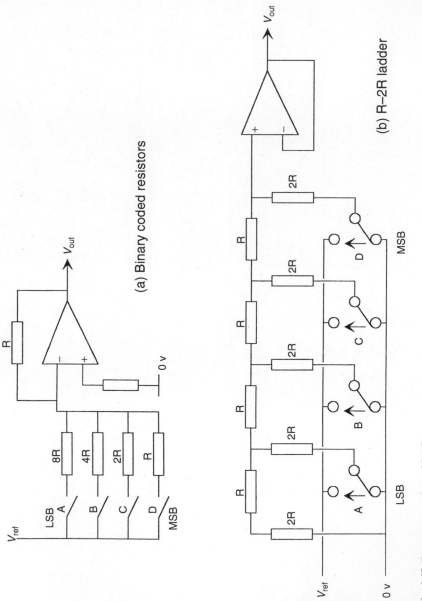

(a) Binary coded resistors

(b) R–2R ladder

Fig. 4.45 Two common types of DAC

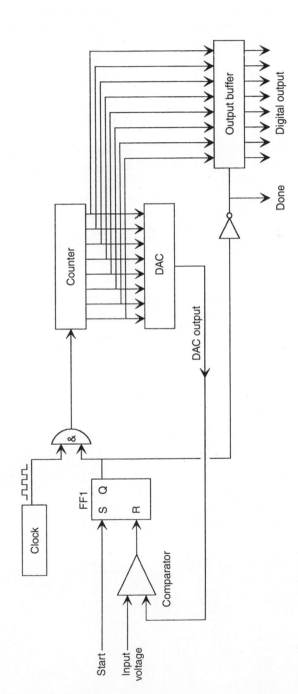

Fig. 4.46 The ramp ADC

is simple and cheap, but relatively slow. Where high speed or high accuracy is required a Successive Approximation ADC is used. The circuit, shown in Fig. 4.47, uses an ordered trial and error process. The sequence starts with the register cleared. The *Most Significant Bit* (MSB) is then set and the comparator output is examined; if the comparator shows the DAC output is less than or equal to V_{in}, the bit is left set. If the DAC output is greater than V_{in}, the bit is reset. Each bit is similarly tested in order, from MSB to *Least Significant Bit* (LSB), causing the DAC output to quickly locate V_{in} as shown. The number of comparisons is equal to the number of bits, so the conversion is much faster than the Ramp ADC. Successive Approximation ADCs are fast and accurate and unlike the Ramp ADC, the conversion time is constant, but they are more complex and expensive.

The conversion from analogue to digital is not instantaneous. Typically signals are read about ten times per second and an analogue input card thus takes regular 'snapshots' of each analogue signal. In Fig. 4.48 the sampling is too slow and a false view of the signal is given; this effect is known as '*aliasing*', so it is important to have a sufficiently fast conversion time. Every analogue signal will have a maximum frequency f_c at which it can change. To obtain a true series of snapshots the signal must

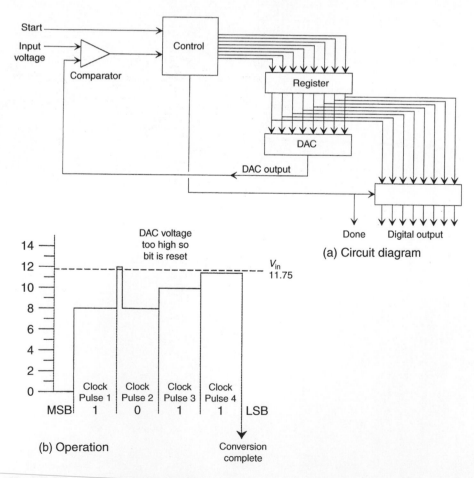

Fig. 4.47 The successive approximation ADC

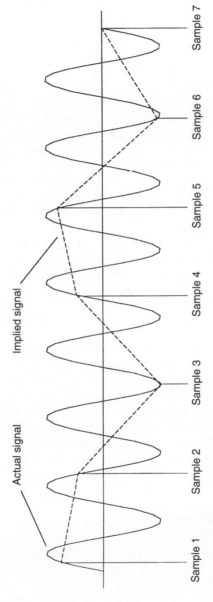

Fig. 4.48 Sampled systems and aliasing

be sampled at least twice the rate of f_c, so if an analogue signal has a maximum frequency of 2 Hz, it must be sampled at least at 4 Hz, or once every 250 ms. This is the basis of *Shannon's Sampling Theorem*. This f_c is rarely known precisely and a scan rate of between $4 f_c$ and $10 f_c$ is normally chosen. It is a good practice to pass the signal through a low-pass filter, known as an *anti-aliasing filter*, before the ADC to ensure frequencies above f_c are removed.

4.12   Installation notes

Measurement systems are generally based on low voltages and are consequently vulnerable to electrical noise. In most plants, a PLC may be controlling 415 V motors at 100 A, and reading thermocouple signals of a few millivolts. Great care must be taken to avoid interference from the higher power signals.

The first precaution is to take care with the earthing layout. A badly laid out system as shown in Fig. 4.49 provides common return paths, and currents from the load returning through the common impedances Z_1 to Z_3 will induce error voltages into the low level analogue circuit. There are at least three distinct types of earth in a system, these being a safety earth (as used for doors, frames etc.), a dirty earth (used for higher voltage and higher current signals), and a clean earth, for low voltage analogue signals. These should meet at only one point.

Screened cable should be used for all analogue signals, foil screening being preferred to braided screen. The screen should be earthed at one point only (ideally the receiving end) as any difference in earth between the two earthing points will cause current to flow in the screen (as in Fig. 4.50) and induce noise into the signal lines. When a screened cable passes through intermediate junction boxes, screen continuity must be maintained, and the screen must be sleeved to prevent it from touching the frame of the junction boxes.

High voltage and low voltage cable should be well separated; most manufacturers suggest at least one metre between 415 V and low voltage cables, but this can be difficult to achieve in practice. It is a good practice to use trunking or conduit for low voltage signals as a way of identifying them for future installers. The same result can also be achieved using cables with differently coloured PVC sleeves. Inevitably high voltage and low voltage cables will have to cross at some points. If adequate spacing cannot be achieved, these crossings should always be at 90°.

Further commentary on interference is given in section 14.2.

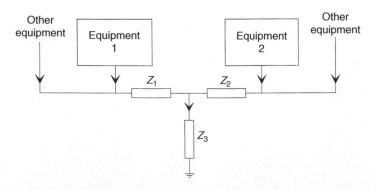

Fig. 4.49 Earthing layouts

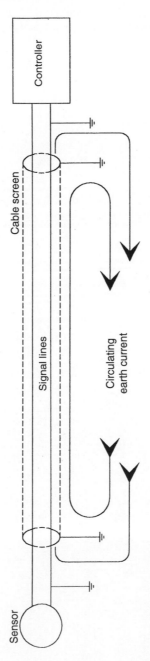

Fig. 4.50 Earth loops in screened cables

Chapter 5

Generators

Dr G.W. McLean
Generac Corporation

5.1 Introduction

Throughout the world there is a need for generators in many different applications. In addition to the underlying need for a public supply of electricity, there are a number of situations in which independent supplies are needed. The applications for generators are categorized broadly as follows:

- public supply networks in which a number of high-power generator sets may operate in parallel
- private or independent generators which may run in parallel with the public supply or isolated from it. Examples of this include:
  - *peak shaving* to reduce the maximum demand of electricity by a user; this can avoid large financial penalties during times of generally high demand on the system
  - *standby emergency generators* to protect the supply to critical circuits such as hospitals or water supplies
  - *temporary supplies* which are needed by the construction industry, or in cases of breakdown
  - *combined heat and power* using the waste heat from the generator engine is used for other purposes such as building heating
- portable supplies, often trailer-mounted, where no alternative supply is available

5.2 Main generator types

The two main types of generator are *'turbo'* or *cylindrical-rotor* and *salient-pole* generators. Both these types are *synchronous* machines in which the rotor turns in exact synchronism with the rotating magnetic field in the stator. Since most generators fall under this class, it forms the basis of most of the chapter.

The largest generators used in major power stations are usually turbo-generators. They operate at high speeds and are usually directly coupled to a steam or gas turbine. The general construction of a turbo-generator is shown in Fig. 5.1. The rotor is made from solid steel for strength, and embedded in slots within the rotor are the field or excitation windings. The outer stator also contains windings which are located in slots, this is again for mechanical strength and so that the teeth between the slots form a good magnetic path. Most of the constructional features are very specialized, such as

(a)

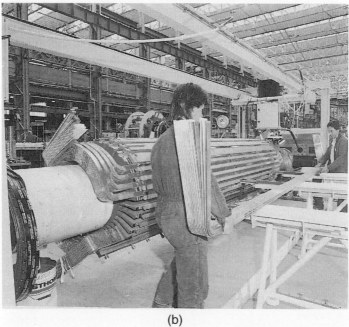

(b)

Fig. 5.1 Turbo-generator construction: (a) stator (b) rotor (c) assembly (courtesy of ABB ALSTOM Power)

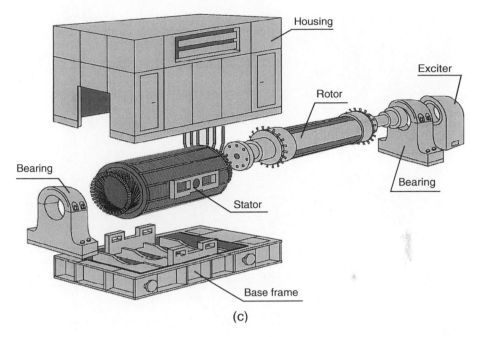

Fig. 5.1 (*Contd*)

hydrogen cooling instead of air, and direct water cooling inside the stator windings, so only passing reference is made to this class of machines in the following descriptions.

More commonly used in smaller and medium power ranges is the salient-pole generator. An example is shown in Fig. 5.2. Here, the rotor windings are wound around the poles which project from the centre of the rotor. The stator construction is similar in form to the turbo-generator stator shown in Fig. 5.1.

Less commonly used are *induction generators* and *inductor alternators*.

Induction generators have a simple form of rotor construction as shown in Fig. 5.3, in which aluminium bars are cast into a stack of laminations. These aluminium bars require no insulation and the rotor is therefore much cheaper to manufacture and much more reliable than the generators shown in Figs 5.1 and 5.2. The machine has characteristics which suit wind turbines very well, and they also provide a low-cost alternative for small portable generators. The basic action of the induction generator will be described later in this chapter, but the operation of the machine is very similar to the induction motor referred to in Chapter 10.

Inductor alternators have laminated rotors with slots, producing a flux pulsation in the stator as the rotor turns. These machines are usually used for specialized applications requiring high frequency.

5.3   Principles of operation

5.3.1   No-load operation

The basic operation of all these generator types can be explained using two simple rules, the first for magnetic circuits and the second for the voltage induced in a conductor when subjected to a varying magnetic field.

Fig. 5.2 Salient-pole generator construction: Top: stator, Bottom: rotor (courtesy of Generac Corporation)

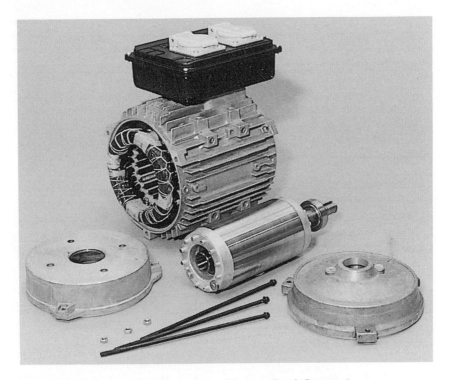

Fig. 5.3 Induction generator construction (courtesy of Invensys Brook Crompton)

The means of producing a magnetic field using a current in an electric circuit have been explained in section 2.2.3, and eqns 2.13 and 2.18 have shown that the flux Φ in a magnetic circuit which has a reluctance R_m is the result of a magneto-motive force (mmf) F_m, which itself is the result of a current I flowing in a coil of N turns.

$$\Phi = F_m/R_m \qquad (5.1)$$

and

$$F_m = IN \qquad (5.2)$$

The main magnetic and electrical parts of a salient-pole generator are shown in Fig. 5.4. In Fig. 5.4(a), dc current is supplied to the rotor coils through brushes and sliprings. The product of the rotor or field current I and the coil turns N results in mmf F_m as in eqn 5.2, and this acts on the reluctance of the magnetic circuit to produce a magnetic flux, the path of which is shown by the broken lines in Fig. 5.4(b). As the rotor turns, the flux pattern created by the mmf F_m turns with it; this is illustrated by the second plot of magnetic flux in Fig. 5.4(b). In section 2.2.3 it has also been explained that when a magnetic flux Φ passes through a magnetic circuit with a cross section A, the resulting flux density B is given by

$$B = \Phi/A \qquad (5.3)$$

Figure 5.4(a) also shows a stator with a single coil with an axial length l. As the rotor turns, its magnetic flux crosses this stator coil with a velocity v, it has been explained

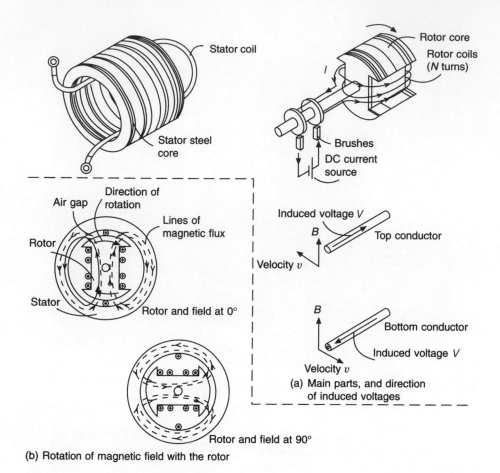

Fig. 5.4 Principles of generator operation

in section 2.2.4 that an electromotive force (emf) V will be generated, where

$$V = Bvl \tag{5.4}$$

The direction of the voltage is given by Fleming's right-hand rule, as shown in Fig. 2.6.

Figure 5.4(b) shows that as the magnetic field rotates, the flux density at the stator coil changes. When the pole face is next to the coil, the air gap flux density B is at its highest, and B falls to zero when the pole is 90° away from the coil. The induced emf or voltage V therefore varies with time (Fig. 5.5) in the same pattern as the flux density varies around the rotor periphery. The waveform is repeated for each revolution of the rotor; if the rotor speed is 3000 rpm (or 50 rev/s) then the voltage will pass through 50 cycles/second (or 50 Hz). This is the way in which the frequency of the electricity supply from the generator is established. The case shown in Fig. 5.4 is a 2-pole rotor, but if a 4-pole rotor were run at 1500 rpm, although the speed is lower, the number of voltage alternations within a revolution is doubled, and a frequency of 50 Hz would also result. The general rule relating the synchronous speed n_s (rpm), number of poles p and the generated frequency f (Hz) is given by

$$f = n_s p / 120 \tag{5.5}$$

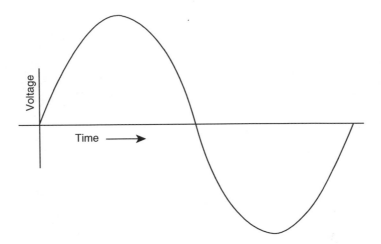

Fig. 5.5 Induced voltage waveform

The simple voltage output shown in Fig. 5.5 could be delivered to the point of use (the 'load') with a pair of wires as a single-phase supply. If more coils are added to the stator as shown in Fig. 5.4(a) and if these are equally spaced, then a three-phase output as shown in Fig. 5.6 can be generated. The three phases are conventionally labelled 'U', 'V' and 'W'. The positive voltage peaks occur equally spaced, one-third of a cycle apart from each other. The nature of single-phase and three-phase circuits has been explained in Chapter 2. The three coils either supply three separate loads, as shown in Fig. 5.7(a) for three electric heating elements, or more usually they are arranged in either 'star' or 'delta' arrangement in a conventional three-phase circuit (Fig. 5.7(b)).

In a practical generator the stator windings are embedded in slots, the induced voltage remaining the same as if the winding is in the gap as shown in Fig. 5.4(b). Also, in a practical machine there will be more than the six slots shown in Fig. 5.6(a). This is arranged by splitting the simple coils shown into several subcoils which occupy

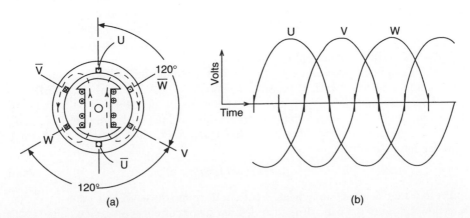

(a)

(b)

Fig. 5.6 Three-phase generation

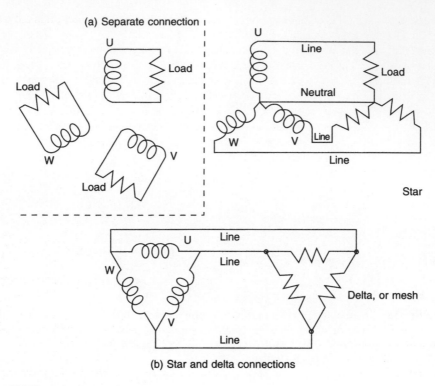

(a) Separate connection

Star

Delta, or mesh

(b) Star and delta connections

Fig. 5.7 Three-phase connections

separate slots, each phase still being connected together to form a continuous winding. Figures 5.1 and 5.2 show the resulting complexity in a complete stator winding.

5.3.2 The effect of load

In the circuits shown in Fig. 5.7 currents flow in each phase, and these currents will have a waveform similar to the voltage waveform shown in Fig. 5.6(b). The concept of a phase shift between voltage and current in ac circuits which contain inductance or capacitance has been explained in section 2.3.2. If an inductive or capacitive load is connected, then the current waveforms will respectively 'lag' or 'lead' the voltage waveforms by 90°. For the inductive load case shown in Fig. 5.6(a), the current in the U phase will be zero, but current will be flowing in V and W phases. It can be seen that the lines of magnetic flux now enclose not only the rotor excitation current, but also the stator currents in the V and W phases. Equations 5.1 and 5.2 show that the flux is the result of the mmf acting on a magnetic circuit, but it can now be seen that the mmf is a combination of the ampere-turns from the rotor and the stator winding. If I_r, I_s, N_r, and N_s are the currents and turns in the stator and rotor windings respectively, then eqns 5.1 and 5.2 combine to give

$$\Phi = (I_r N_r + I_s N_s)/R_m \tag{5.6}$$

In Fig. 5.6 a cross is used to indicate a current flow into the page, and a dot shows current flowing out of the page. It is seen that the stator currents oppose the field current in the rotor and their effect is to reduce the flux, with a corresponding reduction in the generated voltage. This demagnetizing effect is called *'armature reaction'*; it is the way in which Lenz's Law (section 2.2.4) operates in a generator.

The underlined currents in eqn 5.6 indicate that these are vectors and a vector addition is necessary. The armature reaction effect therefore depends on the extent to which the stator currents lag or lead the voltages (often called the 'phase' or 'phase angle'). If, for example, the generator load is capacitive, the currents will lead the voltages by 90°, and they will be opposite in direction to that shown in Fig. 5.6 for an inductive load. The ampere-turns of stator and rotor windings will add in this case and the flux and the generated voltage will be higher. In the case of a resistive load, the ampere-turns of the stator will act at 90° to the rotor poles, tending to concentrate the flux towards the trailing edge of the pole and producing magnetic saturation here when large stator currents flow; this reduces the flux and the output voltage, but not so much as in the inductive load case.

The output voltage is influenced not only by armature reaction, but also by voltage drop within the stator winding. This voltage drop is partly due to the internal resistance of the winding, and partly due to flux which links the stator winding but not the rotor winding; this flux is known as *'leakage flux'* and it appears in the stator electrical circuit as a *leakage inductance*, which also creates a voltage drop. The phase angle between stator currents and voltages will affect this voltage drop, producing a greater drop at lagging currents, and a negative drop (an increase) in voltage at leading currents.

In order to maintain a constant output voltage it is therefore necessary to change the excitation current in the rotor to compensate for the load conditions. The variation in rotor current to do this is shown in Fig. 5.8. To achieve constant output voltage, an *Automatic Voltage Regulator (AVR)* is used on the majority of generators, except for some small self-regulating units (see section 5.4).

The effects described so far can be conveniently summarized in a phasor diagram which combines stator and rotor mmfs as well as the emf drops due to resistance and leakage reactance of the stator winding. Figure 5.9 shows the phasor diagram of a cylindrical-rotor machine. It is assumed that the machine is symmetrically loaded and the diagram is drawn for one phase.

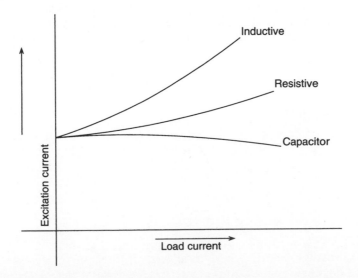

Fig. 5.8 Variation of excitation current with load current to maintain constant output voltage

114

Generators

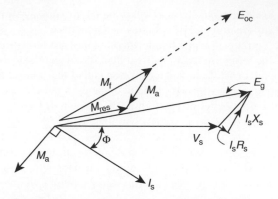

Fig. 5.9 Phasor diagram for a cylindrical-rotor synchronous generator

Phasor M_f represents the field-winding mmf. The output voltage that this would produce with no connected load is given by E_{oc} and its magnitude found by referring to the *open-circuit saturation curve* shown in Fig. 5.10. The reduced slope at high field currents results from saturation of the magnetic circuit. The output voltage V_s in Fig. 5.9 when under load is chosen to lie on the real axis and the output current is lagging by an angle Φ. The armature reaction mmf lags the stator current by 90° and is shown as M_a. When added to the field mmf the resultant mmf acting on the gap is shown as M_{res} which produces the resultant airgap flux resulting in the 'airgap emf' E_g. The magnitude of E_g from M_{res} is again found by using the saturation curve of Fig. 5.10. The output voltage is found by subtracting from E_g the resistive and leakage reactive drops I_sR_s and I_sX_s. The phasor diagram is a convenient way of calculating the steady state performance of a generator at different loads.

A similar diagram can be produced for salient-pole machines. In this case, however, the armature reaction is resolved into two components, one along the centre line of the rotor poles (the *direct* or *d axis*), and the other at an angle of 90° to this (the *quadrature* or *q axis*). These axes have different magnetic reluctance values.

These phasor diagrams are valid for steady state conditions only, and cannot be used to predict transient performance such as a sudden short circuit. A common means

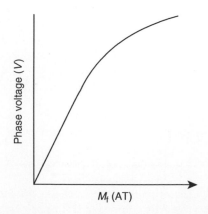

Fig. 5.10 Open-circuit saturation curve

of predicting transient performance is to represent the interaction of the stationary stator circuits and the rotor circuits by a matrix of mutual inductances which will be a
function of rotor position. If this variation is assumed to be sinusoidal, it is possible to
transform the matrix to one with constant mutual inductances by multiplying the above
matrix by a transform matrix, and as a result the stator currents and voltages are trans-
formed to '*d–q values*'. This simple transform, known as a *d–q transform*, allows
the transient performance of a generator to be predicted and is particularly useful in
representing fault conditions in power system networks.

5.3.3   Damping of transients

Transient changes in stator load result in a change of flux in the rotor pole, and if it
can be arranged that this flux change induces a voltage and a flow of current in the pole
face, this current will oppose the change in stator flux. To achieve this, it is normal to
insert into the pole face a set of aluminium or copper 'damper' bars, connected at either
end by a ring or end plate to form a conducting cage in the pole faces.

The *damper cage* has a considerable influence on the transient current flow in the
stator, particularly in the case of a short circuit. In addition, if the load in the three
phases is unbalanced, the induced currents in the damper cage will act to reduce
distortion of the waveform and to reduce asymmetry in the output phase voltages.
A single-phase generator represents a severe case of asymmetry, and this requires very
careful damper cage design because of the high induced currents.

The cage also helps to damp mechanical oscillations of the rotor speed about the
synchronous speed when the generator is connected in parallel with other machines.
These oscillations might otherwise become unstable, leading to the poles 'slipping' in
relation to the frequency set by other generators, and resulting in a loss of synchronism.
Such a condition would be detected immediately by the generator protection circuits
and the generator would then be isolated from the network.

5.3.4   Voltage waveform

The specified voltage waveform for a generator is usually a sine wave with minimum
distortion. A sine wave supply has advantages for many loads because it minimizes
the losses in the equipment; this is especially the case with motors and transformers.
The voltage waveform of a practical generator usually contains some distortion or harmonics, as shown in Fig. 5.11. The distorted waveform shown in Fig. 5.11(a) can be
represented as a series of harmonics, consisting of the fundamental required frequency
and a series of higher frequencies which are multiples of the fundamental frequency.

The harmonic distortion is calculated using Fourier analysis or other means of obtaining the spectrum of harmonics. An example of a spectrum is shown in Fig. 5.11(b).
Distortion is defined by a '*distortion factor*', where

$$\text{Distortion factor} = (\Sigma V_n^2)^{1/2}/V_1 \qquad (5.7)$$

In eqn 5.7, V_n is the magnitude of the nth harmonic, V_1 being the magnitude of the
fundamental one.

There are several ways in which a generator can be designed to produce minimum
distortion factor.

The higher frequency '*slot harmonics*' are due to distortions in the air gap flux density wave, these being created by the stator slot openings. The distortions can be
reduced by skewing the stator slots so that they are no longer parallel to the rotor shaft,

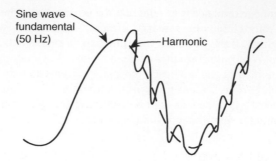

Sine wave fundamental (50 Hz)

Harmonic

Fig. 5.11 Harmonic distortion

but form part of a helix. The slots are often skewed by an amount close to the pitch between one stator slot and the next.

A second step is to use more than one stator slot per pole for each phase winding; Fig. 5.12 shows a winding with three stator slots per pole for each phase. This distributes the effect of the winding better and reduces the harmonics.

Harmonics can also be reduced by *'short-pitching'* the stator winding as shown in Fig. 5.12. Except for the smallest machines, most generators have a double-layer winding in which one side of a coil is laid into a slot above the return side of a different coil. The simplest (fully pitched) winding has all the coil sides of one phase in slots above the return sides of coils in that same phase, but by displacing one layer of the winding with respect to the other, the harmonics are reduced.

A fourth technique is to shape the face of the rotor pole so that the airgap between the rotor and stator is larger at the tips. This prevents a 'flat-topped' shape to the flux wave in the air gap and therefore reduces voltage distortion.

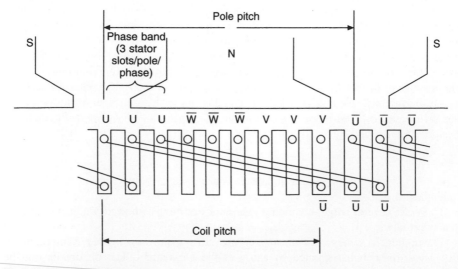

Fig. 5.12 Slot and winding layout

Finally, correct spacing of the damper bars in the pole face and proper choice of the arc length of the pole face reduces the high frequency harmonics which are produced by currents induced in the bars by the passage of the stator slots.

5.3.5   Connecting generators in parallel

Generators are frequently required to operate in parallel, either as part of a larger power system, or in a smaller isolated system. In order to connect in parallel with another generator or a larger system, the generator must produce the same magnitude and frequency of voltage as the other generators on the system. This is achieved by a synchronizing controller which controls the prime mover governor and the AVR field control. The governor must further be controlled to ensure that the phase angle of the voltages at each side of the generator circuit breaker contacts is equal. Once this condition is reached the contacts can be closed with a minimum of transient current. Once the generator is connected to a system, the mode of operation of the governor control must change to provide the required power, and the AVR mode must change to provide the required power factor or VArs. Where two or more generators are connected in *island mode* to an isolated load the governors and AVRs of all the sets must be controlled to share the power in the required ratio and also to ensure the same power factor so that circulating currents between generators are minimized.

5.3.6   Operating limits when in parallel with the mains

When a generator operates in parallel with a system, the power supplied by the generator is controlled by the torque control on the prime mover, for instance the rack of a diesel engine fuel injection pump. The power factor is controlled by the field or excitation current of the generator. By increasing the field current the phase angle of the stator current progressively lags that of the terminal voltage, and conversely a reduction in field current results in the stator current phase angle leading that of the terminal voltage. The power supplied by the generator will remain constant during this process if the prime mover torque is kept constant. The operator or automatic controller must be aware of the limits of the generator otherwise excessive current may occur in the field or stator windings. To this end, an *operating chart* is often provided by the manufacturer (sometimes referred to as the *capability chart*). An example is given in Fig. 5.13.

The vertical axis represents output power and the horizontal axis represents output kVAr. The output voltage is shown as OV and the output current is represented by OB, shown here lagging the voltage by ϕ. The field current is represented by AB. The output power is given by OD (or $VI \cos \phi$), and the output VAr is represented by OG ($VI \sin \phi$). The limits for generator operation are represented by the shaded area, where:

- BH represents the field current temperature limit
- KB represents the power limit of the prime mover
- KJ represents the practical limit of stability. Reducing field current reduces the peak torque capability of the generator, and the rotor load angle moves progressively towards the peak output angle. Any fluctuations in voltage or load can then lead to pole slipping. Automatic relay protection of generators will produce an alarm if the generator operation comes outside this area.

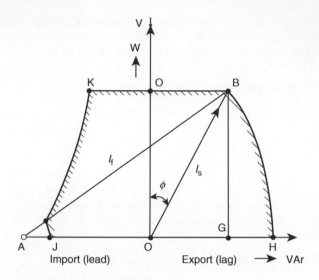

Fig. 5.13 Operating chart for a cylindrical-rotor generator

5.4   The Automatic Voltage Regulator (AVR)

While some small generators have an inherent ability to produce a reasonably constant voltage as the load varies, it is clear from the previous explanations that some form of automatic voltage control is required in the usual form of generator. The Automatic Voltage Regulator (AVR) already referred to in the preceding sections is based on a closed-loop control principle.

The basis of this closed loop control is shown in Fig. 5.14. The output voltage is converted, usually through a transformer or resistor network, to a low voltage dc signal, and this feedback signal is subtracted from a fixed reference voltage to produce an error signal.

The error signal is processed by a compensator before being amplified to drive the rotor excitation current. The change in rotor excitation current produces a variation in output voltage, closing the control loop. If the gain of the control loop is large enough then only a small error is required to produce the necessary change in excitation current, but a high gain can lead to instability in the circuit, with oscillations in the output voltage. The purpose of the compensating circuit is to enable small errors to be

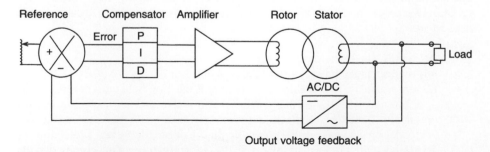

Fig. 5.14 Closed-loop voltage control

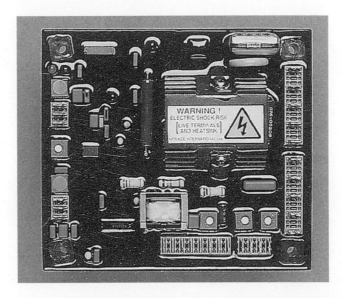

Fig. 5.15 Layout of an automatic voltage regulator (AVR) (courtesy of Newage International)

handled in a stable way. The most common form of compensator is a PID circuit in which the error is amplified proportionately (P), integrated (I) and differentiated (D) in three parallel circuits before being added together. Many AVRs have adjustment potentiometers which allow the gains of each channel to be varied in order to achieve the best performance. The integral term enables compensator output to be achieved at zero error, and this produces the minimum error in output voltage.

The layout of a commercial AVR is shown in Fig. 5.15. Many AVRs are now offered with digital circuitry. The principle of the feedback loop remains the same, but the feedback signal is converted to digital form using an analogue-to-digital converter. The calculations are performed digitally in a microprocessor and the output is on or off, using pulse width modulation (PWM) to vary the average level of dc supplied to the rotor excitation winding. Alternatively, the phase angle of a thyristor bridge can be used to vary the output level; this is known as phase-angle control.

The continuous improvement in power electronic controls and processor power is bringing further advances in voltage and speed control, with more flexible protection of the generator and its connected circuits. An example is the variable-speed constant-frequency generator from Generac Corporation; this is illustrated in Fig. 5.16. This consists of a permanent-magnet generator driven by a variable-speed engine and feeding a power electronic frequency-changer circuit, which delivers output at constant frequency. A microprocessor is used to control the switching of the output devices and to regulate the engine speed depending upon the load applied to the generator. At low power demand the engine speed is reduced to minimize noise, increase efficiency and extend life. The result is a saving in the volume and weight of the generator.

5.5   Brushless excitation

Although some generators are still produced with brushes and sliprings to provide the rotor current as illustrated in Fig. 5.4, most now have a brushless excitation system.

Fig. 5.16 Variable-speed constant-frequency generator (courtesy of Generac Corporation)

The two main techniques for synchronous generators are the separate exciter and capacitor excitation and these are described in the following sections. Also included for convenience here is a brief description of induction generators, since these also provide a brushless system.

5.5.1 Separate exciter

The most common way of supplying dc current to the rotor winding without brushes and sliprings is shown in Fig. 5.17.

The output of the AVR drives a dc current I_f through the pole windings of the *exciter*, which are mounted in a stator frame. The poles produce a stationary field which induces a voltage in the exciter rotor winding as it turns. Figure 5.18 shows that the exciter rotor is mounted on the same shaft as the main generator. The ac voltage produced by the rotor winding of the exciter is converted to dc by a bridge rectifier which is also mounted on the rotor shaft. This rectifier unit is shown clearly at the end of the shaft in Fig. 5.23. The dc output of the rectifier is connected to the main rotor windings by conductors laid in a slot along the rotor shaft. The inductance of the main generator rotor coils is usually sufficient to smooth out the ripple in the bridge rectifier output.

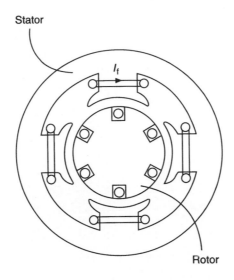

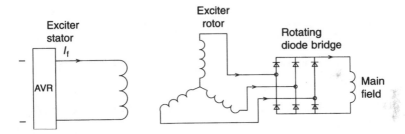

Fig. 5.17 Separate brushless exciter

The power supply to the AVR is either provided by a separate excitation winding in the main generator stator, or by a small permanent-magnet generator mounted on the shaft of the main generator, often referred to as a *'pilot exciter'*. The advantage of the pilot exciter is that the generator has a source of power available once the shaft is turning; the voltage supplied to the AVR is completely independent of generator load and there is no reliance on residual flux in the magnetic circuit of the main generator to start the self-excitation process. The pilot exciter also enables the generator to supply current to a connected network even when a short circuit occurs, enabling the high current to be detected by protection relays which will then disconnect the faulty circuit. If the AVR is supplied from an excitation winding in the main generator stator, the supply voltage is very small when the stator windings experience a short circuit, and the AVR is unable to drive an adequate rotor excitation current.

One manufacturer uses two excitation windings to provide a voltage from the AVR under short-circuit conditions, so that sufficient current is supplied into the fault to trip the protection system. During a short circuit the airgap flux density in these machines shows a pronounced harmonic component. This component induces voltage in coils of one of the excitation windings, which are short pitched and therefore deliver a

Fig. 5.18 Cut-away section of an ac generator (courtesy of Newage International)

voltage to the AVR even under short-circuit conditions. The second excitation winding is fundamental-pitched and provides the major drive for the AVR under normal operating conditions. It is claimed that the performance of this system is comparable to a machine using a permanent-magnet exciter.

Another method used to provide voltage to the AVR under short-circuit conditions is a series transformer driven by the generator output current.

5.5.2 Capacitor excitation

The use of this technique is usually restricted to single phase generators with a rated output less than 10 kW.

A separate excitation winding in the stator has a capacitor connected directly across its output as shown in Fig. 5.19. The rotor is usually of salient-pole construction as described previously, but in this case the rotor winding is shorted through a diode. On starting, the residual flux in the rotor body induces a small voltage in the stator excitation winding and a current flows through the capacitor. This current produces two waves of magnetic flux around the air gap of the generator. One wave travels in the same direction as the rotor, to create the armature reaction described in section 5.3.2. The second wave travels in a direction opposite to the rotor, and induces a voltage in the rotor windings at twice the output frequency. The current circulated in the rotor windings by this induced voltage is rectified by the diode to produce a dc current. This dc current increases the magnetic flux in the machine, which in turn drives more current through the stator excitation winding, which in turn produces more rotor current. This self-excitation process continues until the flux reaches a point at which the

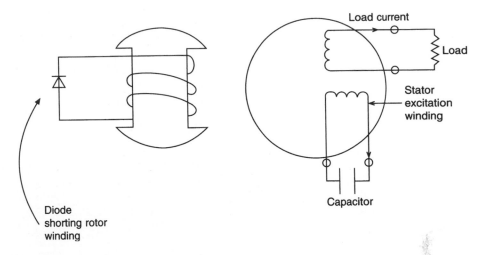

Load current

Load

Stator excitation winding

Diode shorting rotor winding

Capacitor

Fig. 5.19 Capacitor excitation

magnetic circuit is saturated, and a stable voltage results. The process also produces an inherent AVR action, since any load current in the output stator winding induces more rotor current to offset the armature reaction effect.

5.5.3   Induction generator

The principles of the cage induction motor are explained in Chapter 10. If a machine of this type is connected to a supply, it accelerates as a motor up to a speed near its synchronous speed. If the machine is driven faster than the synchronous speed by an engine or other prime mover, the machine torque reverses and electrical power is delivered by the machine (now acting as a generator) into the connected circuit.

A simple form of wind turbine generator uses an induction machine driven by the wind turbine. The induction machine is first connected to the three-phase supply, and acting as a motor it accelerates the turbine up to near the synchronous speed. At this point, the torque delivered by the wind turbine is sufficient to accelerate the unit further, the speed exceeds the synchronous speed and the induction machine becomes a generator.

It is also possible to operate an induction machine as a generator where there is no separate mains supply available. It is necessary in this case to self-excite the machine, and this is done by connecting capacitors across the stator winding as shown in Fig. 5.20(a). The leading current circulating through the capacitor and the winding produces a travelling wave of mmf acting on the magnetic circuit of the machine. This travelling wave induces currents in the rotor cage which in turn produces the travelling flux wave necessary to induce the stator voltage. For this purpose, some machines have an excitation winding in the stator which is separate from the main stator output winding. Figure 5.20(b) shows a single-phase version of the capacitor excitation circuit.

In small sizes, the induction generator can provide a low-cost alternative to the synchronous generator, but it has a relatively poor performance when supplying a low power factor load.

Although induction generators have useful characteristics for use in combination with wind turbines, the magnetizing current must be supplied by other generators

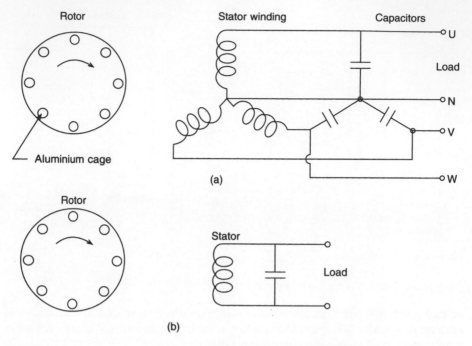

Fig. 5.20 Self-excitation of an induction generator (a) three phase (b) single phase

running in parallel, or capacitors connected across the stator windings. Another problem is that the efficiency of an induction generator drops if its speed differs significantly from the synchronous speed, due to high rotor copper loss in the rotor cage. This can be overcome by using a slipring-fed wound rotor combined with a power electronic converter connected between the stator and rotor windings. Such schemes are often referred to as *slip energy recovery* using a *doubly fed induction generator.*

The slip s, of an induction machine is the per unit difference between the rotor speed and the synchronous speed given by:

$$s = (N_s - N_r)/N_s \tag{5.8}$$

where N_s is the synchronous speed and N_r is the rotor speed. It can be shown that if T_r is the mechanical torque supplied by a turbine to the rotor of the induction generator, the generated electrical stator power transferred across the air gap is given by T_rN_s. Since the input mechanical power to the generator is T_rN_r, the difference $T_rN_r - T_rN_s$ must be the power lost in the rotor, produced mainly by copper loss in the cage. By substitution from eqn 5.8:

$$\text{power transfer to stator} = T_rN_s = \text{rotor loss}/s \tag{5.9}$$

With a simple squirrel cage rotor therefore the slip must be low to avoid high rotor loss with a resultant low efficiency.

If the cage is replaced by a three-phase winding, and sliprings are fitted, the same power balance can be achieved by removing the generated rotor power via the sliprings. This power can then be returned to the stator of the generator via a frequency converter. The rotor generated frequency is given by the stator frequency times slip $(f \times s)$. The circuit illustrating this slip recovery is shown in Fig. 5.21. The advantage

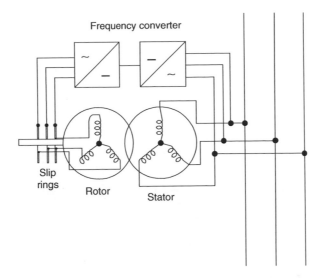

Fig. 5.21 Slip energy recovery induction generator

of this scheme is that the power rating of the electronic converter from eqn 5.9 is (slip × stator power generated). The scheme therefore allows the speed of the wind turbine to be varied to obtain more efficient turbine operation at different wind speeds. The rating of the converter is proportional to the maximum rated slip value, typically about ± 30 per cent, where positive slip applies to subsynchronous rotor speeds. In comparison, a variable-speed, constant-frequency system using a synchronous generator would require a frequency converter rated at full generator power.

The above slip recovery system requires sliprings which can be a disadvantage in marine environments. The sliprings can be eliminated in generators using two mechanically coupled stators and wound rotors with the second stator supplied from the frequency converter. This is known as a *brushless doubly fed* system, but adds considerable cost to the system.

5.5.4 Inductor alternator

From eqn 5.5 it is evident that to achieve high frequency, either the speed or the pole number of the generator must be high. The number of poles that can be considered in a conventional generator design is limited because the slots become too small and the leakage fluxes become too high. A robust construction that overcomes these problems is known as the inductor alternator, and one configuration is shown in Fig. 5.22.

In this quarter cross section the field windings are inserted into the large stator slots and produce a stationary four-pole flux pattern. The armature coils are placed in the smaller slots, each coil enclosing one tooth, and connected in an alternate sense in series to form a phase. The rotor contains no windings and has double the tooth-pitch of the stator. With the rotor in the position shown, the rotor teeth line up with the odd numbered stator slots, and the flux driven by the field coils will flow through this low reluctance path linking the coils round these teeth. The even numbered teeth experience a low flux level. If the rotor is now moved one stator tooth pitch, the high flux will now link the even numbered coils and the stator windings experience an alternating flux linkage and generate an alternating voltage. Unlike the synchronous generator, the stator teeth and air gap experience flux in one direction only. The flux varies

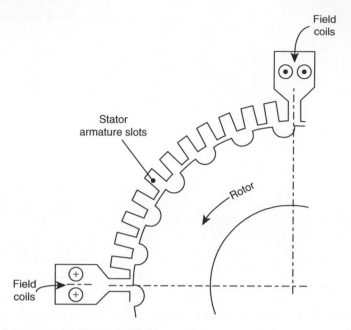

Fig. 5.22 Inductor alternator slotting arrangement

from a maximum to a minimum in one direction, while the synchronous machine experiences a reversal in flux direction. This results in a lower effective magnetic loading, and power density. Numerous slotting arrangements have been devised in order to ease winding and minimize losses and reactances for particular duties.

5.6 Construction

5.6.1 Stator

The stator core is constructed from a stack of thin steel sheets or laminations which are stamped to shape and insulated electrically from each other, either by a thin coating or by an oxide layer which is produced during heat treatment. The steel used has a low silicon content as described in Chapter 3; this increases the resistivity of the steel and therefore reduces the losses due to eddy currents. The steel is carefully processed in order to minimize the hysteresis losses because the whole stator core is subjected to alternating magnetic flux. In large turbogenerators the core is built up in segments and grain-oriented steel is used to reduce the losses further.

The stator windings are located in axial slots in the stator core which are formed by the shape of the laminations. Except in high-voltage machines, the individual coils of the winding are wound with copper wire covered with a layer of polyester/polyamide enamel which is about 0.05 mm thick. The slots are lined with a tough insulating sheet, usually about 0.25–0.5 mm thick; a popular material is a laminate of Mylar and Nomex. The coils are impregnated when in place with a resin to give the winding mechanical strength as well as to improve the heat transfer from the copper to the cooling air. Windings operating at different voltages, such as the three phases, have a further sheet of insulating material separating them in the end-winding region. A general guide to the insulation materials and processes in use is given in section 3.3.

The mounting of the stator core in its frame differs according to the size of the machine, but for a majority of medium-sized generators, the arrangement is as shown in Fig. 5.18. At either end of the frame are bearing housings which locate the rotor shaft. These housings or end-bells are cast in smaller and medium size machines, and fabricated in larger sizes. The generator is often mounted directly onto the engine, and in this case it is usual to eliminate the drive-end bearing, using the rear bearing of the engine to locate the generator shaft.

5.6.2 Rotor

It has already been noted that the construction of a turbogenerator is very specialized and the rotor for these machines are not dealt with here. However, even within the class of salient-pole generators, quite different forms of rotor construction are used, depending upon the size.

Generators rated up to about 500 kW use rotor laminations which are stamped in one piece. In larger machines the poles are made separately from stacks of laminations, and each pole is keyed using a dovetail arrangement onto a spider which is mounted on the rotor shaft. In large high-speed machines the poles can be made from solid steel for extra strength and to reduce mechanical distortion; these solid poles are screwed to the shaft, as shown in the large 4-pole machine in Fig. 5.23.

The nature of the rotor coils also depends upon the size of the machine. Because the ratio of surface area to volume is larger in the coils of small generators, these are

Fig. 5.23 Large salient-pole rotor (courtesy of Brush Electrical Machines)

Fig. 5.24 Layer-wound rotor (courtesy of Generac Corporation)

easier to cool. Generators rated above about 25 kW therefore use a 'layer-wound' coil in which each layer of the coil fits exactly into the grooves formed by the layer below; this is illustrated in Fig. 5.24. Rectangular cross section wire can be used to minimize the coil cross section. The simplest and cheapest way to make the coils, often used in smaller machines, is to wind them in a semi-random way as shown in Fig. 5.25. In either case, the coils are impregnated after winding like the stator windings to give extra mechanical strength and to improve the heat transfer by removing air voids within the coil.

The coils are under considerable centrifugal stress when the rotor turns at full speed, and they are usually restrained at both ends of the pole by bars, and by wedges in the interpole spaces, as shown in Fig. 5.23.

5.6.3 Cooling

Adequate cooling is a vital part of the design and performance of a generator. Forced cooling is needed because of the high loss densities that are necessary to make economic use of the magnetic and electrical materials in the generator.

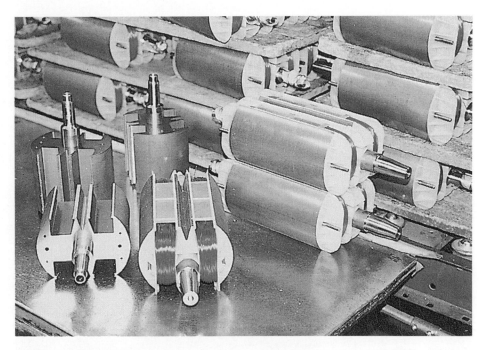

Fig. 5.25 Random-wound rotors (courtesy of Generac Corporation)

The most critical areas in the machine are the windings, and particularly the rotor winding. It is explained in section 3.3 that the life expectancy of an insulation system decreases rapidly if its operating temperature exceeds recommended temperatures. It is crucial for reliability therefore that the cooling system is designed to maintain the winding temperatures within these recommended limits. As shown in Table 3.4, insulation materials and systems are defined by a series of letters according to their temperature capability. As improved insulation with higher temperature capability has become available, this has been adopted in generator windings and so the usual class of insulation has progressed from class A (40–50 years ago) through class E and class B to class F and class H, the latter two being the systems generally in use at present. Class H materials are available and proven, and this system is becoming increasingly accepted. An important part of the generator testing process is to ensure that the cooling system maintains the winding temperature within specified limits, and this is explained later in section 5.8.

In turbogenerators it has already been noted that very complex systems using hydrogen and de-ionized water within the stator coils are used. The cooling of small and medium-sized machines is achieved by a flow of air driven around the machine by a rotor-mounted fan. A typical arrangement is shown in Fig. 5.18. In this case the cooling air is drawn in through ducts; it is then drawn through the air gap of the machine and ducts around the back of the stator core before reaching the centrifugal fan which then expels the air from the machine. There are many variations to the cooling system, particularly for larger generators. In some machines there is a closed circulated air path cooled by a secondary heat exchanger which rejects the heat to the outside atmosphere. This results in a large and often rectangular generator enclosure as shown in Fig. 5.26.

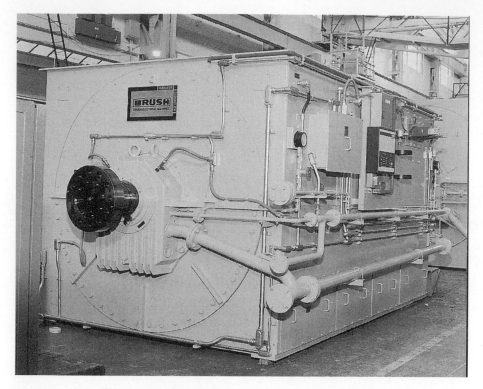

Fig. 5.26 Closed air circuit cooled generator (courtesy of Brush Electrical Machines)

5.7   Rating and specification

In order that a generator can be selected to suit a particular application, manufacturers issue specification data. This can be used and interpreted according to the following sections.

5.7.1   Rated output

The key aspect of the specification is the rated output of the generator, which is normally expressed in terms of the apparent power (VA, kVA or MVA) when supplying the maximum load at the rated power factor, assumed to be 0.8. The rated output is usually based upon continuous operation in a maximum ambient temperature of 40°C. If the machine has a special short-time rating, the nameplate should state the time limits of operation.

The rated output from a given size of generator is related to the size and speed of the machine as follows:

$$\text{Rated output power} = K \times D_g^2 \times L_c \times n_s \qquad (5.10)$$

where D_g is the stator bore diameter, L_c is the stator core length and n_s is the speed. Here, K is a design constant which is proportional to the product of the airgap flux density and the stator current density.

Table 5.1 Typical derating factors for class H insulation in ambient temperatures above 40°C

Ambient temperature (°C)	Derating factor
40	1.00
45	0.97
50	0.94
55	0.91
60	0.88

The rated output of a machine is reduced in ambient temperatures exceeding 40°C and at altitudes above sea level exceeding 1000 m. The latter is because the air density is decreased, and its ability to cool the machine is reduced. Derating factors are applied for these conditions and typical values are summarized in Tables 5.1 and 5.2.

5.7.2 Reactances

The generator can be characterized by several reactances, each being useful in working out the performance and protection requirements under different circumstances. These include the *synchronous reactance*, the *transient reactance*, the *subtransient reactance*, the *Potier reactance* and the *negative- and zero-sequence reactances*.

The subtransient reactance represents the output impedance of the generator within the first few cycles after a short-circuit occurs at the generator terminals. It is used for selecting protective relays for the connected load circuit. The lower the value of the subtransient reactance, the more onerous is the protection requirement.

Transient reactance represents the impedance of the machine over a slightly longer period, and is relevant to the performance of the generator and its AVR under changing load conditions. A low transient reactance is beneficial in responding to load changes.

Associated with the subtransient and transient reactances are time constants which define the rate of decay of these reactances.

The negative and positive sequence reactances influence the performance of the generator when supplying an unbalanced three-phase load.

5.7.3 Main items of specification

In summary, the following items are important when considering an application and specifying the appropriate generator:

- rated output, expressed as apparent power (VA)
- cooling air requirement (m³/s)

Table 5.2 Typical altitude derating factors

Altitude (m)	Derating factor
1000	1.00
1500	0.97
2000	0.97
2500	0.94
3000	0.87
3500	0.82
4000	0.77

Table 5.3 Typical parameters for a 200 kW generator

Line–line voltage, frequency		400 V, 50 Hz
Cooling air flow		0.4 m³/s
Rotor moment of inertia		1.9 kg m²
Weight		600 kg
Efficiency	full load	93%
	3/4 full load	94%
	1/2 full load	94%
	1/4 full load	93%
Stator resistance		0.025 Ω
Rotor resistance		1.9 Ω
Reactances	synchronous, X_d	1.9 pu
	transient, X_d^1	0.2 pu
	subtransient, X_d^{11}	0.1 pu
	positive sequence	0.07 pu
	negative sequence	0.1 pu
	zero sequence	0.07 pu
Time constants	transient, T_{do}^1	0.8 s
	subtransient, T_d^{11}	0.01 s
Short-circuit current		2.5 pu

- moment of inertia of the rotor
- generator weight
- efficiency at full, 3/4, 1/2 and 1/4 load
- stator winding resistance
- reactances and time constants, as listed in section 5.7.2
- maximum short-circuit current delivered by the generator

Typical values for a 200 kW generator are shown in Table 5.3.

5.8 Testing

Type tests are performed by manufacturers in order to confirm that a design meets its specifications, and production tests are done in order to check that each machine as it is manufactured conforms to performance and safety standards.

These tests usually include the following

- *full-load tests* to measure the temperature rise of the machine windings and insulation. The temperature rise is calculated from the change in resistance of the stator windings.
- *tests to determine the excitation current* required to deliver a given output voltage. These are done for open-circuit conditions and also for various load currents and power factors. The resulting curves are usually known as 'saturation' curves.
- *short-circuit tests* to determine the current that can be driven by the generator into a short-circuit fault in the connected load.
- *transient short-circuit tests* to determine the subtransient and transient reactances and time constants.
- *overspeed tests* to confirm that the rotor does not distort or disintegrate. This test is normally performed at 150 per cent of rated speed, at full rated temperature.

- *insulation tests* to confirm that the insulation system is capable of withstanding operating and transient conditions. Typical requirements are that stator windings can withstand ac voltages of 1 kV plus twice rated voltage to the frame. Similarly between stator high-voltage and low-voltage windings, rotor windings are also tested to frame but with voltages typically ten times that of the rated rotor voltage but not less than 1.5 kV.
- *impulse tests* are often used by manufacturers to confirm correct connection of coils and/or shorted turns. This test relies on applying a high-voltage impulse and comparing the waveforms of the transient. Shorted turns will be indicated by travelling wave reflections and non-standard waveforms.

The results from these tests are used to calculate the data described in section 5.7, an example of which has been shown in Table 5.3.

It is now necessary for any generator manufactured or imported into the European Union that relevant EU Directives are met through certification. Strictly it is the manufacturer of the generator set, including the engine and all controls, that is responsible for this certification, but many manufacturers of generators and AVRs will assist in the tests.

5.9 Standards

The leading international, regional and national standards adopted by users and suppliers of generators for manufacturing and testing are shown in Table 5.4.

Table 5.4 International, regional and national standards relating to generators

IEC/ISO	EN/HD	BS	Subject of standard	N. American
IEC 34-1	EN 60034-1	4999-101	Ratings and performance	
IEC 34-2A		4999-102	Losses and efficiency	
IEC 34-4	EN 60034-4	4999-104	Synchronous machine quantities	
IEC 34-6	EN 60034-6	4999-106	Methods of cooling	
IEC 34-8	HD 53.8	4999-108	Terminal markings	
		4999-140	Voltage regulation, parallel operation	
IEC 34-14/				
ISO 2373		4999-142	Vibration	
IEC 529	EN 60529	5490	Degrees of protection	
ISO 8528-3		7698-3	Generators for generating sets	
ISO 8528-8		7698-8	Low power generating sets	
IEC 335-1	EN 60335-1	3456-1	Safety of electrical appliances	
	EN 60742	3535-1	Isolating transformers	
			Motors and generators	NEMA MG 1
			Cylindrical-rotor synchronous generators	IEEE C50.13
			GT-driven cyl-rotor synch generators	IEEE C50.14

Chapter 6

Transformers

Professor D.J. Allan
Merlindesign

6.1 Principles of operation

In simple form, a transformer consists of two windings connected by a magnetic core. One winding is connected to a power supply and the other to a load. The circuit containing the load may operate at a voltage which differs widely from the supply voltage, and the supply voltage is modified through the transformer to match the load voltage.

In a practical transformer there may be more than two windings as well as the magnetic core, and there is the need for an insulation system and leads and bushings to allow connection to different circuits. Larger units are housed within a tank for protection and to contain oil for insulation and cooling.

With no load current flowing, the transformer can be represented by two windings on a common core, as shown in Fig. 6.1. It has been explained in Chapter 2 (eqns 2.25 and 2.26) that the input and output voltages and currents in a transformer are related by the number of turns in these two windings, which are usually called the *primary* and *secondary* windings. These equations are repeated here for convenience.

$$\frac{V_1}{V_2} = \frac{N_1}{N_2} \tag{6.1}$$

$$\frac{I_1}{I_2} = \frac{N_2}{N_1} \tag{6.2}$$

The magnetic flux density in the core is determined by the voltage per turn:

$$\frac{V_1}{N_1} = 4.44 f B_m A \tag{6.3}$$

Equation 6.3 represents a key relationship between the frequency, the number of turns in a winding and the size of the core. If the winding turns are increased, the core cross-sectional area may be reduced. If the frequency is increased (from 50 to 60 Hz), the size of the core can be reduced.

In the no-load case, a small current I_0 flows to supply the magnetomotive force which drives the magnetic flux around the transformer core; this current lags the primary voltage by almost 90°. This I_0 is limited in magnitude by the effective resistance (R_c) and reactance (X_c) of the magnetizing circuit, as shown in Fig. 6.2. The magnetizing current is typically 2–5 per cent of the full load current and it has a power factor in the range 0.1–0.2.

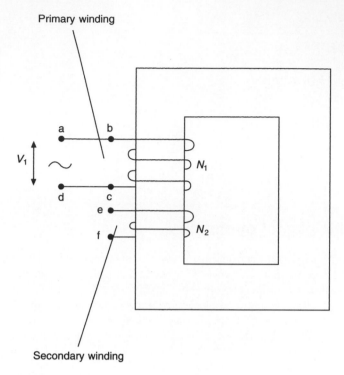

Fig. 6.1 Simple transformer circuit

When the transformer is loaded, there is an internal voltage drop due to the current flowing through each winding. The voltage drops due to the primary and secondary winding resistances (R_1 and R_2) are in phase with the winding voltage, and the voltage drops due to the primary and secondary winding leakage reactances (X_1 and X_2) lag the winding voltage by 90°. The leakage reactances represent those parts of the transformer flux which do not link both the windings; they exist due to the flow of opposing currents in each winding and they are affected strongly by the winding geometry.

The current flow and voltage drops within the windings can be calculated using the equivalent circuit shown in Fig. 6.2. This circuit is valid for frequencies up to 2 kHz.

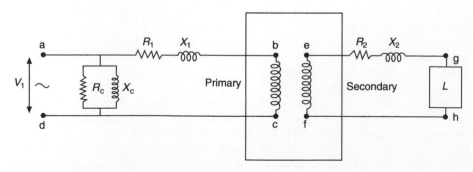

Fig. 6.2 Equivalent circuit of a transformer

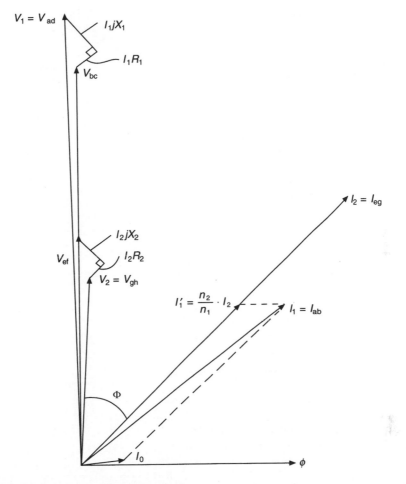

Fig. 6.3 Vector diagram of a transformer on load

A vectorial representation of the voltages and currents is shown in Fig. 6.3 for the case of a load L with a power factor angle Φ.

The decrease in output voltage when a transformer is on load is known as *regulation*. The output voltage is less than the open-circuit voltage calculated according to eqn 6.1 because of voltage drops within the winding when load current flows through the resistive and reactive components shown in Fig. 6.2. The resistive drops are usually much smaller than the reactive voltage drops, especially in large transformers, so the impedance Z of the transformer is predominantly reactive. Regulation is usually expressed as a percentage value relating the vector addition of the internal voltage drops (shown in Fig. 6.3) to the applied voltage.

For many applications, the equivalent circuit shown in Fig. 6.2 can be simplified by ignoring R and X, and referring R_2, X_2 and L to the primary side as shown in Fig. 6.4, where

$$R = R_1 + R_2 \left(\frac{N_1}{N_2} \right)^2 \qquad (6.4)$$

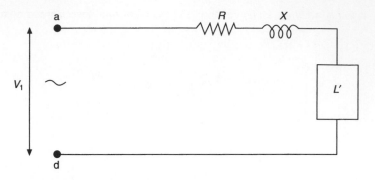

Fig. 6.4 Simplified equivalent circuit

$$X = X_1 + X_2 \left(\frac{N_1}{N_2} \right)^2 \qquad (6.5)$$

and

$$L' = L \left(\frac{N_1}{N_2} \right)^2 \qquad (6.6)$$

Equations 6.4 and 6.5 indicate how resistance and reactance values can be referred from secondary to primary windings (and vice versa) using the square of the turns ratio. Equation 6.6 is used to refer impedances between secondary and primary circuits, allowing circuits to be normalized using the square of the turns ratio, so that simple circuit calculations can be made.

The impedance of the transformer is given by:

$$Z = \left(R^2 + X^2 \right)^{1/2} \qquad (6.7)$$

Values for R_1 and R_2 can be established by measuring the resistance of the windings. The value of X is determined by calculation or by derivation from the total impedance Z, which can be measured with one winding of the transformer short circuited. This Z is given by

$$Z = \frac{V}{I} \qquad (6.8)$$

where V is the voltage necessary to circulate the full-load current I in the windings under short circuit conditions. When V is expressed as a percentage of rated voltage, this gives Z as a percentage value referred to rated power.

When the transformer is energized, but without a load applied, the no-load power loss is due to the magnetic characteristics of the core material used and the flow of eddy currents in the core laminations. The loss due to magnetizing current flowing in the winding is small and can be ignored.

When the transformer is loaded, the no-load loss is combined with a larger component of loss due to the flow of load current through the winding resistance. Additional losses on load are due to eddy currents flowing within the conductors and to circulating currents which flow in metallic structural parts of the transformer. The circulating currents are induced by leakage flux which is generated by the load current flowing in the windings and they are load dependent. These additional losses are known as *stray losses*.

The efficiency of a transformer is expressed as:

$$\text{Efficiency} = \frac{\text{output}}{\text{input}} \times 100\%$$

$$= \frac{(\text{input} - \text{losses})}{\text{input}} \times 100\%$$

$$= \left(1 - \frac{\text{losses}}{\text{input}}\right) \times 100\% \qquad (6.9)$$

where input, output and losses are all expressed in units of power.

The total losses consist of the no-load loss (or iron loss) which is constant with voltage and the load loss (or copper loss) which is proportional to the square of load current. Total losses are usually less than 2 per cent for distribution transformers, and they may be as low as 0.5 per cent in very large transformers.

When losses are considered in kilowatt values, the lowest loss transformer is found to have no-load loss and loss values approximately equal. However, since many distribution and transmission transformers usually operate at only one-half or one-quarter of full load current, the equal losses criterion is an over simplification.

No-load and load losses are often specified as target values by the user for larger transformers, or they may be evaluated by the *capitalization* of losses. The capitalization formula is of the type:

$$C_C = C_T + A \times P_0 + B \times P_K \qquad (6.10)$$

where: C_C = capitalized cost
C_T = tendered price
A = capitalization rate for no-load loss (£/kW)
P_0 = guaranteed no-load loss (kW)
B = capitalization rate for load loss (£/kW)
P_K = guaranteed load loss (kW).

Values assigned to capitalization rates for no-load and load loss vary with the load operating regime of the transformer, the discount rate and forward interest rates. Typical capitalization rates for transmission transformers are £ 4000/kW for no-load loss and £ 1000/kW for load loss. Where a purchaser is responsible for the cost of losses the capitalization formulae are respected, but when a purchaser can pass the cost of losses on to a consumer, transformers are purchased at the lowest capital cost.

6.2 Main features of construction

6.2.1 The core

The magnetic circuit in a transformer is built from sheets or laminations of electrical steel. Hysteresis loss is controlled by selecting a material with appropriate characteristics, often with large grain size. Eddy current loss is controlled by increasing the resistivity of the material using the silicon content, and by rolling it into a very thin sheet. The power loss characteristics of the available materials are shown in Fig. 6.5. Generally speaking the steels with lower power loss are more expensive, so that the 0.3 mm thick CGO steel shown is the cheapest of the range shown in Fig. 6.5, and the very low loss

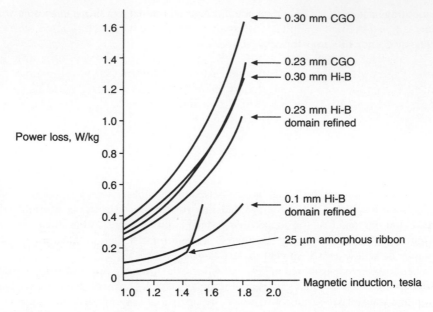

Fig. 6.5 Power loss characteristics of various electrical steels and amorphous materials at 50 Hz

0.23 mm thick Hi-B material with domain control techniques used to modify the apparent crystal size is the most expensive. New materials with thicknesses of 0.1 mm are available on a laboratory basis, but the use of such thin materials presents some production problems. A 25 μm thick amorphous ribbon can be used in very low loss transformers up to a few megavolt ampere and a core with losses of only one-sixth of the conventional 0.3 mm CGO material is possible. The general principles applying to electrical steels are discussed in Chapter 3.

The audible sound radiated by a transformer is generated by magnetostrictive deformation of the core and by electromagnetic forces in the windings, tank walls and magnetic shields. The dominant sound is generated by longitudinal vibrations in the core laminations, which are induced by the magnetic field. The amplitude of the vibrations depends upon the flux density in the core and on the magnetic properties of the core steel. In a large power transformer operating at high flux density, the audible sound level can exceed 100 dB(A) and it may be necessary to use high-quality core material, improved core-joint techniques and perhaps external cladding or a sound enclosure to reduce the sound to an acceptable level.

For many small transformers the cores are assembled using lamination stampings in C and I shapes and coils wound on to the bobbins from thin enameled wire are fitted to the core before closing the circuits.

In small power transformers, the core may take the form of a continuous strip of steel wound into a coil. The windings may be formed directly onto this core using toroidal winding machines, or the core may be cut to allow preformed windings to be fitted, and re-interlaced with the windings in place.

Where transformer weight is critical, it may be advantageous to operate a local power system at high frequency. Equation 6.3 relates voltage per turn to frequency and core cross section. If the transformer is to operate at 400 Hz (a typical requirement for aircraft), it can be seen that the core cross section will be only 12.5 per cent of that which would be necessary at 50 Hz. The clear advantage in cost and weight must be balanced against higher core losses at the increased frequency (although these can be reduced by using thinner laminations) and against higher load losses caused by high-frequency currents in the windings.

A longer established form of construction uses the stacked-core technique. For transformers which are rated in hundreds of watts the laminations may be stamped in E and I shapes, or in C and I shapes, then built into cores and assembled round the windings. At higher ratings in kilovolt ampere or megavolt ampere, the usual construction is to cut laminations to length, and to assemble them in a building berth which includes part of the core-clamping structure. When the core has been built, the remaining part of the clamping structure is fitted and the core is turned upright. If it has been fitted during the stacking stage, the top yoke is then removed, the windings are mounted on the core and the top yoke is then re-interlaced.

Single-phase transformers usually have a three-leg core with high-voltage and low-voltage windings mounted concentrically on the centre leg to reduce leakage flux and minimize the winding impedance. The outer legs form the return path for the magnetizing flux. This type of construction is shown in Fig. 6.6(a). For very high ratings in the region of 500 MVA, it may be economic to use a two-leg construction with two sets of windings connected in parallel and mounted one on each leg, as shown in Fig. 6.6(b). Where a transport height limit applies, a large single-phase transformer may be mounted on a four-leg core where the outer legs are used to return remnant flux, in conjunction with smaller top and bottom yokes. This arrangement is shown in Fig. 6.6(c).

Three-phase transformers are also based usually on a three-leg construction. In this case the high-voltage and low-voltage windings of each phase are mounted concentrically with one phase on each of the three legs. The phase fluxes sum to zero in the top and bottom yokes and no physical return path is necessary. This arrangement is shown in Fig. 6.6(d).

Where transport height is a limitation, a three-phase transformer may be built using a five-leg construction as shown in Fig. 6.6(e). The yoke areas in this case are reduced and the outer legs are included as return flux paths.

Core laminations are compacted and clamped together using yoke clamps connected by *flitch plates* on the outer faces of the legs, or by tie-bars locking the top and bottom yoke clamps together. Figure 6.7 shows a large three-phase, three-leg core in which the yokes are clamped by rectangular section steel frames and insulated straps; flitch plates lock the two clamps together and the core-bands on the legs are temporary, to be removed when the windings are fitted.

6.2.2 Windings

Winding conductors may be of copper or aluminium, and they may be in foil or sheet form, or of round or rectangular section. Foil or sheet conductors are insulated from each other by paper or Nomex interleaves, whereas round or rectangular conductors may be coated with enamel or wrapped with paper or some other solid insulation covering.

Low-voltage windings for transformers with ratings up to about 4 MVA may use foils with the full width of the winding, or round conductors. For higher-power transformers a large cross section may be necessary to carry the current in the low-voltage winding, and it may be economic to use stranded or parallel conductors in order to reduce eddy currents. It may also be necessary to transpose conductors during the winding operation to reduce circulating currents within the winding. For very high current applications, *Continuously Transposed Conductor* (CTC) may be used, where typically 40 or 50 conductors are machine transposed as shown in Fig. 6.8. In a CTC,

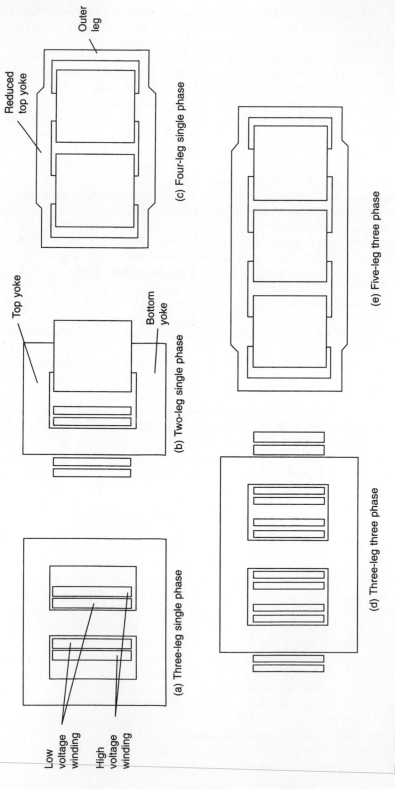

Fig. 6.6 Typical core constructions

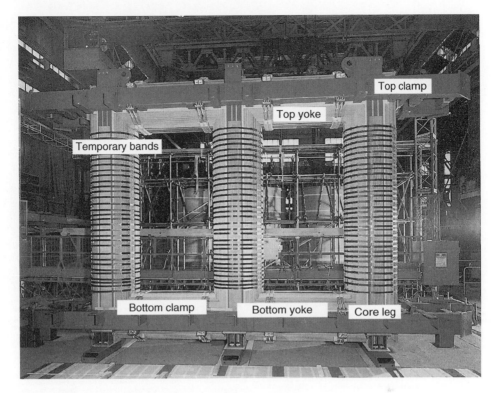

Fig. 6.7 Three-leg core for a three-phase transmission transformer

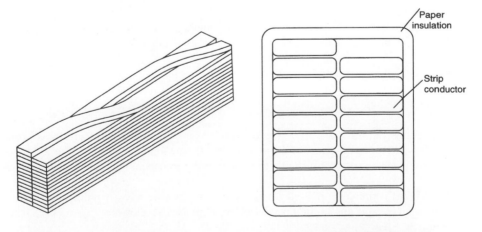

(a) Transposed strip conductor (27 strips in parallel)

(b) Transposed strip conductor in paper envelope

Fig. 6.8 Continuously transposed conductor

each strand is enamel insulated and the cable is enclosed in a paper covering. The conductors for low-voltage windings are usually wound in layers, with cooling ducts and interlayer insulation as part of the construction.

High-voltage windings have more turns and carry less current than the low-voltage windings. They are usually formed from round enamelled wire or paper-wrapped rectangular conductors. As with low-voltage windings, if the winding loss is to be minimized, parallel conductors or CTC may be used. The conductors are wound in layers or in discs, as shown in Figs. 6.9(a) and (b) respectively. The labour cost in a layer winding is high, and disc windings are usually considered to be more stable mechanically under the effects of through-current faults.

Many transformers have three or more windings. These may include regulating windings, tertiary windings to balance harmonic currents or supply auxiliary loads, multiple secondary windings to supply separate load circuits, or multiple primary circuits to connect to power supplies at different voltages or frequencies. The constructional aspects are common between multiple windings, but design aspects are more complicated.

6.2.3   Winding connections

Three-phase transformers are usually operated with the high-voltage and low-voltage windings connected in Y (star), D (delta) or Z (zigzag) connection. The three styles are shown in Fig. 6.10.

In star connection, one end of each of the three-phase windings is joined together at a neutral point N and line voltage is applied at the other end; this is shown in Fig. 6.10(a). The advantages of star connection are:

- it is cheaper for a high-voltage winding
- the neutral point is available
- earthing is possible, either directly or through an impedance
- a reduced insulation level (graded insulation) is possible at the neutral
- winding tappings and tapchanger may be located at the neutral end of each phase, with low voltages to earth and between phases
- single-phase loading is possible, with a neutral current flowing

In delta connection, the ends of the three windings are connected across adjacent phases of the supply as shown in Fig. 6.10(b). The advantages of a delta connection are:

- it is cheaper for a high-current low-voltage winding
- in combination with a star winding, it reduces the zero-sequence impedance of that winding

A delta-connected tertiary winding is often used on large three-phase autotransformers to allow zero-phase sequence currents to circulate within the windings, or to allow triplen-frequency harmonic currents to flow in order to cancel out harmonic fluxes in the core.

The zigzag connection is used for special purposes where two windings are available on each leg and are interconnected between phases as shown in Fig. 6.10(c). The advantages of a zigzag connection are:

- it permits neutral current loading with an inherently low zero-sequence impedance, and it is used in 'earthing transformers' to create an artificial neutral terminal on the system

(a)

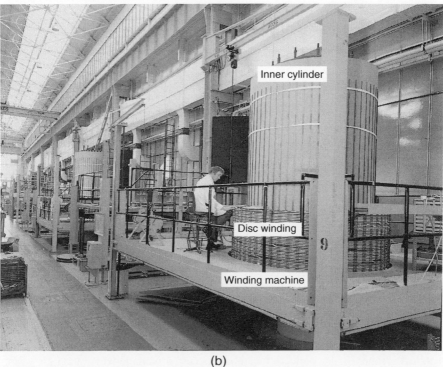

(b)

Fig. 6.9 (a) Layer-type winding on a horizontal winding machine (b) Disc-type winding on a vertical winding machine

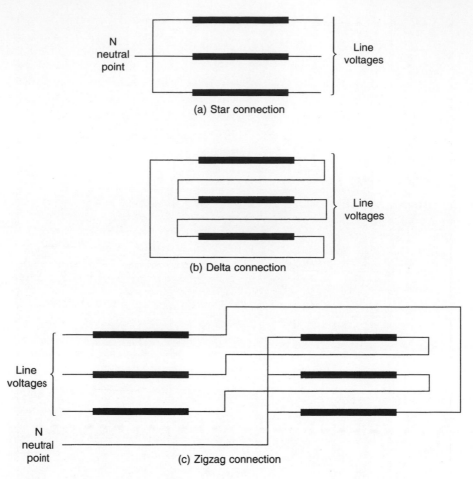

(a) Star connection

(b) Delta connection

(c) Zigzag connection

Fig. 6.10 Three-phase winding connections

- voltage imbalance is reduced in systems where the load is not evenly distributed between phases

In order to define the range of possible connections, a designation has been adopted by IEC in which the letters Y, D, Z and N are assigned to the high-voltage windings and y, d, z and n are assigned to the low-voltage windings. Clock-hour designations 1 to 12 are used to signify in 30° steps the phase displacement between primary and secondary windings. Yy0 is the designation for a star–star connection where primary and secondary voltages are in phase. Yd1 is the designation for a star–delta connection with a 30° phase shift between primary and secondary voltages. The common connections for three-phase transformers are shown in Fig 6.11.

In many transmission transformers, an autotransformer connection is employed. The connection is shown in Fig. 6.12, and unlike the two-winding transformer in which primary and secondary are isolated, it involves a direct connection between two electrical systems. This connection can have a cost advantage where the ratio of input and output voltages is less than 5:1. Current flowing in the 'series' winding corresponds to that in the higher-voltage system only. The 'common' winding carries a current

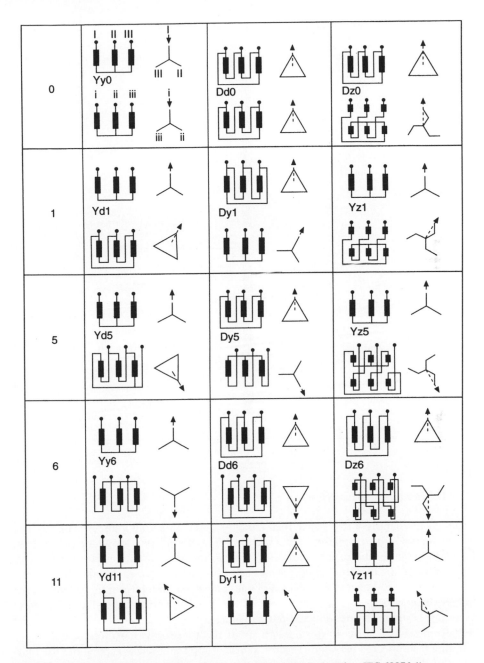

Fig. 6.11 Clock-hour designation of three-phase winding connections (based on IEC 60076-1)

which is the difference between the two systems, and being sized for this lower current it can be significantly cheaper. The autotransformer is more susceptible to damage from lightning impulse voltages and it has a lower strength against through-fault currents; both of these weaknesses can be corrected, but at increased cost.

It may occasionally be necessary to make a three-phase to a two-phase transformation. This might be in order to supply an existing two-phase system from a new three-phase

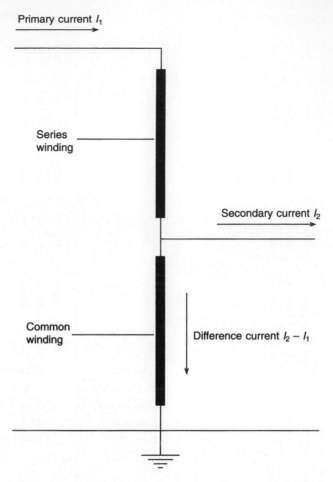

Primary current I_1

Series winding

Secondary current I_2

Common winding

Difference current $I_2 - I_1$

Fig. 6.12 Autotransformer connection

system, or to supply a two-phase load (such as a furnace) from a three-phase system, or to supply a three-phase load (such as a motor) from a two-phase system. In all these cases the usual method of making the transformation is by using two single-phase transformers connected to each of the systems and to each other by the *Scott connection* shown in Fig. 6.13. On one transformer the turns ratio is equal to the transformation ratio, and the mid-point of the winding connected to the three-phase system is brought out for connection to the other single-phase transformer. The second transformer has a turns ratio of 0.866 times the transformation ratio. The primary winding is connected between the third phase of the three-phase supply and the mid-point of the primary winding of the first transformer. The secondary windings of the two transformers are connected to the two-phase supply (or load) using a three- or four-wire connection.

6.2.4   Bushings

Where transformers are enclosed within tanks to contain the insulating and cooling fluid, it is necessary to link the winding connections to the network using through-bushings which penetrate the tank.

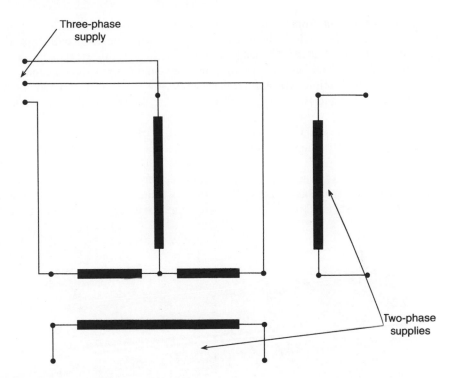

Fig. 6.13 Scott connection of transformers

Low-voltage bushings are generally solid. They are made of porcelain, ceramic or epoxy insulation with sufficient electrical strength to withstand abnormal voltages due to lightning activity or switching operations, and to withstand the service voltage over the lifetime of the transformer. Some low-voltage bushings carry high current and cooling is necessary. In such cases, it is usual to employ a hollow porcelain construction in which the internal connections are cooled by the oil in the tank.

High-voltage bushings must withstand much higher voltage transients. They are usually of composite construction with a core of oil-impregnated or resin-bonded paper in an outer porcelain or epoxy cylinder. This outer cylinder is 'shedded' on the outside to increase its electrical strength under wet conditions. A typical high-voltage bushing with an oil-paper core is shown in Fig. 6.14(a); its internal construction detail is shown in Fig. 6.14(b).

6.2.5 Tapchangers

It has been explained in section 6.1 that when a transformer carries load current there is a variation in output voltage which is known as *regulation*. In order to compensate for this, additional turns are often made available so that the voltage ratio can be changed using a switch mechanism known as a tapchanger.

An off-circuit tapchanger can only be adjusted to switch additional turns in or out of circuit when the transformer is de-energized; it usually has between two and five tapping positions. An on-load tapchanger (OLTC) is designed to increase or decrease the voltage ratio when the load current is flowing, and the OLTC should switch the transformer load current from the tapping in operation to the neighbouring tapping without interruption. The voltage between tapping positions (the step voltage) is normally between 0.8 per cent and 2.5 per cent of the rated voltage of the transformer.

The OLTC mechanisms are based either on a slow-motion reactor principle or a high-speed resistor principle. The former is commonly used in North America on the low-voltage winding, and the latter is normally used in Europe on the high-voltage winding.

The usual design of an OLTC in Europe employs a selector mechanism to make connection to the winding tapping contacts and a diverter mechanism to control current flows while the tapchanging takes place. The selector and diverter mechanisms may be combined or separate, depending upon the power rating. In an OLTC which comprises a diverter switch and a tap selector, the tapchange occurs in two operations. First, the next tap is selected by the tap switch but does not carry load current, then the

Fig. 6.14 (a) High-voltage oil–air bushing

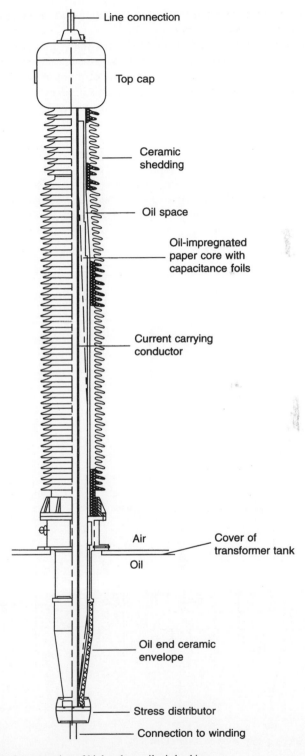

Line connection

Top cap

Ceramic
shedding

Oil space

Oil-impregnated
paper core with
capacitance foils

Current carrying
conductor

Air   Cover of
                       transformer tank

Oil

Oil end ceramic
envelope

Stress distributor

Connection to winding

Fig. 6.14 (b) Internal construction of high-voltage oil–air bushing

diverter switches the load current from the tap in operation to the selected tap. The two operations are shown in seven stages in Fig. 6.15.

The tap selector operates by gearing directly from a motor drive, and at the same time a spring accumulator is tensioned. This spring operates the diverter switch in a very short time (40–60 ms in modern designs), independently of the motion of the motor drive. The gearing ensures that the diverter switch operation always occurs after the tap selection has been completed. During the diverter switch operation shown in Fig. 6.15(d), (e) and (f), transition resistors are inserted; these are loaded for 20–30 ms and since they have only a short-time loading the amount of material required is very low.

The basic arrangement of tapping windings is shown in Fig. 6.16. The linear arrangement in Fig. 6.16(a) is generally used on power transformers with moderate regulating ranges up to 20 per cent. The reversing changeover selector shown in Fig. 6.16(b) enables the voltage of the tapped winding to be added or subtracted from the main winding so that the tapping range may be doubled or the number of taps reduced. The greatest copper losses occur at the position with the minimum number of effective turns.

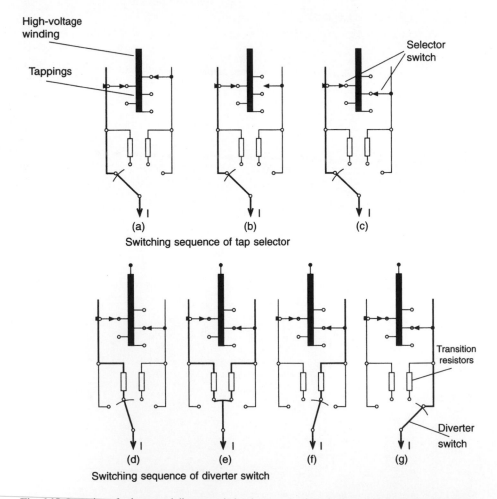

Fig. 6.15 Operation of selector and diverter switches in an on-load tapchanger of high-speed resistor type

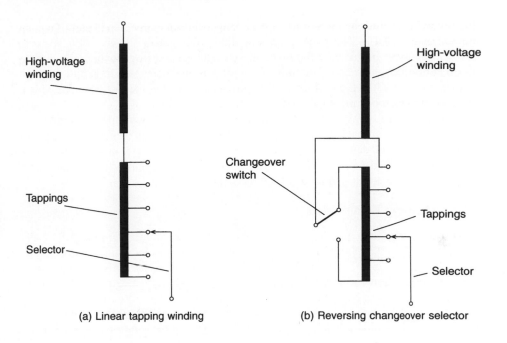

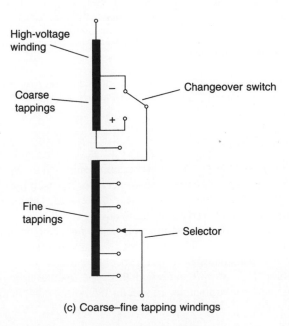

(a) Linear tapping winding

(b) Reversing changeover selector

(c) Coarse–fine tapping windings

Fig. 6.16 Basic arrangements of tapping windings

This reversing operation is achieved with a changeover selector which is part of the tap selector of the OLTC. The two-part coarse–fine arrangement shown in Fig. 6.16(c) may also be used. In this case the reversing changeover selector for the fine winding can be connected to the 'plus' or 'minus' tapping of the coarse winding, and the copper losses are lowest at the position of the lowest number of effective turns. The coarse changeover switch is part of the OLTC.

Regulation is mostly carried out at the neutral point in star windings, resulting in a simple, low-cost, compact OLTC and tapping windings with low insulation strength to earth. Regulation of delta windings requires a three-phase OLTC, in which the three phases are insulated for the highest system voltage which appears between them; alternatively three single-phase OLTCs may be used.

6.2.6   Cooling equipment

Transformers may be naturally cooled, in which case the cooling medium (oil or air) circulates by thermosyphon forces, or they may be forced cooled, with fans or pumps to circulate the air or oil over the core and through the windings.

In oil-cooled transformers a more economic solution is to use directed oil-flow cooling in which oil is pumped directly into the windings. The oil is cooled by passing (or pumping) it through plate radiators, which may be externally cooled by fans, or by using forced air-cooled tubular construction, or by pumping it through watercoolers. Designations for these cooling systems which identify both the cooling fluid and the type of cooling have been assigned by IEC and these are summarized in Table 6.1.

6.3   Main classes of transformer

Transformers are used for a wide variety of purposes, with the complete range of voltage and power ratings as well as many special features for particular applications. The following covers the main types.

6.3.1   Transformers for electronics

Transformers for electronic circuits or for low-voltage power supplies are used to match the supply voltage to the operating voltage of components or accessories, or to match the impedance of a load to a supply in order to maximize power throughput. They may be used to match impedances in primary and secondary circuits using eqn 6.6.

The core is usually constructed in low-power transformers from C- and I-laminations or from E- and I-laminations. The windings are usually of round enamelled wire, and the assembly may be varnished or encapsulated in resin for mechanical consolidation and to prevent ingress of moisture. Increasing numbers of this type operate at high frequencies in the kilohertz range and use laminations of special steel often containing cobalt to reduce the iron losses.

6.3.2   Small transformers

These are used for stationary, portable or hand-held power supply units, as isolating transformers and for special applications, such as burner ignition, shavers, shower heaters, bells and toys. They may be used to supply three-phase power up to 40 kVA at frequencies up to 1 MHz. These transformers are usually air insulated, the smaller units using enamelled windings wires and ring cores and the larger units using C- and I- or E- and I-laminated cores.

Table 6.1 Cooling system designations

For oil immersed transformers a four-letter code is used:

First letter: internal cooling medium in contact with windings:
   O mineral oil or insulating liquid with fire point ≤ 300°C
   K insulating liquid with fire point > 300°C
   L insulating liquid with no measurable fire point

Second letter: circulation mechanism for internal cooling medium:
   N natural thermosyphon flow
   F forced oil circulation, but thermosyphon cooling in windings
   D forced oil circulation, with oil directed into the windings

Third letter: external cooling medium
   A air
   W water

Fourth letter: circulation mechanism for external cooling medium
   N natural convection
   F forced circulation

Examples: ONAN or OFAF

For dry-type transformers a four-letter code is used:

First letter: internal cooling medium in contact with windings:
   A air
   G gas

Second letter: circulation mechanism for internal cooling medium:
   N natural convection
   F forced cooling

Third letter: external cooling medium
   A air
   G gas

Fourth letter: circulation mechanism for external cooling medium
   N natural convection
   F forced cooling

Examples: AN or GNAN

Safety is a major concern for these transformers and they are identified as class I, class II or class III. Class I units are insulated and protected by an earth terminal. Class II transformers have double insulation or reinforced insulation. Class III transformers have outputs at Safety Extra-Low Voltages (SELV) below 50 V ac or 120 V dc (see section 16.2.2.4 for background on SELV).

6.3.3 Distribution transformers

These are used to distribute power to domestic or industrial premises. They may be single-phase or three-phase, pole-mounted or ground-mounted, and they have ratings ranging from 16 kVA up to 2500 kVA.

The windings and core are immersed in mineral oil, with natural cooling, and there are two windings per phase. The primary (high-voltage) winding has a highest voltage ranging from 3.6 kV to 36 kV; the secondary (low-voltage) winding voltage does not exceed 1.1 kV. The high-voltage winding is usually provided with off-circuit tappings of ±2.5 per cent, or +2 × 2.5 per cent, −3 × 2.5 per cent.

The preferred values of rated output are 16, 25, 50, 100, 160, 250, 400, 630, 1000, 1600 and 2500 kVA, and the preferred values of short-circuit impedance are 4 or 6 per cent.

Losses are assigned from lists, for instance from BS 7281-1, or by using a loss–capitalization formula.

The core and windings of a typical distribution transformer rated at 800 kVA, 11 000/440 are shown in Fig. 6.17.

6.3.4 Supply transformers

These are used to supply larger industrial premises or distribution substations. Ratings range from 4 MVA to 30 MVA, with primary windings rated up to 66 kV and secondary windings up to 36 kV.

Transformers in this class are fluid cooled. Most supply transformers use mineral oil; but for applications in residential buildings, oil rigs and some factories, the coolant may be synthetic esters, silicone fluid or some other fluid with a higher fire point than mineral oil.

6.3.5 Transmission (or intertie) transformers

These are among the largest and highest voltage transformers in use. They are used to transmit power between high-voltage networks. Ratings range from 60 MVA to

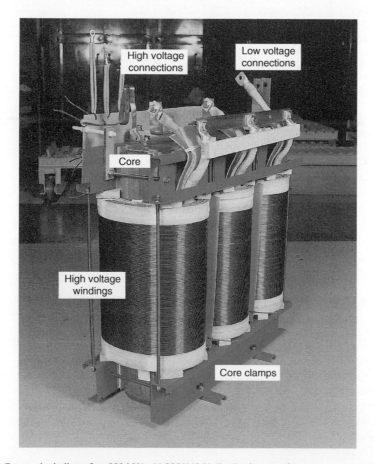

Fig. 6.17 Core and winding of an 800 kVA, 11 000/440 V distribution transformer

1000 MVA and the windings are rated for the networks which they link, such as 33, 66, 132, 275 and 400 kV in the UK, or voltages up to 500 kV or 800 kV in other countries. The impedance of a transmission transformer is usually 18 per cent in the UK, or 8 per cent in continental Europe, but for some system conditions, an impedance of up to 30 per cent is used.

Transmission transformers are oil filled, and are usually fitted with oil pumps and radiator fans to assist cooling of the windings and cores. They are usually fitted with OLTCs, but some networks at 400 kV and 275 kV are linked by transformers without regulating windings. The core and windings of a three-legged transmission transformer rated at 1000 MVA and 400 kV/275 kV/11 kV are shown in Fig. 6.18.

6.3.6   Generator (or step-up) transformers

Power is usually generated in large power stations at typically 18–20 kV, and generator transformers are used to step up this voltage to the system voltage level. These transformers are usually rated at 400, 500, 630, 800 or 1000 MVA.

Generator transformers are usually fitted with regulating windings and OLTCs.

6.3.7   Phase-shifting transformers

Where power is transmitted along two or more parallel transmission lines, the power flow divides between the lines in inverse proportion to the line impedances. Higher power is therefore transmitted through the line with lowest impedance and this can result in overload on that line, when the parallel line is only partly loaded. Phase-shifting

Fig. 6.18 Core and windings of a 1000 MVA, 400/275 kV transmission transformer

transformers are used to link two parallel lines and to control power flow by injecting a voltage 90° out of phase *(in quadrature)* with the system voltage into one line, at either leading or lagging power factor. Where the transformer controls the phase angle but not the voltage, the unit is known as *a quadrature booster*. Where the voltage is also controlled, the unit is known as *a phase-shifting transformer*. Figure 6.19 shows a 2000 MVA, 400 kV quadrature booster transformer on site; the unit is split between two tanks in order to meet construction limitations of size and weight.

6.3.8 Converter transformers

Where power is transmitted through an HVDC system, a converter station is used to change ac power to dc using multiple rectifier bridges. Direct current power is converted back to ac using inverter bridges. Converter transformers handle ac power and power at mixed ac/dc voltages by combining the power flow through 12 phases of rectifier/inverter bridges through dc valve windings.

The insulation structure must withstand all normal and abnormal conditions when ac voltage is mixed with dc voltage of differing polarities over the operating temperature range. The presence of dc currents may also cause dc saturation of the core, leading to abnormal magnetizing currents and variations in sound.

A phase of a three-phase converter transformer bank typically comprises a high-voltage primary winding and two secondary ac/dc valve windings. Three such transformers together form the two secondary three-phase systems; one is connected in delta and the other in star. Each secondary system feeds a six-pulse bridge and the two bridges are connected in series to form a 12-pulse arrangement, as shown schematically in Fig. 6.20. Two such transformer banks are used with the secondary circuits connected in opposite polarity to form a ±215 kV dc transmission system.

Fig. 6.19 2000 MVA 400 kV quadrature booster transformer in two tanks on site

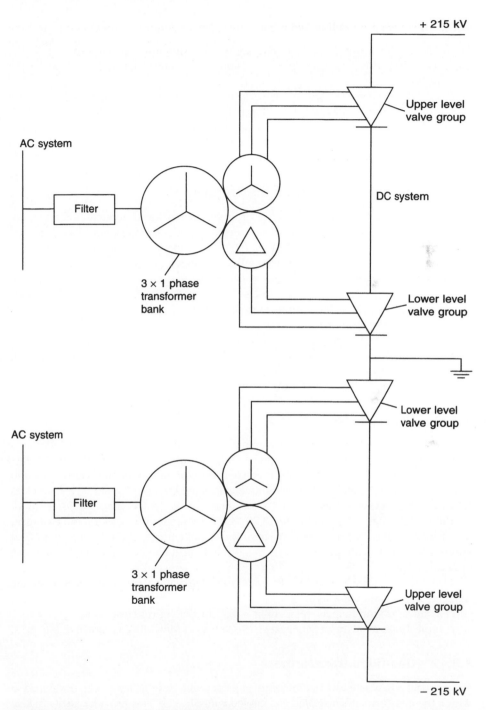

Fig. 6.20 Schematic diagram of ac/dc transmission system

6.3.9   Railway transformers

Transformers for railway applications may be trackside units to supply power to the track, or on-board transformers in the locomotive or under the coaches, to power the drive motors.

Trackside transformers are subjected to uneven loading depending upon the position of the train in the railway system. On-board transformers are designed for the lowest possible weight, resulting in a high-loss performance. Modern train control systems using thyristors, GTOs or IGBTs subject the transformers to severe harmonic currents that require special design consideration.

6.3.10   Rectifier and furnace transformers

Special consideration is needed for transformers in industrial applications involving arc furnaces or heavy-current dc loads in electrochemical plant. The primary windings in such cases are usually rated at 33 kV or 132 kV in the UK, but the secondary windings carry many thousands of amperes and are rated at less than 1 kV. Current sharing between parallel paths in the transformer becomes important because of the magnetic fields created by the high currents. These strong magnetic fields can cause excess heating in magnetic steels if these are used in the structure of the transformer, because of the flow of proximity currents in the steel. To reduce this excess heating, non-magnetic steel is often used to form part of the tank or the cover.

The OLTCs in furnace transformers are subject to a heavy duty; they may perform hundreds of thousands of operating cycles a year, which is more than a lifetime's duty for many transmission transformers.

6.3.11   Dry-type transformers

A dry-type construction is possible where a higher-temperature class of insulation is required than is offered by cellulose and a class 'O' or class 'K' fluid. Dry-type transformers use non-cellulosic solid insulation and the windings may be varnish dipped to provide a class 'C' capability, or vacuum encapsulated in epoxy resin to form a class 'F' or class 'H' system. (See Table 3.4 for details of the temperature classification of insulation.) Ratings are generally up to 30 MVA at voltages up to 36 kV, but cast resin transformers have recently been successfully manufactured at 110 kV using a novel winding design. Overload performance is limited but it can be augmented by the use of cooling fans.

This type is more expensive than a fluid-filled equivalent, and because of the reduced fire risk they are used in special applications where the public are involved, such as underground tunnels, residential blocks of flats or oil-rigs.

A typical cast-resin transformer rated 2500 kVA, 11 000/440 V is shown in Fig. 6.21.

6.3.12   Gas-filled transformers

For applications where low flammability is paramount, designs have been developed in which the transformer is insulated and cooled with SF_6 gas. This provides an alternative to dry-type construction where the risk of fire must be eliminated and the possible contamination of the environment by oil spillage must be avoided. High-voltage SF_6 transformers are available at ratings up to 300 MVA at 275 kV and prototype designs have been tested at up to 500 kV. Gas-filled transformers and reactors are more expensive than oil-filled units but the costs may be justified to eliminate a risk of fire,

Fig. 6.21 Dry-type 2500 kVA, 11 000/440 V transformer with cast resin encapsulation

particularly at a site where the cost of land is high and where the overall 'footprint' of the unit can be reduced by the elimination of fire-fighting equipment.

6.4   Rating principles

6.4.1   Rated power

The rated power of each winding in a transformer refers to a continuous loading and it is a reference value for guarantees and tests concerning load losses and temperature rise. Where a transformer winding has different values of apparent power under different cooling conditions, the highest value is defined as the rated power. A two-winding transformer is given only one value of rated power, which is identical for both windings, but a multi-winding transformer may have reduced levels of rated power on auxiliary windings.

When rated voltage applied to the primary winding and rated current flows through the secondary winding, the transformer is carrying rated power for that set of windings. This definition of rated power, used by IEC, implies that the rated power is the value of power input to the transformer, including its own absorption of active and reactive power. The secondary voltage under full load differs from the rated voltage by the voltage drop in the transformer.

In North America (ANSI/IEEE standard C57.1200) rated power is based on the output that can be delivered at rated secondary voltage, and allowance must be made so that the necessary primary voltage can be applied to the transformer. The difference between basing rated power on loaded voltages (IEC) or open-circuit voltages (ANSI/IEEE) is significant.

6.4.2   Overloading

Although transformers are rated for continuous operation, it is possible to supply over-loads for limited periods. The analysis of overloading profiles is based on the deterioration of cellulose. At temperatures above the rated temperature, cellulose degrades at a faster rate, and the load cycle should therefore include a period of operation at lower temperature to balance effects of these faster ageing rates at higher temperature. In general, the ageing rate doubles for every 8°C rise in temperature.

When the ambient temperature is below the rated ambient temperature it is also permissible to overload the transformer, the limit of the overloading being the operating temperature of the winding. Under severe operational pressures it is sometimes necessary to overload a transformer well beyond its nameplate rating, but such an operation will usually result in shorter life. IEC 354 provides a guide to overloading of oil-filled transformers based on these principles.

A maximum winding temperature of 140°C should not be exceeded even under emergency conditions, because free gas can be produced by cellulose at temperatures above this, and gas bubbles may cause dielectric failure of the windings.

6.4.3   Parallel operation of transformers

For operational reasons it may be necessary to operate transformers connected directly in parallel. For successful parallel operation, the transformers must have:

*   the same voltage ratings and ratio (with some tolerance)
*   the same phase–angle relationship (clock-hour number)
*   the same percentage impedance (with some tolerance)

It is not advisable to parallel-operate transformers with widely different power ratings as the natural impedance for optimal design varies with the rating of the transformer. The power divides between parallel-connected transformers in a relationship which is inversely proportional to their impedances; a low-impedance transformer operated in parallel with a higher-impedance unit will pass the greater part of the power and may be overloaded. A mismatch in loading of up to 10 per cent is normally acceptable.

6.5   Test methods

6.5.1   Specification testing

Transformers for power supply units and isolating transformers are subject to production testing, where compliance is generally checked by inspection and electrical tests. The electrical tests establish earth continuity (for class I), no-load output voltage, dielectric strength between live parts of the input circuit and accessible conducting parts of the transformer, and dielectric strength between input and output circuits.

Tests on larger transformers are carried out as part of the manufacturing process to ensure that transformers meet the characteristics specified by the purchaser. The tests prescribed by IEC are grouped in three categories, these being routine tests, type tests and special tests:

(a) *routine tests,* to be carried out on all transformers
(b) *type tests,* to be carried out on new designs or the first unit of a contract
(c) *special tests,* to be carried out at the specific request of the purchaser

(a) *Routine tests:* These are carried out on all transformers and include,

- measurement of winding resistance using a dc measurement circuit or a bridge. This is the winding resistance used in the equivalent circuit.
- measurement of voltage ratio and check of phase displacement. The voltage ratio is checked on each tap position and the connection symbol of three-phase transformers is confirmed.
- measurement of short-circuit impedance and load loss. The short-circuit imped-ance (including reactance and ac resistance in series) is determined by measur-ing the primary voltage necessary to circulate full-load current with the secondary terminals short-circuited. This provides the value of impedance used in the equivalent circuit. The load loss (copper loss) is measured using the same circuit.
- measurement of no-load loss and magnetizing current, at rated voltage and fre-quency. This is measured on one set of windings with the other windings open-circuit. Because the power factor at no-load may be between 0.1 and 0.2, special low power factor wattmeters are necessary to ensure accuracy. A three-wattmeter connection is used for load loss measurement of three-phase transformers, because the previously used two-wattmeter method is inaccurate.
- dielectric tests, consisting of applied voltage and induced voltage tests, sometimes linked with a switching surge test. The applied voltage test verifies the integrity of the winding insulation to earth and between windings, and the induced voltage test verifies the insulation between turns and between wind-ings. The general requirements set down by IEC are shown in Table 6.2. For transformers rated up to 300 kV, the induced voltage test is for one minute at an overvoltage of between 2.5 and 3.5 times rated voltage, carried out at a higher frequency to avoid core saturation. For rated voltages over 300 kV, the induced test is for 30 min at 1.3 or 1.5 times rated voltage, with an initial short period at a higher voltage to initiate activity in any of the vulnerable partial discharge sites; this is linked with a partial discharge test. This long-term test will not prove the insulation against switching overvoltages, and so for this it must be linked to a switching impulse test.
- tests on OLTCs to confirm the timing of the tapchanger mechanism and to prove the capability on full-voltage and full-current operation.

(b) *Type tests:* These are carried out on new designs, or on the first unit of a contract. They include,

- temperature rise test using the normal cooling equipment to verify the temper-ature rise of the windings under full-load or overload conditions. For smaller transformers it may be possible to carry out the temperature rise test at full volt-age by supplying a suitable load. For larger transformers it is more usual to sup-ply the total losses (load loss and no-load loss) under short-circuit conditions

Table 6.2 Dielectric test values for transformers

Rated withstand voltages for windings of transformers < 300 kV

Highest voltage for equipment kV (rms)	Rated short duration power frequency voltage test kV (rms)	Rated lightning impulse voltage test kV (peak)
≤ 1.1	3	–
3.6	10	20 or 40
7.2	20	40 or 60
12	28	60 or 75
24	50	95 or 125
36	70	145 or 170
72.5	140	325
145	230 or 275	550 or 650
245	325 or 360 or 395	750 or 850 or 950

Rated withstand voltages for windings of transformers ≥ 200 kV, in conjunction with a 30 m test at 1.3 × or 1.5 × rated voltage

Highest voltage for equipment kV (rms)	Rated switching impulse withstand voltage kV (peak)	Rated lightning impulse withstand voltage kV (peak)
300	750 or 850	850 or 950 or 1050
420	950 or 1050	1050 or 1175 or 1300 or 1425
525	1175	1425 or 1550
765	1425 or 1550	1550 or 1800 or 1950

Notes: 1 In general the switching impulse test level is ≈ 83% of the lightning impulse test level.
2 These test levels are not applicable in North America.

and to establish the top oil temperature and the average winding temperature at a lower voltage. Where a number of transformers are available, it is possible to supply them in a 'back-to-back' connection and to make the test at full voltage and full current. The test is carried out over a period of 8 to 24 h to establish steady-state temperatures, and it may be associated with *Dissolved Gas Analysis* (DGA) of oil samples taken during the test to identify possible evidence of micro-deterioration in the insulation at high local temperatures.

- a *lightning impulse voltage test* to prove performance under atmospheric lightning conditions. The specified impulse waveform has a 1 µs front time (time-to-crest) and a 50 µs tail time (time from crest to half-crest value). The winding is tested at a level prescribed by IEC which is linked to the rated voltage. This is set out in Table 6.2. One test application is made at 75 per cent of the test value (*Reduced Full Wave*, or RFW) and this is followed by three test applications at full test value (FW). The test is considered satisfactory if there are no discrepancies between the recorded oscillograms of applied voltage and between the recorded oscillograms of neutral current.

(c) *Special tests:* These are carried out at the specific request of the purchaser. The more usual special tests are,

- *chopped-impulse wave test* the purchaser may wish to simulate the situation when a lightning surge propagated along a transmission line is chopped to earth by a lightning protection rod gap mounted on the bushing. The Chop Wave (CW) is specified as an FW waveform with a chop to earth occurring between 3 and 7 µs after the crest. The chop wave test is combined with the lightning impulse voltage test and the test sequence is 1-RFW, 1-FW, 2-CW and 2-FW test applications, sometimes followed by a further RFW for comparison purposes. A successful test shows no discrepancies between applied voltage oscillograms or between neutral current oscillograms, although if rod gaps are used to perform the chopped-wave test there may be acceptable deviations between neutral current oscillograms after the time of chop. If a triggered gap is used to control time of chop the oscillograms should be identical.

- determination of *zero-sequence impedance*, measured between the line terminals of a star-connected or zigzag-connected winding connected together, and its neutral terminal. This is expressed in ohms per phase, and it is a measure of the impedance presented by the transformer to a three-phase fault to earth.
- short-circuit withstand test to verify the integrity of the transformer under through-fault conditions. IEC recommendations allow the withstand performance of a winding to be demonstrated by calculation or by test. The resource needed to test fully a large power transformer is substantial, and few test facilities are available in the world.
- determination of sound levels, which can be measured in terms of sound power, sound pressure or sound intensity. The latter method was developed especially for measurement of low sound levels in the presence of high sound ambients. In transformers which have been designed with low-flux density to reduce the magnetostriction effect, the predominant sound may be produced by the windings as a result of movements caused by the load currents. This load current generated sound should be measured by the sound intensity method.

6.5.2   In-service testing

Two types of in-service testing are used. Surveillance testing involves periodic checks, and condition monitoring offers a continuous check on transformer performance.

(a) *Surveillance testing – oil samples*

When transformers are in operation, many users carry out surveillance testing to monitor operation. The most simple tests are carried out on oil samples taken on a regular basis. Measurement of oil properties, such as breakdown voltage, water content, acidity, dielectric loss angle, volume resistivity and particle content all give valuable information on the state of the transformer. DGA gives early warning of deterioration due to electrical or thermal causes, particularly sparking, arcing and service overheating. Analysis of the oil by *High-Performance Liquid Chromatography* (HPLC) may detect the presence of furanes or furfuranes which will provide further information on moderate overheating of the insulation.

(b) *On-line condition monitoring*

Sensors can be built into the transformer so that parameters can be monitored on a continuous basis. The parameters which are typically monitored are winding temperature, tank temperature, water content, dissolved hydrogen, partial discharge activity, load current and voltage transients. The data collection system may simply gather and analyse the information, or it may be arranged to operate alarms or actuate disconnections under specified conditions and limits which represent an emergency.

Whereas surveillance testing is carried out on some distribution transformers and almost all larger transformers, the high cost of on-line condition monitoring has limited the application to strategic transformers and those identified as problem units. As the costs of simple monitoring equipment fall, the technique should become more applicable to substation transformers.

6.6   Commissioning, maintenance and repair

6.6.1   Commissioning

Small power transformers can be transported to site complete with oil, bushings, tapchangers and cooling equipment. It is then a relatively simple matter to lift them onto a pole or plinth and connect them into the system.

Large transformers are subject to weight restrictions and size limitations. When they are moved by road or rail and it is necessary to remove the oil, bushings, cooling equipment and other accessories to meet these limitations. Very large transformers are usually carried on custom-built transporters, such as the 112-wheel transporter shown in Fig. 6.22; this road transporter is carrying a 1000 MVA, 400 kV transmission transformer that has been dismantled to meet a 400 tonne road weight restriction. Once a transformer of this size arrives on site, it must be lifted or jacked onto its plinth for re-erection. In some cases with restricted space it may be necessary to use special techniques, such as water skates to manoeuvre the transformer into position.

When the transformer has been erected and the oil filled and reprocessed, it is necessary to carry out commissioning tests to check that all electrical connections have been correctly made and that no deterioration has occurred in the insulation system. These commissioning tests are selected from the routine tests and usually include

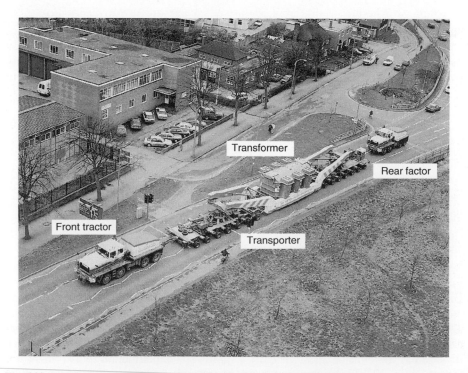

Fig. 6.22 Road transporter carrying a 1000 MVA, 400 kV transmission transformer

winding resistance and ratio, magnetizing current at 440 V, and analysis of oil samples to establish breakdown strength, water content and total gas content. If oil samples indicate high water content then it may be necessary to dry the oil using methods addressed in the following section.

6.6.2   Maintenance

Transformers require little maintenance in service, apart from regular inspection and servicing of the OLTC mechanism. The diverter contacts experience significant wear due to arcing, and they must be replaced at regular intervals which are determined by the operating regime. For furnace transformers it may be advisable to filter the oil regularly in a diverter compartment in order to remove carbon particles and maintain the electrical strength.

The usual method of protecting the oil breather system in small transformers is to use silicone gel breathers to dry incoming air; in larger transformers refrigerated breathers continuously dry the air in a conservator. Regular maintenance (at least once a month) is necessary to maintain a silica gel breather in efficient working order.

If oil samples indicate high water content then it may be necessary to dry the oil using a heating–vacuum process. This also indicates high water content in the paper insulation and it may be necessary to redry the windings by applying a heating and vacuum cycle on site, or to return the transformer to the manufacturer for reprocessing or refurbishment. An alternative procedure is to pass the oil continuously through a molecular sieve filter. Molecular sieves absorb up to 40 per cent of their weight of water.

6.6.3   Diagnostics and repair

In the event of a failure, the user must first decide whether to repair or replace the transformer. Where small transformers are involved, it is usually more economic to replace the unit. In order to reach a decision, it is usually necessary to carry out diagnostic tests to identify the number of faults and their location. Diagnostic tests may include the surveillance tests referred to in section 6.5.2, and it may also be decided to use acoustic location devices to identify a sparking site, low-voltage impulse tests to identify a winding fault and frequency response analysis of a winding to an applied square wave to detect winding conductor displacement.

If the fault is in a winding, it usually requires either replacement of the winding in a repair workshop or rewinding by the manufacturer, but many faults external to the windings, such as connection or core faults can be corrected on site.

Where a repair can be undertaken on site it is essential to maintain dry conditions in the transformer by continual purging using dry air. Any material taken into the tank must be fully processed and a careful log should be maintained of all materials taken into and brought out of the tank.

When a repair is completed, the transformer must be re-dried and re-impregnated, and the necessary tests carried out to verify that the transformer can be returned to service in good condition.

6.7   Standards

The performance requirements of power transformers are covered by a range of international, regional and national standards. The main standards in IEC, CENELEC and BSI are shown in Table 6.3, together with their inter-relationship. Where appropriate, the equivalent ANSI/IEEE standard is also referenced.

Table 6.3 Comparison of international, regional and national standards for transformers

IEC	EN/HD	BS	Subject of standard	ANSI/IEEE
60076-1	EN 60076-1	BSEN 60076-1	Power transformers: general	C57.12.00
60076-2	EN 60076-2	BSEN 60076-2	Power transformers: temperature rise	C57.12.90
60076-3	EN 60076-3	BSEN 60076-3	Power transformers: dielectric tests	C57.12.90
60076-4	EN 60076-4	BSEN 60076-4	Power transformers: guide to lightning impulse testing	C57.98
60076-5	EN 60076-5	BSEN 60076-5	Power transformers: ability to withstand short circuit	C57.12.90
60076-8		BS 5953	Power transformers: application guide	
60076-10	EN 60076-10	BSEN 60076-10	Power transformers: determination of sound levels	C57.12.90
60076-10-1	EN 60076-10-1	BSEN 60076-10-1	Power transformers: determination of sound levels: user guide	
60137	EN 60137	BSEN 60137	Insulated bushings for ac voltages above 1 kV	C57.19.00
	HD 506	BS 7616	Bushings for liquid-filled transformers 1 kV to 36 kV	
60214-1	EN 60214-1	BSEN 60214-1	Tapchangers: performance requirements and test methods	C57.131
60214-2	EN 60214-2	BSEN 60214-2	Tapchangers: application guide	
60289	EN 60289	BSEN 60289	Reactors	C57.21
60354		BS 7735	Loading guide for oil-immersed power transformers	C57.92
TR 60616			Terminal and tapping markings for power transformers	C57.12.70
60726	EN 60726	BSEN 60726	Dry-type power transformers	C57.12.01
60905			Loading guide for dry-type power transformers	C57.96
61248			Transformers and inductors for electronics and telecommunications	
61558			Safety of power transformers, power supply units and similar	
61378-1	EN 61378-1	BSEN 61378-1	Convertor transformers: transformers for industrial applications	
61738-2	EN 61738-2	BSEN 61738-2	Convertor transformers: transformers for HVDC applications	
61738-3	EN 61738-3	BSEN 61738-3	Convertor transformers: user guide	
	HD 428-1	BS 7281-1	Oil filled distribution transformers: general to 24 kV	C57.12.22
	HD 428-2	BS 7281-2	Oil-filled distribution transformers: transformers with cable boxes	
	HD 428-3	BS 7281-3	Oil-filled distribution transformers: transformers for 36 kV	
	HD 428-4	BS 7281-4	Oil-filled distribution transformers: harmonic currents	
	HD 538-1	BS 7844-1	Dry-type distribution transformers: general to 24 kV	C57.12.51
	HD 538-2	BS 7844-2	Dry-type distribution transformers: transformers for 36 kV	
		BS 3535	Safety requirements for transformers that may be stationary or portable, single phase or polyphase	

Chapter 7

Switchgear

N.P. Allen
T. Harris

C.J. Jones
VA TECH Transmission & Distribution Limited

S. Lane
Whipp & Bourne

7.1 Introduction

Switchgear is used to connect and disconnect electric power supplies and systems. It is a general term which covers the switching device and its combination with associated control, measuring, protective and regulating equipment, together with accessories, enclosures and supporting structures.

Switchgear is applied in electrical circuits and systems from low voltage, such as domestic 220/240 V applications, right up to transmission networks up to 1100 kV. A wide range of technology is needed, and this chapter is split into the four main subdivisions of low voltage, medium voltage (distribution), high voltage (transmission) and dc.

The main classes of equipment are as follows:

- disconnectors, or isolators
- switches
- fuse–switch combinations
- circuit breakers
- earthing switches

A *disconnector* is a mechanical switching device which in the open position provides a safe working gap in the electrical system which withstands normal working system voltage and any overvoltages which may occur. It is able to open or close a circuit if a negligible current is switched, or if no significant change occurs in the voltage between the terminals of the poles. Currents can be carried for specified times in normal operation and under abnormal conditions.

A *switch* is a mechanical device which is able to make, carry and interrupt current occurring under normal conditions in a system, and to close a circuit safely, even if a fault is present. It must therefore be able to close satisfactorily carrying a peak current corresponding to the short-circuit fault level, and it must be able to carry this fault current for a specified period, usually one or three seconds.

A fuse and a switch can be used in combination with ratings chosen so that the fuse operates at currents in excess of the rated interrupting or breaking capacity of the switch.

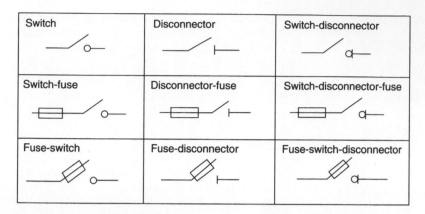

Switch	Disconnector	Switch-disconnector
Switch-fuse	Disconnector-fuse	Switch-disconnector-fuse
Fuse-switch	Fuse-disconnector	Fuse-switch-disconnector

Fig. 7.1 Summary of equipment definitions

Such a device is known as a *'fuse switch'* if the fuseholder is also used as part of the main moving contact assembly, or a *'switch fuse'* if the fuse is a separate and static part of an assembly which includes the switch connected in series.

Figure 7.1 shows schematically the various combinations of fuse, switch and disconnector that are available.

A *circuit breaker* is a mechanical switching device which is not only able to make, carry and interrupt currents occurring in the system under normal conditions, but also to carry for a specified time and to make and interrupt currents arising in the system under defined abnormal conditions, such as short circuits. It experiences the most onerous of all the switching duties and is a key device in many switching and protection systems.

An *earthing switch* is a mechanical device for the earthing and short-circuiting of circuits. It is able to withstand currents for a specified time under abnormal conditions, but it is not required to carry normal service current. An earthing switch may also have a short-circuit making capacity.

7.2   Principles of operation

Since they perform the most arduous duty, circuit breakers are used here for the following brief description of switchgear operating principles.

The heart of a circuit breaker is the contact system which comprises a set of moving contacts, a set of fixed contacts, their current carrying conductors or leads and an opening mechanism which is often spring-loaded.

The fixed and moving contacts are normally made of copper, with sufficient size and cross section to carry the rated current continuously. Attached or plated onto the copper are the contact tips or faces which are of silver or an alloy. The contact tip or face is the point at which the fixed and moving contacts touch, enabling the current to flow. It is a requirement that a low contact junction resistance is maintained, and that the contacts are not welded, destroyed or unduly eroded by the high thermal and dynamic stresses of a short circuit.

As the device closes, the contact faces are forced together by spring or other kinds of pressure generated from the mechanism. This pressure is required in order to reduce the contact junction resistance and therefore the ohmic heating at the contacts; it also assists in the destruction or compaction of foreign material such as oxides which may contaminate the contact faces. The closing process is further assisted by arranging the geometry to give a wiping action as the contacts come together; the wiping helps to ensure that points of purely metallic contact are formed.

The breaking or interruption action is made particularly difficult by the formation of an arc as the contacts separate. The arc is normally extinguished as the current reaches a natural zero in the ac cycle; this mechanism is assisted by drawing the arc out to maximum length, therefore increasing its resistance and limiting the arc current. Various techniques are adopted to extend the arc; these differ according to size, rating and application and this is covered in more detail in the following sections.

The interruption of a resistive load current is not usually a problem. In this case the power factor is close to unity. When the circuit breaker contacts open, draw an arc and interrupt the current, the voltage rises slowly from its zero to its peak following its natural 50 Hz or 60 Hz shape. The build up of voltage across the opening contacts is therefore relatively gentle, and it can be sustained as the contact gap increases to its fully open point.

However, in many circuits the inductive component of current is much higher than the resistive component. If a short circuit occurs near the circuit breaker, not only is the fault impedance very low, and the fault current at its highest, but also the power factor may be very low, often below 0.1, and the current and voltage are nearly 90° out of phase. As the contacts open and the current is extinguished at its zero point, the voltage tends instantaneously to rise to its peak value. This is illustrated in Fig. 7.2. This results in a high rate of rise of voltage across the contacts, aiming for a peak

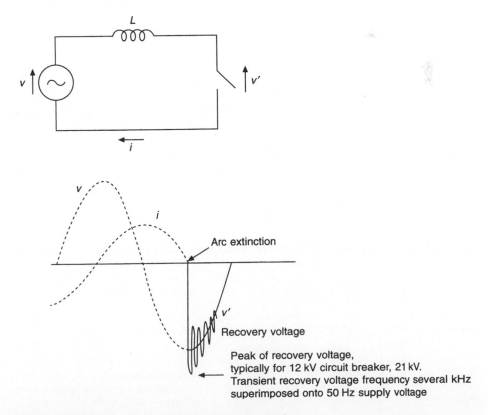

Fig. 7.2 Circuit breaker recovery voltage in an inductive circuit

transient recovery voltage which is considerbly higher than the normal peak system voltage. There is a risk under these circumstances that the arc will re-strike even though the contacts are separating, and the design of the circuit breaker must take this into account.

7.3   Low-voltage switchgear

7.3.1   Switches, disconnectors, switch disconnectors and fuse combination units

7.3.1.1   Construction and operation

These switches are categorized by the ability to make and break current to particular duties which are listed for ac applications in Table 7.1. The category depends upon a multiple of the normal operating current and an associated power factor.

In all these devices, the contacts are usually made of silver-plated copper; it is unusual to find the alloy contact tips used in circuit breakers. The contact system, connecting copperwork and terminals, is usually in a housing of thermoset plastic or of a thermoplastic with a high temperature capability. The mechanism of the switch may be located in a number of positions, but it is usual to mount it at the side of or below the contact system.

The switching devices may be fitted with auxiliaries which normally take the form of small switches attached to the main device. These auxiliary switches provide means of indication and monitoring of the position of the switch.

There are switching devices that are motor-assisted to open and close the contacts, and some have tripping devices incorporated into the mechanism to open the switch under circuit conditions outside the protective characteristics of the fuse.

Many types of fuselink can be accommodated on a switch fuse or a fuse switch. In the UK, the bolted type is probably the most common, followed by the 'clip-in' type. There are many designs available to hold the clip-in fuselink within the switch moulding. In continental Europe the DIN type fuselink is the most common; here the fuse tags are formed as knife blades to fit into spring-tensioned clips.

The construction of a *fuse switch* is illustrated generally in Fig. 7.3. It consists of a moving carriage containing the fuselinks, which is switched into and out of a fixed system containing the contacts and mechanism. The moving carriage is shown in Fig. 7.4; it consists of a plastic and steel framework with copper contacts and terminals onto which are mounted the fuselinks. This is normally a three-pole or four-pole construction. It is usual to be able to extract the moving carriage to facilitate the changing of fuselinks outside the switch. Removal of the carriage also ensures total

Table 7.1 Categories of duty and overload current ratings for switches, disconnectors, switch disconnectors and fuse combination units

Category of duty	% of full-load current rating		Application
	making	breaking	
AC21	150%	150%	resistive loads
AC22	300%	300%	resistive and inductive loads
AC23	1000%	800%	high inductive loads, stalled motor conditions

Fig. 7.3 Main features of a fuse switch

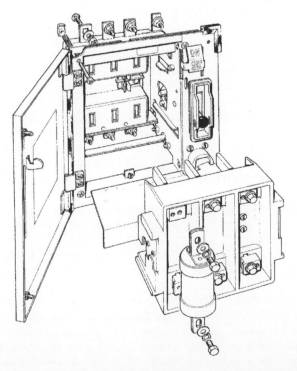

Fig. 7.4 Moving carriage in a fuse switch; Bill Sovereign fuse switch disconnector (courtesy of Bill Switchgear)

isolation of the load from the supply. The moving carriage is moved in and out by a spring-charged mechanism that ensures a quick make and break of the contacts.

In a *switch fuse* the fuselinks remain static; the simplest form is a knife blade contact in series with a fuselink. In the modern form of switch fuse the fuselinks are mounted on a plastic enclosure which contains the contacts and the mechanism to operate these contacts. The contacts are normally arranged on both load-side and supply-side of the fuselink in order to ensure that the fuse is isolated when the switch is in the off position. The contact system is driven by a spring-assisted mechanism which is usually designed to provide quick make and break operation independent of the speed at which the operator moves the on–off handle.

In its simplest form, the switch is a knife blade contact, but the modern switch would follow the form of the fuse switch without the fuselinks. Many manufacturers use a fuse switch or a switch fuse with solid copper links in place of the fuselinks.

The *disconnector* also follows the construction of the switch, but it must be capable of providing complete isolation when the contacts are open. This usually requires the provision of adequate clearance and creepage distances between live parts.

The *switch-disconnector* combines the properties of a switch and a disconnector; this is in most cases what a manufacturer would build for sale.

7.3.1.2 Standards and testing

Switches, disconnectors, switch disconnectors and fuse combination units are designed to comply with the requirements of IEC 947-3 or EN 60947-3. This requires a combination of type tests and routine tests.

A sequence of type tests is required to prove compliance with the standard. This sequence includes the following key tests:

- the making and breaking switching capacity test, which shows that the device is suitable for extreme overload conditions of resistive, inductive or highly inductive loads. The categories of duty and the current overload ratings have been shown in Table 7.1. The test is carried out by operating the device a number of times at the assigned rating.
- temperature-rise verification. This is carried out at the highest current rating of the device to prove that under full-load conditions in service the device will not damage cables, terminals and insulating materials or put operators at risk through contact with hot accessible parts. The limits of acceptable temperature rise are stipulated in the standard.
- the operational performance test. This is conducted to prove the mechanical and electrical durability of the device. A number of on-load and off-load switching operations are made, depending upon the make/break duty assigned by the manufacturer.
- dielectric verification to prove that the device has completed the sequence without damage to its insulation system.

A fuse switch or switch fuse is also proven by a fuse-protected short-circuit withstand (breaking) test. In this test, the fuselinks used have to be of the maximum rating, and with a breaking capacity assigned by the manufacturer. No damage such as welding of the contacts must occur to the switch as a result of this test. A fuse-protected making test is also carried out in which the switch is closed onto the declared rated short-circuit current.

A disconnector has to provide isolation properties and additional type tests are performed to prove these. In particular, a leakage current test is conducted after the main test sequence; maximum levels of acceptable leakage current are specified. In addition, the isolator handle is subjected to a force of three times the normal operation force necessary to switch the device off (within given minimum and maximum limits), the contacts of one phase being artificially locked in the off position and the test force being applied to open the switch. The on–off indicators of the disconnector must not give a false indication during and after this test.

Routine tests to be applied to all switch devices include an operational check in which each device is operated five times to check mechanical integrity. Also a dielectric test is carried out at a voltage which depends upon the rated voltage of the device. For a rated operating voltage of 380/415 V the dielectric test voltage is 2500 V.

7.3.2 Air circuit breakers and moulded case circuit breakers

7.3.2.1 Construction and operation

Both the air circuit breaker and the moulded case circuit breaker (mccb) comprise the following features:

- a contact system with arc-quenching and current-limiting means
- a mechanism to open and close the contacts
- auxiliaries which provide additional means of protection and indication of the switch positions

The modern air circuit breaker is generally used as an incoming device on the supply side of a low voltage switchboard, and it represents the first line of protection on the load side of the transformer. An example is shown in Fig. 7.5. In addition to the above features, it also includes:

- a tripping and protection system to open the circuit breaker under fault conditions (if required)
- a means of isolating the device from the busbars
- usually an open construction, or the contact system housed in a plastic moulding
- current ratings from 400 A to 6300 A

The mccb may be used as an incoming device, but it is more generally used as an outgoing device on the load side of a switchboard. It is normally mounted into a low-voltage switchboard or a purpose-designed panel board. In addition to the three features listed at the start of this section, it also includes:

- an electronic or thermal/electromagnetic trip sensing system to operate through the tripping mechanism and open the circuit breaker under overload or fault conditions
- all parts housed within a plastic moulded housing made in two halves
- current ratings usually from 10 A to 1600 A

The basis of the main contact system has been explained in section 7.3.1.1. The fixed contacts are usually mounted on a back panel or within a plastic moulding, and

Fig. 7.5 Air circuit breaker (courtesy of Terasaki (Europe) Ltd)

the moving contacts are usually supported on an insulated bar or within an insulated carriage.

In addition to these main contacts, arcing contacts are provided. The arcing contacts generally comprise a silver–tungsten alloy which provides high electrical conductivity with excellent wear against arc erosion and good anti-welding properties. The arc contacts are positioned in such a way that they make first and break last during opening and closing of the contacts. Their purpose is to ensure that any arcing takes place on the arcing contacts before the main contacts touch; the arcing contacts can then break once the main contacts have been forced together. During opening, the sequence is reversed. This action protects the main contacts from damage by arcing.

Extinguishing of the arc is an important feature in all low-voltage switchgear. All air circuit breakers have devices to extinguish the arc as quickly as possible. It is well known that an arc, once formed, will tend to move away from its point of origin; by forming the contact system carefully, the magnetic field generated by the current flow is used to move the arc into an arcing chamber where its extinction is aided.

Within the arcing chamber are a number of metal arc-splitter plates which normally have a slot or 'V' shape cut into them in order to encourage the arc to run into the arcing chamber. A typical arrangement is shown in Fig. 7.6. The plates are used to split the arc into a number of smaller arcs, having the effect of lengthening and cooling the arc.

The arc is then extinguished when the voltage drop along it is equal to the voltage across the open contacts.

If the arc can be extinguished very quickly before the full prospective fault current is reached, the maximum current passed through the circuit breaker can be limited. This current limiting is achieved if the arc is extinguished within 10 ms of the start of the fault, before the peak of the sinusoidal current waveform is reached. Many designs have been tried in an attempt to achieve current limiting. Many of these designs are based on forming the current carrying conductors within the circuit breaker in such a way that the electromagnetic forces created by the currents force the arc into the arcing chamber very quickly. A particular development of this is to form 'blow-out' coils from the current-carrying conductors.

The opening and closing mechanism of an air circuit breaker generally uses the energy developed in releasing a charged spring. The spring may be charged manually or with the aid of a winding mechanism attached to a motor. The charge is usually held until a release mechanism is activated, and once the spring energy is released to the main mechanism, linkages are moved into an over-toggle mode to close the moving contacts onto the fixed contacts. The force of the main spring is also used to transfer energy into compressing the moving contact springs, and it is these springs which provide the energy to open the contacts when required.

Attached to and interfacing with the main mechanism is the tripping mechanism which is used to open the contacts. The means of tripping an air circuit breaker is a coil, which may be a shunt trip coil, an undervoltage coil or a polarized trip coil. The shunt trip coil is current-operated and consists of a winding on a bobbin with a moving core at its centre. When a current flows, the magnetic flux causes the core to move and this

Fig. 7.6 Contact system and arc chute in an air circuit breaker

core operates the trip mechanism. An undervoltage coil is similar in construction, but the moving core is held in place by the magnetic flux against a spring; if the voltage across the winding falls this allows the spring to release the core and operate the trip mechanism. In the polarizing trip coil, the output of the coil is used to nullify the magnetic field from a permanent magnet which is set into the coil; when the coil is energized, the permanent magnetic field is overcome and using the spring energy, a moving core is released to operate the trip mechanism.

The release coil receives its signal from the overcurrent detection device, which is usually an electronic tripping unit. This tripping unit has the capability to detect overloads and short circuits, and it may be able to detect earth fault currents. The overload protection device detects an overload current and relates this to the time for which the current has flowed before initiating a trip signal; the time–current characteristic which is followed here is determined by the manufacturer within the guidelines of the standard. The general principles of time–current protection are explained more fully in Chapter 8, but a typical time–current characteristic is shown for convenience in Fig. 7.7. The short-circuit protection enables a high fault current to be detected and the circuit breaker to be opened in less than 20 ms. The short-circuit protection characteristics

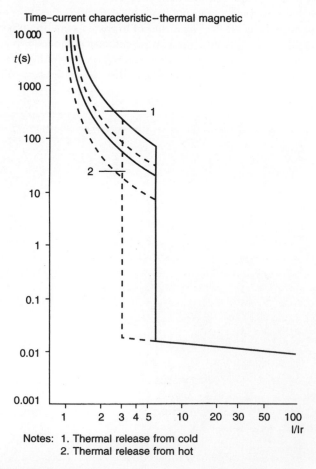

Time–current characteristic–thermal magnetic

Notes: 1. Thermal release from cold
        2. Thermal release from hot

Fig. 7.7 Typical time–current characteristic curves

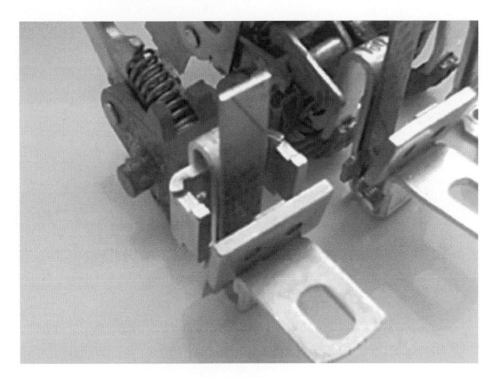

Fig. 7.8 Electromechanical and electromagnetic trips used in a mccb

are usually adjustable in multiples of the full-load rated current, and they may incorporate a time interval which would usually be less than a second.

The mccb also incorporates an electronic trip device similar to that described here, but this may not include all of the sophistication of the air circuit breaker system. Some mccbs will include an electromechanical thermal trip and electromagnetic trip in place of the electronic trip unit. These trips are shown in Fig. 7.8. The electromechanical thermal trip consists of a heater and bimetal, the heater being part of the main current-carrying circuit. In overload conditions the heater causes deflection of the bimetal which is time-related depending upon the current flow through the heater. The bimetal deflection results in operation of the trip mechanism and hence a time–current characteristic can be drawn for the circuit breaker. An electromagnetic trip usually comprises a 'U'-shaped coil around the current-carrying circuit of the mccb, together with a moving pole piece. When a fault current passes, the magnetic flux within the 'U'-shaped coil pulls in the pole piece and this movement is used to trip the circuit breaker.

7.3.2.2   Testing and standards

Low-voltage air circuit breakers and mccbs are designed and tested according to the requirements of IEC 947-2, or EN 60947-2.

A series of type tests are carried out to prove compliance with the standard. This series of tests includes the following:

- *a short-circuit capability test.* This determines the maximum short circuit current the device can withstand, while after the test, retaining the ability to pass full-load current without overheating, and to allow the circuit breaker limited use. This test is carried out in a short-circuit test laboratory able to deliver exactly the current specified in the standard; large generators and transformers are necessary for this work, making it expensive.

- *a short-time withstand test.* This establishes that the circuit breaker can carry a short-circuit current for a period of time; in many cases this would be 50 kA for 1 s. A circuit breaker with this capability can be used as an incoming device to a distribution system. Based on the principles of co-ordination explained in Chapter 8, this allows other devices within the distribution system to open before the incoming circuit breaker, thereby not necessarily shutting off all of that distribution system.
- *an endurance test.* This determines the number of open–close operations the circuit breaker can withstand before some failure occurs which means that the device will no longer carry its rated current safely. The tests are carried out with and without current flowing through the contacts.
- *opening during overload or fault conditions.* This test verifies that the device opens according to the parameters set by the standard and those given by the manufacturer. It determines the time–current characteristic of the circuit breaker, and example of which has been shown in Fig. 7.7.
- *a temperature rise test.* This is carried out after all the above type tests. The standard specifies a maximum temperature rise permissible at the terminals and it ensures that the device will be safe to use after short circuits have been cleared and after a long period in service.

7.3.3   Miniature circuit breakers

Like the mccb, the miniature circuit breaker (mcb) has a contact system and means of arc quenching, a mechanism and tripping and protection system to open the circuit breaker under fault conditions.

The mcb has advanced considerably in the past 30 years. Early devices were generally of the 'zero-cutting' type, and during a short circuit the current had to pass through a zero before the arc was extinguished; this provided a short-circuit breaking capacity of about 3 kA. Most of these early mcbs were housed in a bakelite moulding. The modern mcb is a much smaller and more sophisticated device. All the recent developments associated with moulded case circuit breakers have been incorporated into mcbs to improve their performance, and with breaking capacities of 10 kA to 16 kA now available, mcbs are used in all areas of commerce and industry as a reliable means of protection.

Most mcbs are of single-pole construction for use in single-phase circuits. The complete working system is housed within a plastic moulding, typical external appearance being shown in Fig. 7.9. A section showing the principal parts of the mcb is shown in Fig. 7.10. The contact system comprises a fixed and a moving contact, and attached to each is a contact tip which provides a low-resilience contact junction to resist welding.

Modern mcbs are fitted with arc chutes consisting of metal plates which are held in position by insulating material. The arc chute does not necessarily surround the contact; in some designs arc runners are provided to pull the arc into the arc chute.

The tripping mechanism usually consists of a thermal-magnetic arrangement. The thermal action is provided by a bimetal with, in some cases, a heater. For ratings in the

Fig. 7.9 Miniature circuit breaker, external view

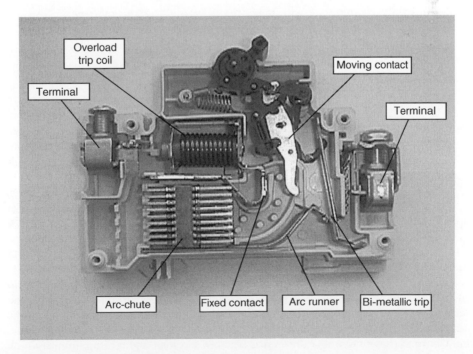

Fig. 7.10 Principal parts of a miniature circuit breaker

Table 7.2 Magnetic trip settings for mcbs

MCB Type	Minimum trip current	Maximum trip current	Application
B	$3\,I_n$	$5\,I_n$	Domestic: resistive or small inductive loads such as lighting and socket outlets
C	$5\,I_n$	$10\,I_n$	Light industrial: inductive loads such as fluorescent lighting and motors
D	$10\,I_n$	$20\,I_n$	Very inductive loads such as welding machines

range 6–63 A the bimetal forms part of the current path, the heat generated within the bimetal itself being sufficient to cause deflection. The deflection is then used to activate the tripping mechanism. The characteristics of the bimetal are chosen to provide particular delays under certain overload or fault currents according to the required time–current characteristic. A high-resistance bimetal is used for low-current devices and a lower resistance bimetal for high-current devices. In very low-current mcbs a heater may be incorporated around the bimetal in order to generate sufficient heat to deflect it.

The magnetic tripping element usually consists of a coil which is wrapped around a tube, there being a spring-loaded slug within the tube. Movement of the slug operates the tripping mechanism to open the mcb. It can also be used to assist in opening the contacts by locating the coil close to the moving contact. When a fault current flows, the high magnetic field generated by the coil overcomes the spring force holding the slug in position; the slug then moves to actuate the tripping mechanism and forces the contacts apart by striking the moving contact arm. For low mcb ratings the coil is formed from thin wire with many turns; for higher ratings the wire is thicker, with fewer turns. The magnetic trip is set by the manufacturer according to the required characteristics. These characteristics are defined in the standard and form 'types' which are shown in Table 7.2.

MCBs are designed and tested according to the requirements of IEC 898.

7.3.4 Residual current devices

The residual current device (rcd) is used to detect earth fault currents and to interrupt supply if an earth current flows. The main application is to prevent electrocution but rcds can also be used to protect equipment, especially against fire. The earth fault currents that operate an rcd can range from 5 mA up to many amperes. For typical domestic applications the typical trip current would be 30 mA.

The rcd can be opened and closed manually to switch normal load currents, and it opens automatically when an earth fault current flows which is about 50 per cent or more of the rated tripping current.

The main features of an rcd are shown in Fig. 7.11. The key component is a toroidal transformer, upon which the load current (live) and return current (neutral) conductors are wound in opposite directions. The toroid also carries a detecting winding. If no earth fault current is flowing, then the load and return currents are equal. In this case the mmfs generated by the load and return current windings are equal; there is no resultant flux in the toroid and the detecting winding does not generate any current. When a fault current flows there is a difference between the load and return currents which generates a resultant flux in the toroid and induces a current in the detecting winding.

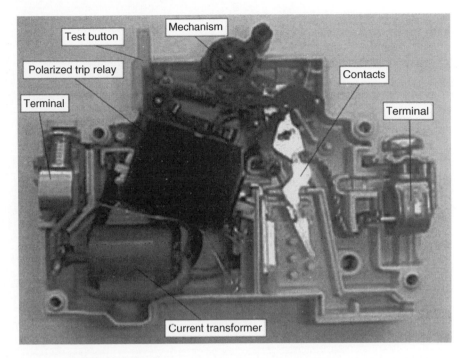

Fig. 7.11 Residual current device

The current generated in the detecting winding operates a relay which opens the main contacts of the rcd.

The detecting winding has to produce from a very small output, sufficient power to operate the tripping mechanism. Two alternative methods are used. In the first method, the output signal from the detecting coil is electronically amplified and the second method uses a polarized relay operating on a sensitive mechanical trip mechanism. The operation of a polarized trip relay has already been described for circuit breakers in section 7.3.2.1; it is based on the magnetic output of a small coil nullifying the field from a permanent magnet, causing the release of an armature. The basic operation is illustrated in Fig. 7.12.

The operation of an rcd has here been described for single-phase operation, but it may also be applied in a three-phase application where typically it might be used in a light industrial system for protection against fire. There are two arrangements of a three-phase rcd. Either the three phases are wound around a current transformer, or the three phases and the neutral are wound onto a balancing transformer.

The rcd has only limited breaking capacity and it is not a replacement for overcurrent protection devices such as the mcb. The residual current breaker with overcurrent (rcbo) is now available; this is an rcd with an overcurrent tripping mechanism and enhanced contacts to cope with interruption of fault conditions.

RCDs are designed and tested according to the requirements of IEC 1008 and IEC 1009.

7.3.5 Standards

The leading international standards adopted by for low-voltage switchgear are shown in Table 7.3.

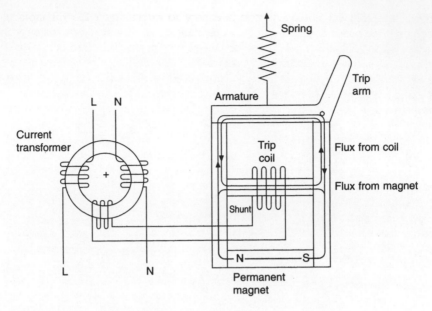

Fig. 7.12 Operation of polarized trip relay

7.4 Medium-voltage (distribution) switchgear

Distribution switchgear, also commonly referred to as medium-voltage switchgear, is generally acknowledged to cover the range 1 kV to 36 kV; 72.5 kV and even 132 kV can be considered as distribution voltages rather than transmission voltages, and the equipment overlaps with high-voltage or transmission switchgear in this range.

Switchgear in the distribution voltage range differs considerably from low-voltage switchgear. There are similarities with aspects of transmission switchgear, but many functions and practices are different.

7.4.1 Types of circuit breaker

In the past, oil- and air-break devices have predominated, but now several alternative methods of arc interruption are used in distribution-voltage circuit breakers.

Early *oil designs* featured plain-break contacts in a tank of oil capable of withstanding the considerable pressure built up from large quantities of gas generated by long arcs. During the 1920s and 1930s various designs of arc control device were introduced to improve performance. These were designed such that the arc created

Table 7.3 International standards relating to low-voltage switchgear

IEC	EN/HD	Subject of standard
IEC 898	EN 60898	Miniature circuit breakers (mcbs)
IEC 947-2	EN 60947-2	Air circuit breakers (acbs) and moulded case circuit breakers (mccbs)
IEC 947-3	EN 60947-3	Switches – fuse combination switches
IEC 1008	EN 61008	Residual current devices (rcds)

between the contacts produces enough energy to break down the oil molecules, generating gases and vapours which by the cooling and de-ionizing of the arc resulted in successful clearance at current zero. During interruption, the arc control device encloses the contacts, the arc, a gas bubble and oil. Carefully designed vents allow the gas to escape as the arc is lengthened and cooled. Figure 7.13 shows an axial-blast arc control device. The use of oil switchgear is reducing significantly in most areas of the world because of the need for regular maintenance and the risk of fire in the event of failure.

Sulphur hexafluoride (SF_6) is the most effective gas for the provision of insulation and arc interruption. It was initially introduced and used predominantly in transmission switchgear, but it became popular at distribution voltages from the late 1970s.

The main early types of SF_6 circuit breaker are known as 'puffer' types. These use a piston and cylinder arrangement, driven by an operating mechanism. On opening, the cylinder containing SF_6 gas is compressed against the piston, increasing the gas pressure; this forces the gas through an annular nozzle in the contact giving a 'puff' of gas in the arc area. The gas helps to cool and de-ionize the arc, resulting in arc interruption at current zero. The dielectric properties of SF_6 are quickly re-established

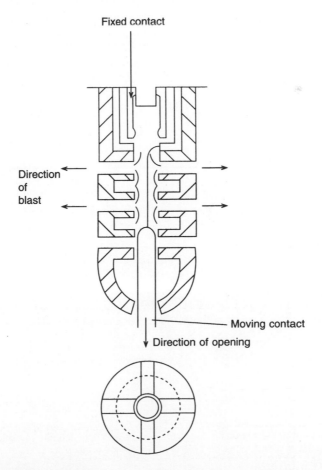

Fig. 7.13 Axial-blast arc control device

in the heat of the arc, giving a rapid increase in dielectric strength as the voltage across the contacts begins to rise. The energy required to drive the puffer type of circuit breaker is relatively high, resulting in the need for a powerful mechanism.

Other designs of SF_6 circuit breaker started to be introduced in the 1980s. These make more use of the arc energy to aid interruption, allowing the use of a lighter and cheaper operating mechanism. Whilst in the puffer design the compressed SF_6 is made to flow over an arc which is basically stationary, these newer designs move the arc through the SF_6 to aid interruption. In the rotating arc SF_6 circuit breaker, a coil is wound around the fixed contact. As the moving contact withdraws from the fixed contact the current is transferred to an arcing contact and passes through the coil. The current in the coil sets up an axial magnetic field with the arc path at right angles to the magnetic flux, and rotation of the arc is induced according to Fleming's left-hand rule (see Chapter 2, eqn 2.16). The arc rotation is proportional to the magnitude of the current. Careful design is required in order to ensure an adequate phase shift between the flux and the current to maintain adequate movement of the arc as the current falls towards its zero crossing. The operation of a rotating-arc interrupter is illustrated in Figs 7.14 and 7.15.

Other types of self-extinguishing circuit breaker use an insulating nozzle which is similar to that in a puffer breaker, but with a coil to produce a magnetic field for moving the arc. The thermal energy in the arc is used to increase the local pressure. This results in arc movement which again is proportional to the current, and successful interruption follows. In some designs a small puffer is also used to ensure there is enough gas flow to interrupt low levels of current, where the thermal energy in the arc is relatively low.

The various types of SF_6 switchgear are referred to with respect to high-voltage switchgear in section 7.5.2, and Fig. 7.24 shows the main configurations.

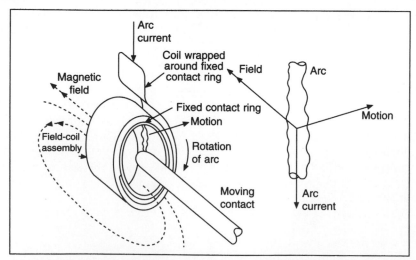

The arc has been extended towards the coil axis by transverse movement of the contact bar. It is shown after transfer to the arcing tube and consequent production of the magnetic field. It is thus in the best plane to commence rotation.

Fig. 7.14 Operation of a rotating-arc SF_6 interrupter

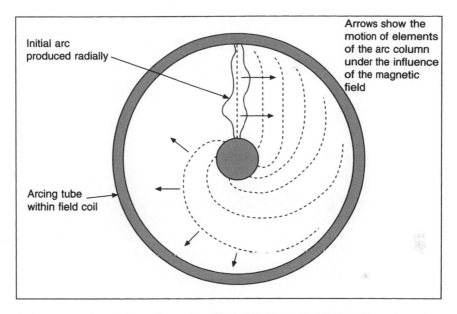

Development of a spiral arc. The arc is motivated to move sideways but this tendency is limited by the shape of the electrodes and it develops quickly into a spiral with each element tending to move sideways as shown by the arrows.

Fig. 7.15 Operation of a rotating-arc SF_6 interrupter

The main alternative to the SF_6 circuit breaker in the medium-voltage range is the *vacuum interrupter*. This is a simple device, comprising only a fixed and a moving contact located in a vacuum vessel, but it has proved by far the most difficult to develop. Although early work started in the 1920s, it was not until the 1960s that the first vacuum interrupters capable of breaking large currents were developed, and commercial circuit breakers followed about a decade later.

The principle of operation of a vacuum interrupter is that the arc is not supported by an ionized gas, but is a metallic vapour caused by vaporization of some of the contact metal. At zero current, the collapse of ionization and vapour condensation is very fast, and the extremely high rate of recovery of dielectric strength in the vacuum ensures a very effective interrupting performance. The features of a vacuum interrupter which are key to its performance are the contact material, the contact geometry and ensuring that the envelope (a glass or ceramic tube with welded steel ends) remains vacuum-tight throughout a working life in excess of 20 years. Even today, specialized manufacturing techniques are necessary to achieve this. A typical vacuum interrupter is shown in Fig. 7.16.

The circuit breaker is located within a switchgear housing, some types of which are described later. The main insulation in the housing is usually air, although some designs now have totally sealed units filled with SF_6 gas. Structural isolation is required to support current-carrying conductors; this is normally some type of cast resin. Thermoplastic materials which can be injection-moulded are often used for smaller components, but larger items such as bushings which are insulation-covered are usually made from thermosetting materials such as polyurethane or epoxy resin mixed with filler to improve its mechanical and dielectric properties.

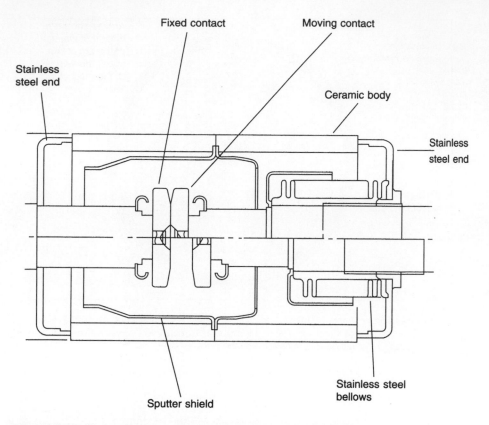

Fig. 7.16 Typical vacuum interrupter (courtesy Meidensha Corporation, Japan)

7.4.2 Main classes of equipment

7.4.2.1 Primary switchgear

Primary substations up to 36 kV are usually indoor, although they can be outdoor where space is not a problem. Indoor equipment is usually metal-enclosed, and it can be subdivided into metalclad, compartmented and cubicle types. In the metalclad type, the main switching device, the busbar section and cable terminations are segregated by metal partitions. In compartmented types the components are housed in separate compartments but the partitions are non-metallic. The third category covers any form of metal-enclosed switchgear other than the two described.

The vertically isolated, horizontally withdrawn circuit breaker is the traditional arrangement in the UK or UK-influenced areas. This originates from the bulk-oil circuit breaker, for which there was a need for easy removal for frequent maintenance, but manufacturers have utilized the already available housings to offer vacuum or SF$_6$ circuit breakers as replacements for oil circuit breakers, or even for new types to this design. The design also provides convenient earthing of the circuit or busbars by alternative positions of the circuit breaker in the housing; this is achieved by disconnecting the circuit breaker, lowering it from its service position, moving it horizontally and then raising and reconnecting the connections between the busbar or cable and earth. A typical section is shown in Fig. 7.17.

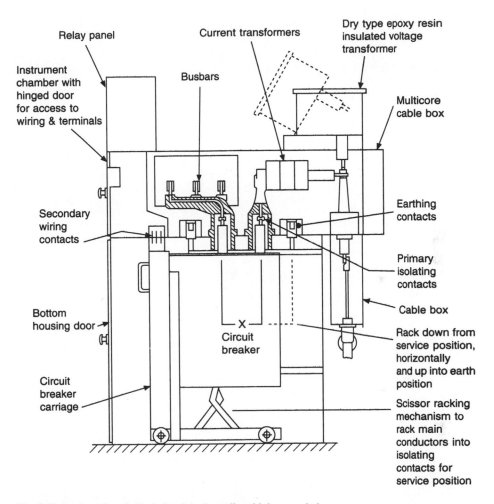

Fig. 7.17 Section of vertically isolated, horizontally withdrawn switchgear

Horizontally isolated, horizontally withdrawn equipment has been used extensively in mainland Europe, with air-break, small oil volume and latterly with SF_6 and vacuum circuit breaker truck designs. A section of this type of switchgear is shown in Fig. 7.18. Since the circuit breaker cannot be used as the earthing device in this design, integral earthing switches or portable earthing devices are required.

Fixed-position circuit breakers were introduced in the 1970s; these depart from the withdrawable arrangements and are based on a 'sealed-for-life' concept. Due to the early problems with vacuum interrupters, some users were reluctant to accept fixed-position vacuum switchgear, but now that excellent long-term reliability of vacuum interrupters has been established, newer designs have been introduced again recently. Several arrangements in which the housing is completely sealed and filled with SF_6 as a dielectric medium are now available and an example is shown in Fig. 7.19. Because of the compact dimensions that SF_6 permits these are gaining popularity, particularly at 36 kV, where there can be a cost advantage not only in the price, but also in the reduced size of the equipment and the substation required.

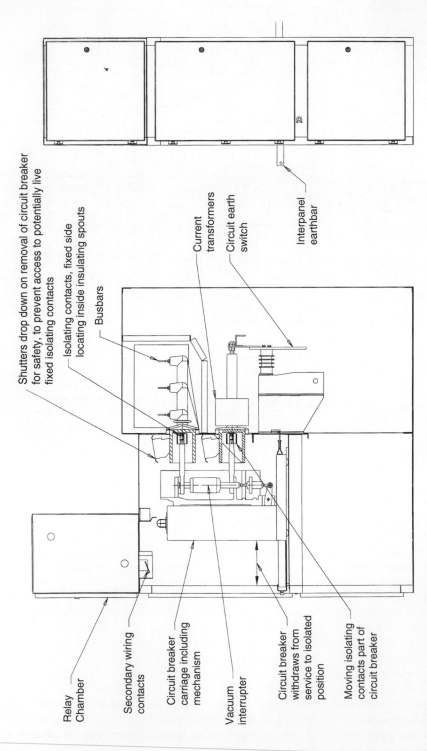

Shutters drop down on removal of circuit breaker for safety, to prevent access to potentially live fixed isolating contacts

Isolating contacts, fixed side locating inside insulating spouts

Busbars

Current transformers

Circuit earth switch

Interpanel earthbar

Relay Chamber

Secondary wiring contacts

Circuit breaker carriage including mechanism

Vacuum interrupter

Circuit breaker withdraws from service to isolated position

Moving isolating contacts part of circuit breaker

Fig. 7.18 Section of horizontally isolated, horizontally withdrawn switchgear

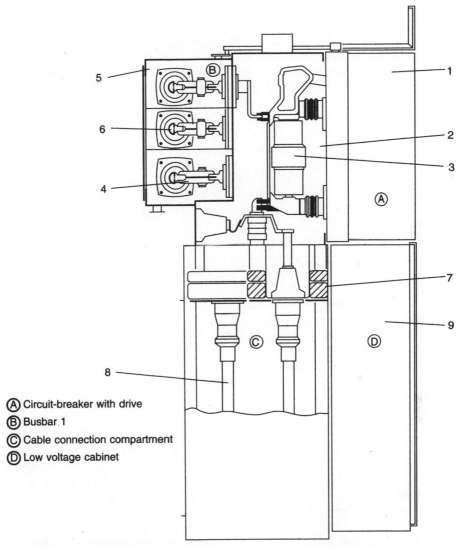

A Circuit-breaker with drive

B Busbar 1

C Cable connection compartment

D Low voltage cabinet

1 Drive for circuit-breaker and disconnector/ earthing switch with control panel	5 Busbar compartment
2 Circuit-breaker compartment	6 Busbar
3 Vacuum interrupter	7 Current transformer
4 Disconnector/earthing switch blades (in position "Disconnector MADE")	8 Connecting cable
	9 Cabinet for control and auxiliary devices (low-voltage cabinet)

Fig. 7.19 Section of 36 kV SF$_6$ gas-insulated switchgear (courtesy AREVA T & D)

7.4.2.2 Secondary switchgear

Secondary distribution switchgear is that which is connected directly to the electrical utility transformers which provide low-voltage supplies to customers. The systems distributing power from the primary substations can be conveniently divided into overhead circuits and underground cable networks.

Most faults on overhead systems are caused by lightning strikes, branches brushing the conductors, clashing of conductors in high wind, or large birds bridging the lines; they are usually transient in nature. In most cases the fault duration is short and the circuit breakers used in these circuits, known as *reclosers*, are programmed to close again a very short time after they have opened. This allows the fault to be cleared with a minimum of disruption to consumers. It is normal to programme the recloser to open, then close up to four times in order to allow time for the fault to disappear. If after this time the fault is still present, it is assumed that the fault is not transient; the recloser then locks out and the faulty section of the line is isolated. A recloser normally controls several circuits. An 11 kV pole-mounted auto-recloser is shown in Fig. 7.20. Reclosers are available with SF_6 and vacuum interrupters; the equipment used in conjunction with reclosers are switches and sectionalizers, which can be air-break switches, vacuum or SF_6 devices expulsion fuses and off-load switching devices.

Most urban areas in UK and Europe are supplied by underground cable. In this case the step-down transformer (for example 11 000/415 V) which supplies a consumer circuit is connected to a protective device such as a fuse switch or circuit breaker in a ring. A simplified distribution network with the main components is shown in Fig. 7.21. The 'ring main unit' usually has two load-break switches with the control protective device combined into a single unit. In UK this type of unit is ruggedly designed for outdoor use, but in continental Europe indoor designs are more normal. The traditional oil-filled unit has been replaced largely by SF_6 equipment over recent years. The ring main unit has normally been manually operated with a spring-operated mechanism to ensure that their operation is not dependent on the force applied by the operator. However, as utilities press to improve the reliability of supplies to consumers and to reduce the time lost through outages caused by faults, the use of automatically operated motor-driven spring mechanisms is now on the increase.

7.4.3   Rating principles

The main consideration in the choice of switchgear for a particular system voltage is the current it is capable of carrying continuously without overheating (the rated normal current) and the maximum current it can withstand, interrupt and make onto under fault conditions (the rated short-circuit current). Studies and calculations based on the circuits of interest will quantify these parameters. The switchgear rating can then be chosen from tables of preferred ratings in the appropriate standard, such as IEC 694 and IEC 56. Table 7.4 shows figures taken from IEC 56 showing the co-ordination of rated values for circuit breakers. Ratings vary commonly from less than 400 A in a secondary system to 2000 A or more in a primary substation, with fault currents of 6 kA in some overhead line systems up to 40 kA and above in some primary circuits.

The standards also specify overvoltages and impulse voltages which the switchgear must withstand. Normal overvoltages which occur during switching or which may be transferred due to lightning striking exposed circuits must be withstood. The switchgear also needs to provide sufficient isolating distances to allow personnel to work safely on a part of the system that has been disconnected. The lightning impulse withstand voltage and the normal power frequency withstand voltage are specified relative to the normal system voltage in the standards.

7.4.4   Test methods

Type tests are performed on a single unit of the particular type and rating in question. They are performed to demonstrate that the equipment is capable of performing the

Fig. 7.20 11 kV pole-mounted auto-recloser (courtesy FKI plc – Engineering Group)

rated switching and withstand duties without damage, and that it will provide a satisfactory service life within the limits of specified maintenance. The main type tests are
as follows:

- *dielectric tests.* Lightning impulse is simulated using a standard waveshape
  having a very fast rise time of 1.2 μs and a fall to half value of 50 μs. Power
  frequency tests are performed at 50 Hz or 60 Hz. The method of performing the
  tests is detailed in standards, for instance IEC 60, 'Guide to high voltage testing techniques'. This requires the switchgear to be arranged as it would be in
  service. If outdoor-rated, the tests need to be carried out in dry and wet conditions,

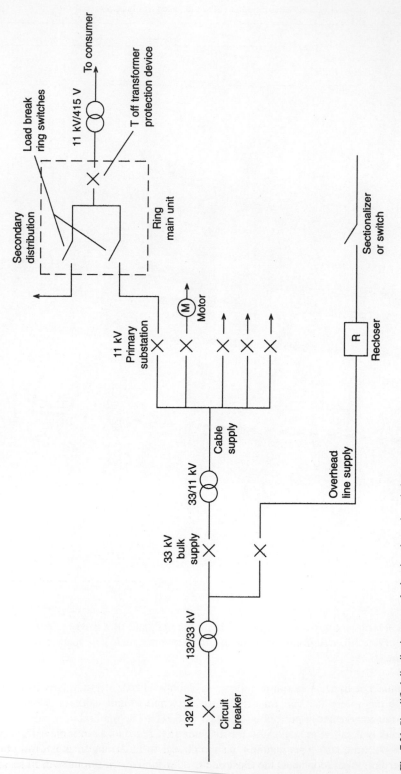

Fig. 7.21 Simplified distribution network showing the main switchgear components

Table 7.4 Co-ordination of rated values for circuit breakers, taken from IEC 56

Rated voltage U (kV)	Rated short-circuit breaking current Isc (kA)	Rated normal current I_n (A)							
3.6	10	400							
	16		630		1250				
	25				1250	1600		2500	
	40				1250	1600		2500	4000
7.2	8	400							
	12.5	400	630		1250				
	16		630		1250	1600			
	25		630		1250	1600		2500	
	40				1250	1600		2500	4000
12	8	400							
	12.5	400	630		1250				
	16		630		1250	1600			
	25		630		1250	1600		2500	
	40				1250	1600		2500	4000
	50				1250	1600		2500	4000
17.5	8	400	630		1250				
	12.5		630		1250				
	16		630		1250				
	25				1250				
	40				1250	1600		2500	
24	8	400	630		1250				
	12.5		630		1250				
	16		630		1250				
	25				1250	1600		2500	
	40					1600		2500	4000
36	8		630						
	12.5		630		1250				
	16		630		1250	1600			
	25				1250	1600		2500	
	40					1600		2500	4000
52	8			800					
	12.5				1250				
	20				1250	1600	2000		
72.5	12.5			800	1250				
	16			800	1250				
	20				1250	1600	2000		
	31.5				1250	1600	2000		

the latter using water of specified conductivity which is sprayed onto the equipment at a controlled rate during application of the high voltage.

- *temperature rise tests.* Again the switchgear is arranged in its service condition, and the normal continuous current is applied until the temperature of the main components has stabilized. The final temperature, measured by thermocouples, must not exceed values stated in the specifications; these values have been chosen to ensure that no deterioration of metal or insulation is caused by continuous operation.
- *short circuit and switching tests.* These are conducted over a range of currents from 10 per cent to 100 per cent of the short circuit rating. The tests are at low power factor, which represents the most onerous switching conditions seen in service.

The different levels represent short circuits at various points in the system, from close to the circuit breaker terminals to a long distance along the cable or line. Whilst the current to interrupt varies with these positions, another major factor is the rate of rise of the transient recovery voltage appearing across the circuit breaker contacts; this can have a significant influence on the circuit breaker performance and it plays a key part in the design. The ability to interrupt the short-circuit current with a decaying dc component superimposed on the power frequency must also be demonstrated. The decrement or rate of decay of the dc voltage is specified in the standards, but the dc component which is present when the contacts open depends upon the opening time of the circuit breaker (it being reduced if the opening time is long). Figure 7.22 shows the determination of short-circuit making and breaking currents and of the dc component. Circuit breakers must also be tested when closing onto a fault condition,

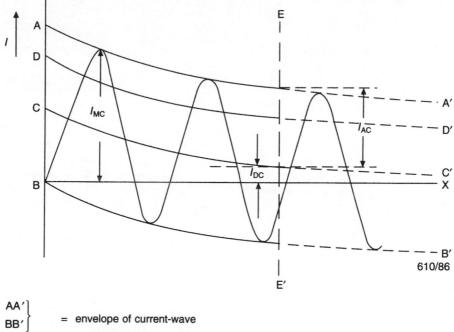

AA' BB' }	= envelope of current-wave
BX	= normal zero line
CC'	= displacement of current-wave zero-line at any instant, the dc component
DD'	= rms value of the ac component of current at any instant, measured from CC'
EE'	= instant of contact separation (initiation of the arc), with dc component I_{DC} indicated
I_{MC}	= making current
I_{AC}	= peak value of ac component of current at instant EE'
$\dfrac{I_{AC}}{\sqrt{2}}$	= rms value of the ac component of current at instant EE'
I_{DC}	= dc component of current at contact separation point
$\dfrac{I_{DC} \times 100}{I_{AC}}$	= percentage value of the dc component

Fig. 7.22 Determination of short-circuit making and breaking currents and of dc component

and then carrying the full short-circuit current for a specified period, usually one or three seconds.

- *mechanical endurance test.* To demonstrate that the equipment has a satisfactory life, this test is performed over a specified number of operations which depends upon the type of equipment and duty; the number of open–close operations can vary from 1000 to 10 000. The tests are carried out without electrical load, but at the rated output of the mechanism of the device.

Routine tests are performed by the manufacturer on every unit of switchgear to ensure that the construction is satisfactory and that the operating parameters are similar to the unit which was subjected to the type test. These tests are specified in the standards, and the minimum performed will include the following:

- power frequency voltage withstand dry tests on the main circuit
- voltage tests on the control and auxiliary circuits
- measurement of the resistance of the main circuit
- mechanical operating tests

7.4.5 Commissioning and maintenance

Switchgear is usually transported in sections which need to be carefully erected and connected together at the substation.

Commissioning tests are performed after installation; these are necessary to ensure that the connections are sound and that the equipment is functioning satisfactorily. These tests will normally include checks of current and voltage transformers, checking of the protection scheme, circuit breaker operation and high-voltage withstand tests.

With the demise of oil-filled switchgear the requirement for maintenance has been significantly reduced. SF_6 and vacuum circuit breakers which are 'sealed for life' should require no servicing of the interrupting device. Manufacturers provide instructions on simple inspection and servicing of other parts of the circuit breaker; this is usually limited to cleaning, adjustment and lubrication of the mechanism.

7.4.6 Standards

Some of the commonly used standards and specifications for distribution switchgear are given in Table 7.5.

7.5 High-voltage (transmission) switchgear

7.5.1 System considerations

The transmission of high powers over long distances necessitates the use of High (HV), Extra High (EHV) or Ultra High (UHV) voltages. Historically, these voltages were classed as 72.5 kV and above. The lower voltages were introduced first, and as the technologies have developed these have increased so that now the highest transmission voltages being used are 1100 kV. For long transmission distances it is also possible to use DC systems; these have particular advantages when two systems are to be connected, but they are beyond the scope of this section, which considers only ac transmission switchgear.

In UK the 132 kV system was first established in 1932, and subsequent growth has required expansion of the system up to 275 kV (1955) and 420 kV (1965). Other countries

Table 7.5 Standards and specifications for distribution switchgear

IEC/BS/EN	Electricity Association	Subject	North American
62271-100		High voltage ac circuit breakers	
62271-102		AC disconnectors and earth switches	
62271-103		High voltage switches	
62271-200		AC metal enclosed switchgear and controlgear for rated voltages above 1 kV up to and including 52 kV	
62271-105		High voltage ac switch fuse combinations	
62271-1		Common specifications for high voltage switchgear and controlgear standards	
	41–26	Distribution switchgear for service up to 36 kV (cable connected)	
	41–27	Distribution switchgear for service up to 36 kV (overhead line connected)	
		Overhead, pad mounted, dry vault, and submersible automatic circuit reclosers and fault interrupters for ac systems	ANSI/IEEEC37.60
		Overhead, pad mounted, dry vault, and submersible automatic line sectionalisers for ac systems	ANSI/IEEEC37.63

have developed similarly, so a multiple voltage transmission system exists in virtually every major electricity-using nation.

In recent years there has been some harmonization of the rated voltages of transmission systems, the agreed levels set down in IEC 694 being 72.5 kV, 100 kV, 123 kV, 145 kV, 170 kV, 245 kV, 300 kV, 362 kV, 420 kV, 550 kV and 800 kV. Systems will be operated at a variety of normal voltage levels such as 66 kV, 132 kV, 275 kV and 400 kV, but the maximum system operating voltage may be higher than these nominal ratings, so the next highest standard value will be selected as the rated voltage of the system.

In order to connect, control and protect these transmission systems it is necessary to use switchgear of various types and ratings.

A switchgear installation which allows the connection and disconnection of the interconnecting parts of a transmission system is referred to as a substation. A substation will include not only the switchgear, but also transformers and the connections to overhead lines or cable circuits. The main functional elements of the switchgear installation in a substation are circuit breakers, switches, disconnectors and earthing switches. Some of the most common configurations for substations are illustrated in Fig. 7.23.

Switching conditions on transmission systems are similar in principle to those on distribution systems, with a requirement to interrupt normal load currents, fault currents and to undertake off-load operations. Because of the different circuit parameters in transmission and distribution networks, the switching duties impose particular requirements which have to be taken into account during the design and type testing of switchgear equipment. A particular requirement for transmission is the ability to deal with the onerous transient recovery voltages which are generated during interruption of faults on overhead lines when the position of the fault is close to the circuit breaker terminals; this is defined in IEC 62271-100 as the Short Line Fault condition. A further requirement is the energization of long overhead lines, which can generate large transient overvoltages on the system unless counter-measures are taken.

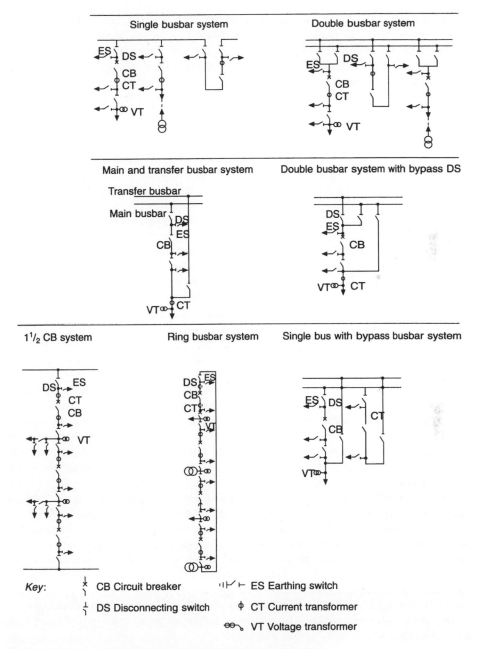

Fig. 7.23 Examples of busbar configurations for high-voltage substations

7.5.2 Types of circuit breaker

The first designs of transmission circuit breaker used minimum-volume oil interrupter technology similar to that which had been developed for distribution switchgear.

The 1940s saw the development and introduction of air-blast circuit breakers throughout the world. These devices used compressed air at pressures up to 1200 psi, not only to separate the contacts but also to cool and de-ionize the arc drawn between the contacts. Early air-blast designs used interrupters which were not permanently pressurized, and reclosed when the blast was shut off. Isolation was achieved separately by a series-connected switch which interrupted any residual resistive and/or capacitive grading current, and had the full rated fault-making capacity.

The increase of system voltages to 400 kV in the late 1950s coincided with the discovery, investigation and specification of the kilometric or short line fault condition. Air-blast interrupters are particularly susceptible to high Rates of Rise of Recovery Voltage (RRRV), and parallel-connected resistors are needed to assist interruption. At the increased short-circuit current levels of up to 60 kA which were now required, parallel resistors of up to a few hundred ohms per phase were required, and the duty of breaking the resistor current could no longer be left to the series-connected air switch. To deal with this, permanently pressurized interruptors incorporating parallel-resistor switching contacts were used. The original pressurized-head circuit breakers used multiple interrupters per phase in order to achieve the required fault-switching capacity at high transmission voltages; 10 or 12 interrupters were used for 400 kV. These were proved by direct unit testing at short-circuit testing stations, each having a direct output of 3 GVA. It was in fact the ability to test which determined the number of interrupters used in the design of a circuit breaker.

The introduction of synthetic testing methods increased the effective capacity of short-circuit testing stations by an order of magnitude, and this allowed the proving of circuit breakers having fewer interrupters. So, 400 kV air-blast circuit breakers with only four or six interrupters per phase were being supplied and installed up to the late 1970s.

The next step-change in technology occurred with the introduction of SF_6 circuit breakers. The merits of electronegative gases such as SF_6 had long been recognized, and freon had been used in the late 1930s to provide the primary insulating medium in 33 kV metalclad switchgear installed in two UK power stations. The reliability of seals was a problem, and it was only during the nuclear reactor programme that sufficiently reliable gas seals were developed to enable gas-insulated equipment to be reconsidered.

The first practical application of SF_6 in switchgear was as an insulating medium for instrument transformers in transmission substations in the late 1950s and early 1960s. These were quickly followed by the use of SF_6 not only as an insulating medium but also as the means of extinguishing arcs in circuit breakers. The first generation of SF_6 circuit breakers were double-pressure devices which were based on air-blast designs, but using the advantages of SF_6. These double-pressure designs were quickly followed by puffer circuit breakers which have already been described in section 7.4.1. Puffer circuit breakers, now in the form of second or third generation of development, form the basis of most present-day transmission circuit breaker designs. Figure 7.24 illustrates the various forms of SF_6 interrupter.

7.5.3 Main classes of equipment

Transmission substations may be built indoors or outdoors but because of the space required it is usually difficult to justify an indoor substation. The strongest justification for an indoor substation is the effect of harsh environments on the design, operation and maintenance of the equipment.

The two main types of transmission switchgear currently available are conventional or Air-Insulated Switchgear (AIS) and gas-insulated metal-enclosed switchgear.

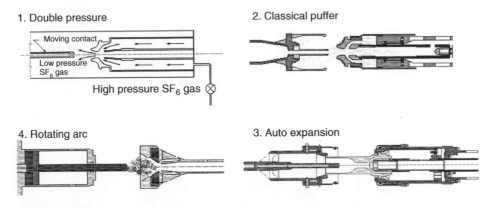

1. Double pressure

Moving contact

Low pressure SF$_6$ gas

High pressure SF$_6$ gas ⊗

2. Classical puffer

4. Rotating arc

3. Auto expansion

Fig. 7.24 Various forms of SF$_6$ interrupter

7.5.3.1 Conventional or Air-Insulated Switchgear (AIS)

Here the primary insulating medium is atmospheric air, and in the majority of AIS the insulating elements are of either solid or hollow porcelain.

In circuit elements such as disconnectors and earthing switches, where the switching is done in air, the insulators are solid, and provide not only physical support and insulation, but also the mechanical drive for the elements. Examples of disconnectors and earth switch arrangements are shown in Fig. 7.25. The choice of disconnector design is dependent on the substation layout and performance requirements.

Equipment such as circuit breakers which need an additional insulating or interrupting medium use hollow porcelain insulators which contain the active circuit elements in the appropriate insulating medium. As previously outlined, initial AIS installations used oil as the insulating and interrupting medium, which in many cases was replaced by the use of compressed air, and modern designs almost always use SF$_6$ gas.

The layout of a substation is governed by the need to ensure that air clearances meet the dielectric design requirements, including conditions when access is required to the substation for maintenance. Rules for the minimum clearances and the requirements for creepage distances on the external porcelain insulators are contained in the standards. For locations with atmospheric pollution, it is necessary to provide extended creepage distances on insulators. Guidelines for the selection of creepage distance are given in IEC 815.

A recent development in AIS technology is the use of composite insulators for functional elements in place of the traditional porcelain. There is a risk of explosion with porcelain insulators when they are filled with insulating gas if a mechanical fault occurs with the equipment. A composite insulator consists of silicone rubber sheds on a filament glass-fibre tube; it is not brittle and will not explode if the insulator is ruptured. Composite insulators are now being used on instrument transformers and circuit breakers for AIS, and their use is likely to become more common.

An example of an outdoor AIS substation is shown in Fig. 7.26.

7.5.3.2 Gas-insulated metal-enclosed switchgear

In metal-enclosed switchgear there is a an external metal enclosure which is intended to be earthed; it is complete except for external connections. The earliest transmission metal-enclosed equipment used oil as the insulating medium, but since the introduction of SF$_6$, all transmission metal-enclosed switchgear is based on SF$_6$ as the insulating medium.

The main significance of the Gas-Insulated Substation (GIS) principle is the dramatic reduction of insulating clearances made possible by SF$_6$, resulting in much more compact substation layouts. GIS are easier and more economic to accommodate where land use is critical, such as in many urban areas. It also offers considerable advantages in applications where environmental pollution is severe, such as in coastal,

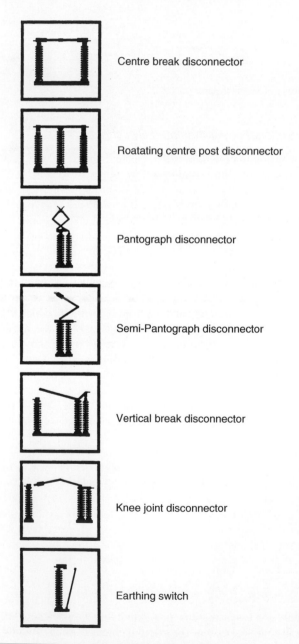

Centre break disconnector

Roatating centre post disconnector

Pantograph disconnector

Semi-Pantograph disconnector

Vertical break disconnector

Knee joint disconnector

Earthing switch

Fig. 7.25 Examples of disconnectors and earth switch arrangements

Fig. 7.26 Example of an outdoor AIS substation

desert or heavily industrialized areas. Environmental conditions can also influence the need for maintenance of the switchgear and the implementation of that maintenance; GIS is so compact that transmission substations can be indoor, reducing the maintenance problems considerably. The greater reliability of GIS is another factor which is encouraging its increased application.

GIS switchgear is generally of modular design. All individual elements such as busbars, disconnectors, instrument transformers and circuit breakers are contained in interconnecting earthed enclosures filled with SF_6 at up to 7 bar. At the highest voltage levels, the three phases are housed in separate phase-segregated enclosures. At the lower voltages it is common for all three phases to be housed in the same enclosure and with some designs this philosophy is extended up to 300kV. Because of the possibility of heating effects from induced currents it is normal for phase-segregated designs to use non-magnetic materials (typically aluminium) for the enclosures, but for phase-integrated designs it is possible to use magnetic steels for some ratings without overheating constraints.

An example of a GIS substation is shown in Fig. 7.27.

7.5.3.3 Hybrid substations

A substation where there is a mix of GIS and AIS technologies is often referred to as a hybrid substation. This can be a substation that consists of bays where some are of AIS components only and some are a mix of AIS and GIS technology, or where some are of GIS components only. Alternatively, and more commonly, elements of AIS and GIS technology are typically mixed in the same bay of equipment and this mixture is a hybrid of the two technologies and is applied across the complete substation.

Fig. 7.27 Example of a GIS substation

There are exceptions to this definition:

- if the only component in a GIS substation utilizing AIS technology is the interface connection (bushing or cable termination) to overhead line, cable or transformer, the substation is still considered as a GIS substation
- in an AIS substation where only one component is provided in GIS technology, or where additional elements have a mixture of air and gas insulation but the primary insulation to earth is still air (as in a dead tank arrangement) then these substations are still considered as AIS

Any other combinations are considered as hybrid, for instance where only busbars are SF_6 insulated, or where an enclosed gas-insulated circuit breaker contains additional equipment such as instrument transformers or earthing switches.

An example of a hybrid substation is shown in Fig. 7.28.

7.5.4   Rating principles

As for distribution switchgear, the main consideration in the choice of rating for a particular system is the rated normal current and the rated short-circuit current. Power system studies using load flow and transient analysis techniques will establish the values of the parameters for individual systems. It is also necessary to select lightning impulse rated values from the range of standard values specified in standards (as for distribution switchgear) and to specify for transmission systems with a rated voltage of 300 kV and above a rated switching impulse withstand voltage. The rated switching surge voltage level is defined as the peak voltage of a unipolar standard

Fig. 7.28 420 kV hybrid-GIS switchgear with AIS busbars

250/2500 μs waveform; this represents the transient voltages which are generated by the operation of circuit breakers when switching overhead lines, shunt reactors and other circuits.

An important consideration in the choice of equipment ratings for transmission systems is insulation co-ordination. Insulation co-ordination is the selection of the electric strength of equipment and its application in relation to the voltages which can appear on the system taking into account the characteristics of any protective devices, so as to reduce to an economically and operationally acceptable level the probability that the resulting voltage stresses imposed on the equipment will cause damage to the equipment or affect continuity of service. For system voltages up to 300 kV, experience has demonstrated that the most important factor in determining system design is the stress due to lightning; for voltages over 300 kV the switching over-voltage increases in importance. It is now common to use Metal Oxide surge Arrestors (MOA) to limit lightning over-voltage levels on transmission systems. In addition, the MOA can in many cases provide adequate limitation of switching over-voltages, but for longer overhead lines it may be required to use a circuit breaker with controlled (or point-on-wave) switching in order to keep over-voltages at acceptable levels. The traditional technique of using circuit breakers with parallel pre-insertion resistors is now being replaced by the use of the MOA and controlled switching.

7.5.5 Test methods

International standards require that switchgear must be so designed and manufactured that it satisfies test specifications with regard to its insulating capacity, switching performance, protection against contact, current-carrying capacity and mechanical function. Evidence of this is obtained by type testing a prototype or sample of the switchgear. In addition, routine tests are performed on each individual item of switchgear manufactured, either on completed or sub-assembled units in the factory or on site.

Type tests are performed on transmission switchgear in a similar manner to those applied to distribution switchgear (see section 7.4.4), but because of the difference in rated values, the necessary test values, the test equipment and techniques differ in detail. Some tests which are specific to transmission switchgear are as follows:

- *dielectric tests*. In addition to lightning impulse and power frequency tests it is necessary to perform switching impulse tests for equipment with a rated voltage above 300 kV.
- *short-circuit and switching tests*. The test parameters for transmission equipment involve much higher energy requirements than for distribution equipment, so special techniques have been developed over the years. To provide the high voltage and high currents needed for transmission switchgear, it is usual to provide a 'synthetic' test source where the high voltage and high currents are provided in the same test from different sources.

Routine tests are performed on all switchgear units or sub-assemblies to ensure that the performance of the factory-built unit will match that of the type-tested unit. The AIS is normally fully assembled in the factory and routine tested as a complete functioning unit. Routine tests will include mechanical operations to check operating characteristics, resistance checks of current-carrying paths, tests on control and auxiliary circuits and power-frequency withstand tests to verify insulation quality. For GIS, particularly at higher voltage levels, it is not practical to assemble fully a complete installation in the factory. Therefore sub-assemblies of GIS or transportable assemblies are individually tested prior to shipment; for GIS at 145 kV or 245 kV this might involve a complete 'bay' or circuit of equipment, but at 420 kV or 550 kV the size of the equipment dictates that only sub-assemblies can be transported to site as a unit. Tests on transportable assemblies include power-frequency withstand with partial discharge detection, verification of gas tightness, mechanical operations and main circuit resistance.

7.5.6 Commissioning and maintenance

Due to the size of transmission switchgear it is necessary to transport the equipment to site in sub-assemblies and then to re-assemble to complete switchgear on site. This procedure is more extensive than for distribution switchgear and it results in more elaborate commissioning tests.

After routine testing at the factory, AIS is dismantled into sub-assemblies for transport to site and then re-assembled at the substation. A short series of operational and functional checks are performed before the equipment is put into service.

Site commissioning procedures for GIS are more elaborate than for AIS, since GIS can be particularly sensitive to assembly defects or to particulate contamination; because of the small electrical clearances made possible by SF_6, it is essential that all particulate contamination down to the size of about 1 mm is excluded from the equipment. For these reasons extensive power-frequency withstand tests are performed on site on the completed installation. The requirements for commissioning of GIS are covered in IEC 62271-203.

Modern designs of transmission switchgear are becoming simpler and inherently more reliable, with reduced maintenance requirements. The traditional preventive maintenance approach, with maintenance performed at prescribed intervals irrespective of the condition of the switchgear, is being replaced by predictive maintenance in

Table 7.6 Standards and specifications for transmission switchgear

IEC/BS/EN	Subject	North American ANSI
60071-1	Insulation co-ordination. Pt 1 – Definitions, principles and rules	C92.1
62271-1	Specifications for high voltage switchgear and controlgear standards	C37.09
62271-100	High voltage ac circuit breakers	C37.04
		C37.06
62271-101	Synthetic testing of high voltage ac circuit breakers	C37.081
62271-102	AC disconnectors and earthing switches	C37.32
62271-103	High voltage switches for rated voltages of 52 kV and above	
62271-203	Gas-insulated metal-enclosed switchgear rated 72.5 kV and above	C37.55
62271-300	Seismic qualification of high voltage ac circuit breakers	C37.81

which maintenance is needed only when the condition of the equipment warrants intervention or where operational duties are severe. Diagnostic techniques and condition monitoring to support this are now becoming available; in particular, it is very difficult to apply conventional partial discharge tests to a complete GIS installation and special diagnostic techniques have been developed to monitor its structural integrity.

Many items of transmission switchgear are still in service after 30, 40, or even 50 years of satisfactory service. A typical design life for modern transmission switchgear is 40 years, but this is conservative in comparison with the figures being achieved on the older equipment still in service today.

7.5.7 Standards

Some of the commonly used standards and specifications for transmission switchgear are given in Table 7.6.

7.6 DC switchgear

7.6.1 Technology and applications

The most obvious difficulty in interrupting dc current is that there is no natural current zero. The current has to be forced to zero by the circuit breaker by generating an arc voltage across the break contacts which is greater than the system voltage. Since the load and the fault current is highly inductive the circuit breaker interrupting device must be capable of dissipating all the energy in the circuit until arc extinction.

Until the early 1970s the main method of arc extinction was to stretch and cool the arc in an arcing chamber under the influence of the magnetic field produced by a series connected *magnetic blow-out coil*. A voltage drop in dc arcs of about 1 V/mm has typically been accepted, and although this is dependent on arcing conditions, it is evident that for any system voltage above a few hundred volts the arc must be fairly long. This is fairly difficult to achieve, and it promotes the chance of the arc re-striking to other components within the switchgear.

The uses for dc switchgear include generator field control, deep mine winding gear and superconducting toroidal coils, but by far the most common requirement is in urban transit systems where dc supplies are ideally suited for light to medium power requirements. The majority of systems operate on supply voltages ranging from 650 V to 1500 V with a few inter-city routes using 3000 V.

Urban dc rail systems can be categorized as *rapid transit systems* or *light rail systems*.

An underground or suburban railway generally using large rolling stock such as the London Underground is typically described as a rapid transit system. Such systems require high normal current, high fault-rated dc circuit breakers because of the large load currents drawn on start-up of the rolling stock and the low impedance of the track power supply connections. In order to limit the let-through current during a fault, and thereby protect the ac/dc conversion equipment, high-speed dc circuit breakers are nearly always specified for this type of system. Typical ratings for use in rapid transit systems are 750 V, 6000 A rectifier circuit breakers and 4000 A track feeder circuit breakers with a maximum fault interrupting rating of 125 kA at 900 V.

A light rail system is typically a tram system, with or without track, using an overhead catenary–pantograph supply. Switchgear ratings are lower with typical ratings of 750 V, 2000 A for rectifier breakers and 1250 A for track feeder breakers, with a maximum fault interrupt rating of 50 kA, 900 V. The circuit breakers used are generally defined as semi-high speed.

7.6.2 Semiconductor circuit breakers

Semiconductors have been considered as a potential advance in the field of dc circuit breakers for many years. The speed of operation far exceeds the capability of conventional mechanical circuit breakers and if the short-circuit current can be detected rapidly a semiconductor circuit breaker could be capable of switching the circuit in fractions of a millisecond. This would limit the let-through fault current to a smaller value with less possibility of damage to equipment. Semiconductor switches are capable of interrupting large short circuit currents on traction dc power supplies, but they are unable alone to absorb and dissipate all the fault energy.

One method involves a forced commutation circuit in parallel with the semiconductor, providing a low impedance path into which the fault current is temporarily diverted. This creates a current zero in the semiconductor switch circuit allowing it to turn off. Other methods such as series-resonant commutation can be used but in each case the high reverse voltage caused by rapid turn off of a highly inductive circuit must be limited, otherwise dc equipment will be damaged. This can be achieved using a surge arrester or capacitor within the circuit breaker to absorb the energy or by circulating the energy around the fault circuit using a free-wheel diode.

Semiconductor circuit breakers are not widely used for three main reasons:

- high forward conduction losses due to forward volt drop of the semiconductor junction and the high current requirement of dc circuit breakers
- isolation when the circuit breaker is switched off is reliant upon semiconductor dielectric strength
- the limitation to the speed of operation is the detection time of the protection device. It is not possible to devise a protection relay that can operate rapidly enough to reap the full benefits of the semiconductor switch and to reliably discriminate between all types of evolving faults and normal transient load conditions.

The use of a mechanical switch in parallel with the semiconductor eliminates the forward conduction loss and a series isolating switch operating after fault clearance provides a physical isolation distance. However, in order to achieve the benefit of high-speed fault clearance the mechanical components have to be capable of closing and opening in a few hundred microseconds. The result is a more complex and expensive switch in parallel with a sophisticated and complex electronic device that will not normally compete with a conventional circuit breaker.

7.6.3 Air circuit breakers

The cold cathode arc chute is the preferred method of arc interruption in modern dc circuit breakers. It is ideally suited to the interruption of dc since it provides a fairly fixed arc voltage irrespective of the arc current. The cold cathode arc chutes are arranged as shown in Fig. 7.29 and the current and voltage characteristics of an interruption can be seen in Fig. 7.30. An arc is formed as the contacts part and at this stage the voltage across that arc will be very low, in the order of tens of volts. This is dependent on arcing current, method of arc movement, contact separation and contact materials. The arc then moves towards the splitter plates along the arc runners, being stretched as it moves and generating an arc voltage of a few hundred volts and it is then split into the series arcs by the splitter plates when a sharp jump in the arc voltage can be seen. The difference between the arc voltage forced in the arc chute and the system voltage describes the arcing time within plates.

There are a number of techniques that are used to force the arc into the splitter plates. By careful arrangement of the contact and runner system it is possible to generate a tight current loop which, by its nature, forces the arc into a larger loop. This however does not provide strong enough magnetic fields at the lower currents to move the arc column away from the contact area. A mechanical puffer driven from the opening mechanism is one technique that is used to move the arc column away from

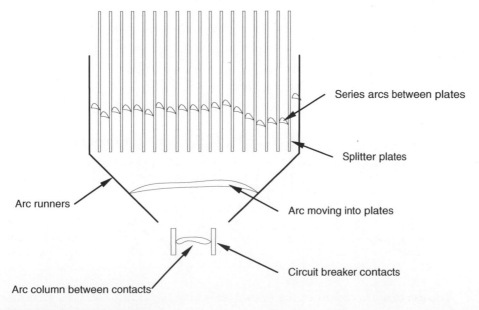

Fig. 7.29 Cold cathode arc chute

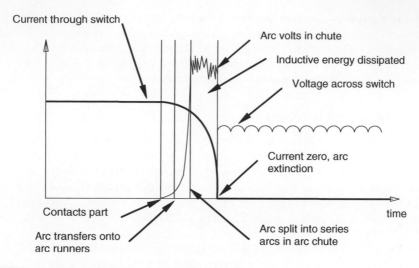

Current through switch

Arc volts in chute

Inductive energy dissipated

Voltage across switch

Current zero, arc extinction

Contacts part

time

Arc transfers onto arc runners

Arc split into series arcs in arc chute

Fig. 7.30 Interruption characterstics of a cold cathode arc chute

the contact area, applying a blast of air between the contacts. This is independent of arcing current but it adds complexity and is only available for a short period of time.

A magnetic field orthogonal to the arc is an effective method to move the arc rapidly from the contact gap into the splitter plates and this has been successfully used on many ac and dc air circuit breakers. The magnetic field can be generated by permanent magnets, secondary coils or by the current itself. If the circuit breaker is to interrupt current in both directions, it can be seen in Fig. 7.31 that the field must be reversible in order to have the same resultant force upwards on the arc and with secondary coils or permanent magnets, it can be difficult to reverse the field especially

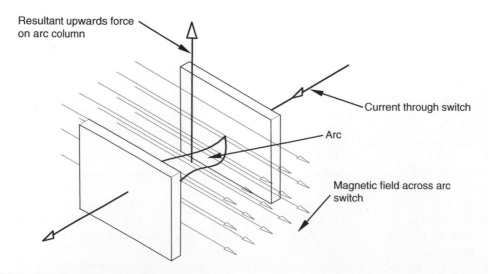

Resultant upwards force on arc column

Current through switch

Arc

Magnetic field across arc switch

Fig. 7.31 Moving the arc from the contact

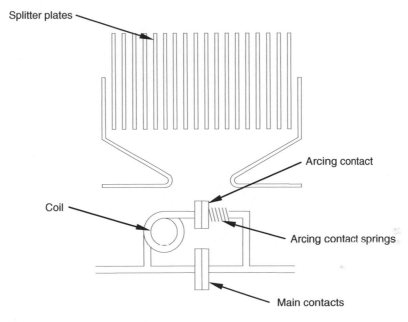

Fig. 7.32 Typical arcing contact arrangement

at low currents. Connecting secondary coils across the contact gap gives the correct field orientation of field, but these have to be switched after the arc has been extinguished and this can lead to problems in isolation. Using the main current to provide the field inherently gives the correct orientation of force on the arc, but in order to develop a uniform field with a high enough flux density across the contacts it is necessary to link the flux from several turns of main conductor. This can be cumbersome because of the large-sectional copper required for the relatively high rated currents of the switch.

Often arcing contacts are used to solve this problem. The main contacts carry normal current and part slightly earlier than the arcing contacts, so they can be made from a less arc-resistant material, which reduces the heat generated in the switch under normal operating conditions. This also enables the addition of a coil in the arcing contact arrangement, as shown in Fig. 7.32, providing a magnetic field across the contacts. It is imperative that the main contacts break earlier than the arcing contacts even when the arcing contacts are badly worn, and that the inductance of the coil is very low to assist in current transfer to the arcing contacts with dI/dt values of potentially tens of thousands of amps per millisecond.

Another method now in use is to link the flux from the main current with a ferrous coil or open core around the main current. The *open-core magnetic blow-out* shown in Fig. 7.33 provides a magnetic field in the correct orientation while the main current is flowing, with enough flux linkage to interrupt dc arcs at very low currents. The avoidance of large coils carrying the main current allows for a very compact passive device with no moving parts that has been proven to work over the full range of duties required.

Without assistance from a device such as the open-core magnetic blow-out, the arc may not transfer into the arc chute at low currents (up to 200 A) as its magnetic field

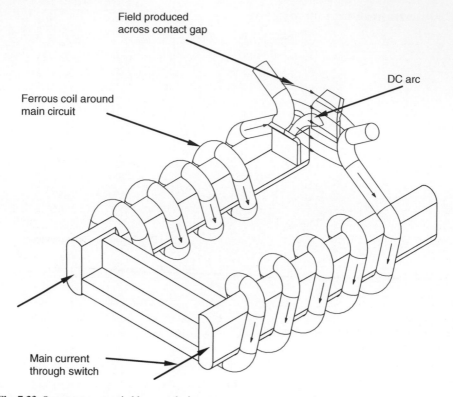

Field produced
across contact gap

DC arc

Ferrous coil around
main circuit

Main current
through switch

Fig. 7.33 Open core magnetic blow-out device

would be insufficient to force it off the contacts. At very low currents (in the order of tens of amperes, depending upon contact gap, system voltage, contact material, circuit time constant etc.) the arc can be extinguished across the contact gap without transferring onto the runners and into the chute. *Critical currents* can occur on some circuit breakers when the current is too small to transfer correctly into the chute and too large to be interrupted across either the contacts or on the arc runners. The result can be very long and inconsistent arcing times or in the worst case failure to interrupt. The position becomes worse when the circuit breaker is required to interrupt a current in its critical area in the reverse direction following the clearance of a relatively large current in the forward direction. The forward current interruption can magnetize the steel plates in the arc chute (and possibly other mechanism components) and the remanent magnetic field in the arc chute is then in the opposite direction to that required to draw the arc into the chute in the reverse direction. If the reverse current is small the magnetic field cannot be reversed and the arc cannot be interrupted.

A type test is specified in the BSEN 50123-2 standard to search for forward and reverse critical currents. These are defined as the current for which the particular circuit breaker has the longest arcing time. These low currents are inductive and cover the range for which it is most likely that air circuit breakers will have difficulty interrupting. The specification also calls for a reverse test at the forward critical current after a large forward interruption.

7.6.4   DC traction fault duties

DC circuit breakers are classified according to their characteristics during fault inter-ruption. Some common requirements are highlighted in the circuit breaker designations:

- very high speed current limiting circuit breakers, V
- high speed current limiting circuit breakers, H
- semi-high speed circuit breakers, S

Type V and type H circuit breakers are also described as current-limiting circuit breakers because they have a break time which is sufficiently short to prevent the short-circuit current reaching the peak value it would have attained without interruption. Very high speed circuit breakers have a break time no greater than 4ms, irrespective of other parameters of the circuit. This speed of operation is limited to semiconductor circuit breakers and hybrid circuit breakers that are a combination of semiconductor and mechanical switches. High speed circuit breakers have an opening time no greater than 5ms and a total break time not greater than 20ms when the current to be inter-rupted has a sustained value of at least 7 times the circuit breaker setting and dI/dt is greater than or equal to 5kA/ms. Type S circuit breakers provide assured current inter-ruption, but current limitation may not take place. The circuit breaker in most common use is the type H because of its ability to operate quickly and limit damaging let-through fault current at an affordable price.

The specification of the circuit breaker type is decided by the fault levels the system can generate, and the types of fault are characterized by their location in the traction system.

The functions for which the dc circuit breaker can be used within a standard trac-tion system are:

- interconnector circuit breaker, I (also called bus-section or section circuit breaker)
- line feeder circuit breaker, L
- rectifier circuit breaker, R

These can have different requirements, and current interruption direction can also be specified, as well as direct-acting overload release direction. It can be advantageous for discrimination purposes to specify the direction in which the release operates, to ensure that only the breaker feeding a fault opens and that supply is maintained elsewhere. Current interruption in both directions is necessary unless it can be shown that there are no circumstances under which a circuit breaker can be opened when the current is flowing in an opposite direction.

7.6.5   DC traction protection

The power supplied through the switchgear fluctuates as trains start, accelerate and regenerate, and one or more trains can be travelling or transferring onto different sections of track. To discriminate between a fault and such a variety of load conditions requires a sophisticated and well co-ordinated protection system. Depending upon the point in the system at which the fault occurs, it is quite common for the magnitude of the current to be less than that flowing due to the traction load.

While short-circuit currents on the dc traction supply can be large and damaging to equipment on the dc supply, the value of the short-circuit current on the ac supply due to dc faults could be fairly small. Also, ac switchgear protection operating and fault clearance times may be too long to protect the ac/dc conversion equipment effectively. The dc circuit breakers and the protection system must therefore be fast-operating. Electronic and/or electromagnetic protection relays are always used on modern dc switchboards, the current being sensed by a shunt or other transducer and the circuit breaker being tripped by the shunt trip coil. As the last line of defence, it is common practice to specify a high-set direct acting instantaneous series overcurrent release within the circuit breaker.

In order to minimize the risk of an undetected fault condition the following types of protection have been developed.

- *Direct-acting overcurrent protection:* This is incorporated into the circuit breaker, the device directly tripping the breaker through a mechanical linkage or magnetic circuit. The purpose is to provide protection for overload conditions and high-speed operation under high-current faults. In most networks the device will not provide discrimination between normal load conditions and low level or distant faults. Line feeder and rectifier circuit breakers utilise this form of protection.

- *Rate of rise protection (dI/dt):* This is now the most commonly used software-driven protection system in traction applications. As it measures the rate of rise of current through the switchgear it is able to react rapidly during the initial increase of current at the on-set of a fault before the current reaches a large, damaging value. When properly developed it is effective in discriminating between distant faults and normal train starting and accelerating currents, but a detailed knowledge of the system is imperative for correct setting of the relay. The settings are chosen to ensure that the device will not trip on the most severe train starting condition but will trip under distant fault conditions. An extension of this principle is *DELTA I (ΔI) protection*, in which the initial dI/dt triggers a current measurement. After a set time, the change in current is measured and if it exceeds the set value (determined from the most onerous train starting current) a trip is issued. The dI/dt setting can be lower than that for standard rate of rise protection, offering better sensitivity for distant fault detection. In addition the time delay setting can be reduced to give more sensitive protection for less distant faults.

- *Undervoltage protection:* Relays can be used to monitor a reduced track voltage under fault or load conditions at the substation and at points in between. The system may be used for both single and double-end fed systems, pilot wires being used to trip the appropriate circuit breaker or breakers. In order to provide discrimination and to allow circuit breakers to be closed onto tracks at a reduced voltage, undervoltage protection system needs to incorporate a time delay. This must be of the order of 100 ms, which renders this type of protection inadequate for close-up faults. As the system measures only voltage it cannot be used to provide directional discrimination.

- *Falling-voltage impedance protection:* Impedance protection is an extension of the falling-voltage principle in which track current is also monitored. The impedance of a track section seen from the circuit breaker is the ratio of track voltage to the current flowing through the circuit breaker. An impedance value, normally equal to the loop impedance of the whole track section provides the setting value. The presence of a fault closer to the substation will lead to the *V/I* being reduced and if it falls below the set impedance value, a trip command can be issued. This system is used extensively by Network Rail in the UK and on systems utilizing additional switching stations located between the rectifier substations.

- *Instantaneous overcurrent:* This offers the same protection as circuit breaker direct-acting overload protection, but since modern digital relays can be used to give more accurate and faster tripping times, it may assist in the scheme co-ordination. Like direct-acting overload protection it must always be set above the level required during trains starting and accelerating and therefore it can not always detect distant faults.

7.6.6 Safety features and interlocking

Modern dc switchgear is almost exclusively of the horizontally withdrawable type to facilitate full isolation, maintenance and testing. Figure 7.34 shows a modern withdrawable dc circuit breaker, cutaway to show the shutter and venting arrangement. It is important that interlocks prevent an incorrect sequence of actions from causing unsafe operation. The circuit breaker compartment usually caters for two positions, a test/isolated position, and a service position. The circuit breaker can be operated in either mode through the secondary control and auxiliary connections. The interlocking principles should follow the following philosophy:

- When the circuit breaker is inserted into the cubicle, the earth connection is made before all other connections. This earth connection should be suitably rated for the system earth fault current.

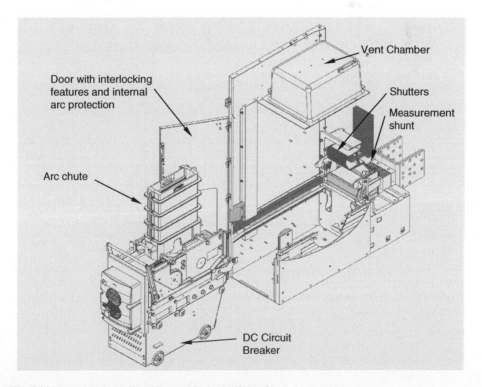

Fig. 7.34 Typical horizontally withdrawable dc circuit breaker

- The circuit breaker can be operated in the test position, and must remain isolated from the primary circuit.
- The main compartment door must be closed before the circuit breaker moves from the isolated to the service position.
- The shutters should not be operated by the circuit breaker mechanism while the door is open, and are independent in operation.
- The door cannot be opened once the circuit breaker has been moved from the isolated position.
- The circuit breaker cannot be moved from isolated or service position until it is open.
- If the auxiliary connections are not made, there will be no possibility of unsafe operation when the breaker is put into the service position.

7.6.7 Applicable standards and testing

Until relatively recently it was the case that most dc network operators would purchase their electrical equipment with only limited third party test evidence. Indeed it was not unusual for equipment either to be put into trial service on the railway or for entire test tracks to be constructed. Such trial work with equipment is extremely expensive and can be disruptive to the operation of the railway network as a whole, and most testing of dc switchgear now takes place at high power testing laboratories using various synthetic test duties to simulate railway conditions.

Testing or specifying equipment can be a difficult task because of the numerous variations between railway network requirements and manufactured equipment, and most customers and manufacturers normally use the various national and international standards that specify dc switchgear, shown in Table 7.7. Of the standards currently in use the EN 50123:2003 series (also known as IEC 61992) and ANSI C37.16-2000 are commonly quoted, the EN series being the more up to date and complete.

The EN 50123 series requires that dc switchgear is subjected to a series of tests which normally includes a range of short circuit duties, electrical endurance testing,

Table 7.7 International standards for dc switchgear

EN	Equipment specified	ANSI
EN 50123-1	Railway applications – Fixed installations – DC Switchgear	
EN 50123-2	Railway applications – Fixed installations – DC circuit breakers	
EN 50123-3	Railway applications – Fixed installations – Indoor DC disconnectors, switch-disconnectors and earthing switches	
EN 50123-4	Railway applications – Fixed installations – Outdoor DC disconnectors, switch-disconnectors and earthing switches	
EN 50123-5	Railway applications – Fixed installations – Surge arresters and low voltage limiters for specific use in DC systems	
EN 50123-6	Railway applications – Fixed installations – Measurement, control and protection devices for specific use in DC traction systems	
EN 50123-7	Railway applications – Fixed installations – DC circuit breakers	
EN 50124-1	Railway applications – Insulation coordination	
EN 50163	Railway applications– Supply voltages of traction systems	
	IEEE Standard for low-voltage DC power circuit breakers used in enclosures	C37.14
	Low-Voltage power circuit breakers – Preferred ratings, related requirements and application recommendations	C37.16

temperature rise testing, dielectric testing, ingress protection and mechanical endurance testing. Each duty or type of test is subject to variation depending on the type of circuit breaker to be tested, the system voltage and the characteristics of the circuit that is being switched. In particular a considerable amount of testing is undertaken to assess performance under the various fault conditions that may arise at low or reverse currents.

From an operator's perspective, testing at maximum energy levels is particularly important because the test can be used to simulate extended electrical endurance causing considerable wear to both the circuit breaker contacts and arc chute.

As with much railway equipment dc switchgear is typically located in remote locations of restricted size, requiring special consideration of mechanical reliability and effect of temperature rise within the equipment. Circuit breakers can be subjected to extended mechanical endurance tests of over 50 000 operations without any maintenance, often including variations in temperature, humidity and control system voltage. Temperature rise testing is normally undertaken on complete switchgear assemblies with conditions as close to those in service as possible, particularly if ventilation is restricted. Such testing sometimes includes operator-defined overload cycles to simulate the effects of increased numbers of trains on the railway.

References

7A.   Ryan, H.M. (ed), *High voltage engineering and testing,* Peter Peregrinus, 1994.
7B.   Ryan, H.M. and Jones, G.R. *SF$_6$ switchgear,* Peter Peregrinus, 1989.
7C.   Flurscheim, C.H. (ed), *Power circuit breaker theory and design,* Peter Peregrinus, 1985.

Chapter 8

Fuses and protection relays

Dr D.J.A. Williams

With amendments by
G. Newbery
Technical Consultant to Cooper-Bussmann

8.1 Protection and co-ordination

Fuse and protection relays are specialized devices for ensuring the safety of personnel working with electrical systems and for preventing damage due to various types of faults. Common applications include protection against overcurrents, short circuits, overvoltage and undervoltage.

The main hazard arising from sustained overcurrent is damage to conductors, equipment or the source of supply by overheating, possibly leading to fire. A short circuit may melt a conductor, resulting in arcing and the possibility of fire; the high electromechanical forces associated with a short circuit also cause mechanical stresses which can result in severe damage. A heavy short circuit may also cause an explosion. Rapid disconnection of overcurrents and short circuits is therefore vital. An important parameter in the design and selection of protective devices is the *prospective current*; this is the current which would flow at a particular point in an electrical system if a short circuit of negligible impedance were applied. The prospective current can be determined by calculation if the system impedance or *fault capacity* at that point is known.

In addition, personnel working with electrical equipment and systems must be protected from electric shock. A shock hazard exists when a dangerous voltage difference is sustained between two exposed conducting surfaces which could be touched simultaneously by different parts of the body. This voltage normally arises between earth and metalwork which is unexpectedly made live and if contact is made between the two through the body, a current to earth is caused. Fuses can provide protection where there is a low-resistance path to earth, because a high current flows and blows the fuse rapidly. To detect a wide range of currents flowing to earth it is necessary to use current transformers and core-balance systems; these operate a protective device, such as a circuit breaker such as the rcd described in section 7.3.4 for low-voltage systems.

When designing an electrical protection system it is also necessary to consider co-ordination so that when a fault occurs, the minimum section of the system around the fault is disconnected. This is particularly important where disconnection has safety implications, for instance in a hospital. An illustration of co-ordination is shown in Fig. 8.1.

Protective devices are described by a *time–current characteristic*. In order to achieve co-ordination between protective devices, their time–current characteristics must be sufficiently separated, as shown in Fig. 8.2, so that a fault downstream of

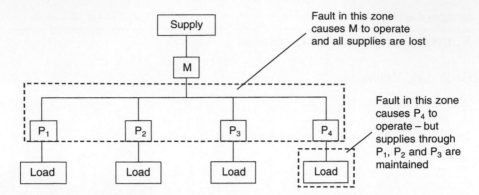

Fig. 8.1 Example of co-ordination

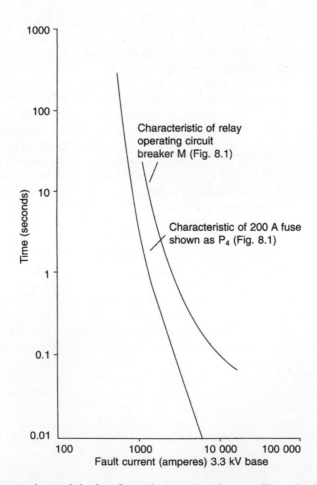

Fig. 8.2 Time–current characteristics for a fuse and a protection relay controlling a circuit breaker

both of them operates only the device nearest to the fault. A variety of shapes of time–current characteristic for both fuses and protection relays are available for different applications.

Another consideration when designing electrical protection is back-up. In some circuits it is desirable to have a device, such as an mcb (see section 7.3.3) which disconnects lower overcurrents and can be reset. Higher overcurrents, which the mcb cannot disconnect without being damaged, are disconnected by an upstream fuse in series with the mcb. This arrangement has the advantage that re-closure onto a severe fault is less likely, because replacement of the fuse would be necessary in this case as well as re-closure of the mcb. It is also possible for a protective device to fail, for example because of mishandling. The effects of a failure can be minimized or avoided by a back-up protective device which operates under such conditions. In Fig. 8.1, protective device M backs up each of the protective devices P. However, if M operates as a back-up device, co-ordination is lost because all four branches of the circuit lose supply, not just the faulty branch.

8.2 Fuses

8.2.1 Principles of design and operation

A fuse consists of a replaceable part (the fuselink) and a fuse holder. Examples of fuse holders are shown in Fig. 8.3.

The simplest fuselink is a length of wire. It is mounted by screw connections in a holder which partly encloses it. When an overcurrent or short-circuit current flows, the wire starts to melt and arcing commences at various positions along it. The arc voltage causes the current to fall and once it has fallen to zero, the arcs are extinguished. The larger the wire cross section, the larger is the current that the fuselink will carry without operating. In the UK, fuses of this type are specified for use at voltages up to 250 V and currents up to 100 A. They are known as *semi-enclosed* or *rewireable* fuses.

The most common fuselink is the *cartridge* type. This consists of a barrel (usually of ceramic) containing one or more elements which are connected at each end to caps fitted over the ends of the barrel. The arrangement is shown in Figs 8.4 and 8.5. If a high current breaking capacity is required, the cartridge is filled with sand of high chemical purity and controlled grain size. The entire fuselink is replaced after the fuse has operated and a fault has been disconnected. Cartridge fuses are used for a much wider range of voltages and currents than semi-enclosed fuses.

Fuselinks can be divided into *current-limiting* and *non-current-limiting* types. A sand-filled cartridge fuselink is of the current-limiting type; when it operates, it limits the peak current to a value which is substantially lower than the prospective current. A non-current-limiting fuse, such as a semi-enclosed fuse, does not limit the current significantly.

The element shown in Fig. 8.4 is a notched tape. Melting occurs first at the notches when an overcurrent flows and this results in a number of controlled arcs in series. The voltage across each arc contributes to the total voltage across the fuse, and this total voltage results in the current falling to zero. Because the number of arcs is limited, the fuselink voltage should not be high enough to cause damage elsewhere in the circuit. The characteristic development of current and voltage during the operation of a fuse is shown in Fig. 8.6.

The function of the sand is to absorb energy from the arcs and to assist in quenching them; when a high current is disconnected, the sand around the arcs is melted.

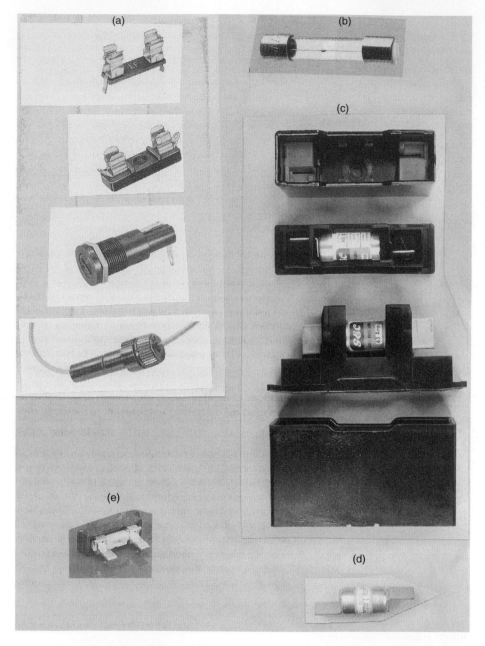

Fig. 8.3 Fuse holders for miniature and compact low-voltage fuses: (a) fuse holders for miniature fuselinks (b) miniature fuselink (c) fuse holders for compact low-voltage fuselinks (d) compact low-voltage fuselink (e) rewirable fuse

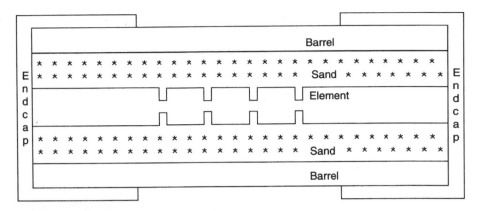

Fig. 8.4 Cross section through a low-voltage cartridge fuselink

The element is usually of silver because of its resistance to oxidation. Oxidation of the element in service would affect the current that could be carried without melting, because the effective cross section of the element is changed. Silver-plated copper elements are also used.

Many elements include an *m-effect blob*, which can be deposited on wire (Fig. 8.3(b)) or notched tape. The blob is of solder-type alloy which has a much lower melting point than the element. If a current flows which is large enough to melt only the m-effect blob, the solder diffuses into the silver. This creates a higher local resistance in the element and the fuse operates at a lower current than it would have done in the absence of the blob.

Other types include the *expulsion fuse* which is used at high voltage, and the *universal modular fuse (UMF)* which is used on Printed Circuit Boards (PCBs).

Fuses offer long life without deterioration in their characteristics or performance, and cartridge fuses have the particular advantage that they contain the arc products completely.

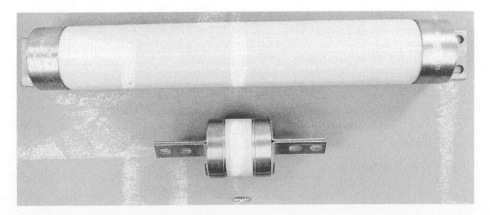

Fig. 8.5 High-voltage and low-voltage cartridge fuselinks (high-voltage fuse (top) and two low-voltage fuses showing a range of fuse sizes)

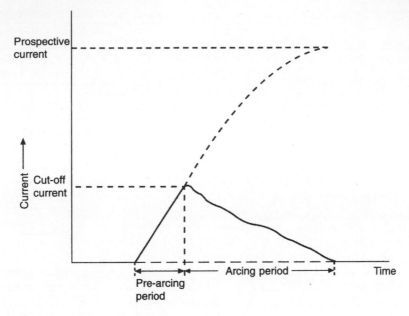

Fig. 8.6 Current and voltage during the operation of a fuse

8.2.2 Rating principles and properties

(a) *Current ratings (IEC)*

The rated current of a fuse is the maximum current that a fuselink will carry indefinitely without deterioration. In the case of ac ratings, an rms symmetrical value is given.

The current rating printed on a fuselink applies only at temperatures below a particular value. Derating may be necessary at high ambient temperatures and where fuses are mounted in hot locations, such as an enclosure with other heat-generating equipment.

It is obviously undesirable to have a proliferation of current ratings and therefore particular or preferred ratings are specified in standards. There is a general tendency, particularly in the IEC standards, to follow the R10 series or, if necessary, the R20 series may be used. For example, using the R10 series fuselinks with ratings from 10 A to 100 A are produced for 10 A, 12 A, 16 A, 20 A, 25 A, 32 A, 40 A, 50 A, 63 A, 80 A and 100 A. In the USA, more traditional ratings including 15 A, 30 A and 60 A are used.

(b) *Voltage ratings (IEC)*

The rated voltage of a fuse is the nominal voltage for which it was designed. Fuselinks will perform satisfactorily at lower voltages, but at much lower voltages, the reduction in current caused by the resistance of the fuselink should be considered. In the case of ac ratings, the rms symmetrical value is given, and for dc ratings the mean value, including ripple, is given.

IEC recommendations are moving towards harmonized low-voltage ac supplies of 230 V, 400 V and 690 V, but although the nominal voltage is being changed in many countries it will be possible for the voltage to remain at its previous non-harmonized level for several years. In Europe, the nominal voltage is 230 V, and the permitted variations will allow supplies to remain at 240 V and 220 V. Fuselinks marked 230 V may have been designed originally for use with higher or lower voltages, and problems may therefore arise when replacing fuselinks because a device manufactured for use at 220 V would not be safe to use on a 240 V system. A fuselink designed for 240 V could safely be used at 220 V. Similar considerations apply where the voltage is changed from 415 V to 400 V, or 660 V to 690 V.

(c) Variations in rating principles

The IEC rating principles are used worldwide, except in North America, where UL (Underwriters Laboratory) standards apply.

The rated current to a UL standard is the minimum current required to operate the fuse after many hours, and the current that it will carry indefinitely (the IEC rated current) is approximately 80 per cent of this rating.

The voltage rating marked on a UL fuselink is the *maximum* voltage at which it can be used, whereas that marked on an IEC fuselink is the *nominal* voltage.

These differences must be considered when replacing fuselinks, particularly in the case of miniature cartridge fuselinks which are interchangeable. In general, it is preferable to replace a fuselink with one of the same rating from the same manufacturer; this ensures that its characteristics are as similar as possible to those of the previous fuselink.

The IEC and UL ratings of fuse holders also differ. The IEC rating is the highest rated current of a fuselink with which it is intended to be used. A higher rating may be given in North America, this being related to the maximum current that does not cause overheating when a link of negligible resistance is used.

(d) Frequency ratings

Fuses are most commonly used in ac circuits with frequencies of 50 Hz or 60 Hz and a fuse designed for one of these frequencies will generally operate satisfactorily at the other. If the arc extinguishes at current zero, then the maximum arcing time on a symmetrical fault will be 10 ms at 50 Hz and 8 ms at 60 Hz.

Fuse manufacturers should be consulted about the suitability of fuses for other frequencies, which may include 17.67 Hz for some railway supplies, 400 Hz for aircraft and higher frequencies for some electronic circuits.

In dc circuits there is no current zero in the normal waveform and the fuselinks designed for ac may not operate satisfactorily. Separate current and voltage ratings are given for fuselinks tested for use in dc circuits. DC circuits can be more inductive for a given current than ac systems, and since the energy in the inductance is dissipated in the fuse, it is necessary for the dc voltage rating to be reduced as the time constant (L/R) of a circuit increases.

(e) Breaking capacity

The breaking capacity of a fuse is the current which can be interrupted at the rated voltage. The required breaking capacity will depend upon the position of the fuse in the supply system. For instance, 6 kA may be suitable for domestic and commercial applications, but 80 kA is necessary at the secondary of a distribution transformer.

The power factor of a short circuit affects the breaking capacity, and appropriate values are used when testing fuses.

(f) Time–current characteristics

The time–current characteristic of a fuse is a graph showing the dependence upon current of the time before arcing starts (the pre-arcing time); an example has been shown in Fig. 8.2. The total operating time of a fuse consists of the pre-arcing time and the arcing time. When pre-arcing times are longer than 100 ms and the arc is then extinguished at its first current zero (that is, an arcing time of less than 10 ms on a 50 Hz supply) then the time–current characteristic can be taken to represent the total operating time.

The *conventional time*, the *conventional fusing current* and the *conventional non-fusing current* are often shown on time–current characteristics. These values are defined in the standards. All fuses must operate within the conventional time when carrying the conventional fusing current; when carrying the conventional non-fusing current they must not operate within the conventional time.

Specific characteristic curves have not been standardized because the constraints which might be imposed by doing so could stifle future development and prevent the introduction of new fuses with slightly different time–current characteristics. The modern trend is to specify a number of points which form gates through which the actual time–current characteristics of all manufacturers' fuselinks must pass if they are to comply with the appropriate standard. Such gates apply to current IEC miniature fuse standards. In addition to gates, some of the IEC standards on low-voltage fuses specify zones within which all characteristics must lie. The zones are specified so that all fuselinks of a given rating would operate in a shorter time at any current than any fuselink of at least 1.6 times the rating (two steps in the R10 series).

(g) I^2t

I^2t is defined as the integral of the square of the current let through by a fuse over a period of time. Values are given by manufacturers for pre-arcing I^2t and total let-through I^2t.

Table 8.1 shows typical values of I^2t for low-voltage cartridge fuses of selected current ratings; values differ between manufacturers.

The heat generated in a circuit in a short circuit or fault condition before the fuse disconnects is given by the product of I^2t and the circuit resistance. As the let-through I^2t becomes constant above a particular level of fault current, the heat generated does not increase for prospective currents above this value, unless the breaking capacity is exceeded.

Table 8.1 Example of the variation of I^2t with current rating

Current rating (A)	Pre-arcing I^2t (kA²s)	Total I^2t (kA²s)
16	0.3	0.8
40	3.0	8.0
100	30	80
250	300	800
630	3000	8000

(h) *Power dissipation*

The resistance of a fuse will result in dissipation of power in the protected circuit when normal currents are flowing. This should be considered when designing the layout of a protection system.

(i) *Cut-off current*

A current-limiting fuse prevents a fault current from rising above a level known as the *cut-off current*. This is illustrated in Fig. 8.6. The cut-off current is approximately proportional to the cube root of the prospective current, and the maximum current is therefore very much lower than it would be if a non-current-limiting protection device were used.

8.2.3   Main classes of equipment

Fuses are produced in many shapes and sizes, and various types are illustrated in Figs 8.5, 8.7, 8.8 and 8.9. The main three categories are:

- miniature (up to 250 V)
- low voltage (up to 1000 V ac or 1500 V dc)
- high voltage (greater than 1000 V ac)

All three categories include current-limiting and non-current-limiting types.

(a) *Miniature fuses*

Cartridge fuses have in the past been the most common form of miniature fuse, but the UMF (see Fig. 8.7) is becoming increasingly used on PCBs. A UMF is much smaller than a cartridge fuse, and it is mounted directly on the PCB, whereas a cartridge fuse is mounted in a holder. Subminiature fuses have pins for mounting on PCBs.

Miniature cartridge fuses and subminiature fuses are rated for use at 125 V or 250 V. The UMFs have additional voltage ratings of 32 V and 63 V which make them more suitable for many types of electronic circuit. Miniature fuses are available with current ratings from 2 mA to 10 A. The maximum sustained power dissipation which is permitted in cartridge fuses ranges from 1.6 W to 4 W.

Miniature fuses may have a low, intermediate or high breaking capacity. All three ranges are available for UMFs, and these are given in Table 8.2.

Cartridge fuses are available with low or high breaking capacity. Low breaking capacity types have glass barrels without sand-filler and a visual check can therefore be made on whether or not the fuse has operated. High breaking capacity cartridge fuselinks are generally sand-filled and have ceramic barrels; they can interrupt currents of up to 1500 A.

A range of speeds of operation are available. Time-lag (surge-proof) fuses are required in circuits where there is an inrush current pulse, for instance when capacitors are charged or when motors or transformers are magnetized. The fuse must not be operated by these normal-operation surges, which must not cause deterioration of the fuse. The five categories of time-lag are medium time-lag (M), time-lag (T), long time-lag (TT), very quick-acting (FF) and quick-acting (F). They are available as cartridge fuses and the last two are used in the protection of electronic circuits. The alphabets shown in parentheses are marked on the endcaps.

Fig. 8.7 Examples of UMFs

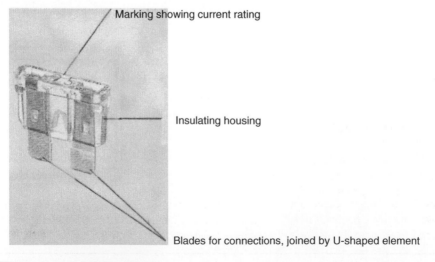

Fig. 8.8 Blade-type automotive fuselink

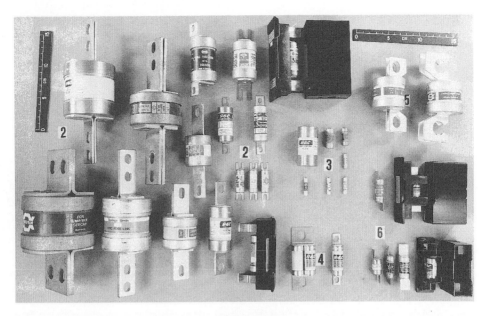

Fig. 8.9 A range of low-voltage fuselink types: (2) general purpose industrial fuselinks (3) fuselinks for domestic purposes (4) fuselinks for protecting semiconductors (5) fuselinks for use in UK electricity supply networks and (6) compact fuselinks for industrial applications

The UMFs are available in similar categories, which are super quick-acting (R), quick-acting (F), time-lag (T) and super time-lag (S).

The time–current characteristics of miniature fuses of the same type but with different ratings are similar in shape. Time can therefore be plotted against multiples of rated current and it is unnecessary to show separate characteristics for each current rating. Examples of time–current characteristics are shown in Fig. 8.10.

Cartridge fuses have various types of elements. Fast-acting types have a straight wire, and time-delay types use wire with an m-effect blob (Fig. 8.3(b)), helical elements on a heat-absorbing former or short elements with springs connecting them to the endcaps.

In addition to the most common types which have been described, miniature fuses are produced in a wide range of shapes and sizes. As an example of this, a blade-type

Table 8.2 Breaking capacity of UMFs

Voltage rating (V)	Breaking capacity (A)		Maximum overvoltage (kV)
32	Low	35*	0.33
63	Low	35*	0.5
125	Low	35*	0.8
250	Low	100	1.5
250	Intermediate	500	2.5
250	High	1500	4.0

*or 10 × rated current, whichever is the greater.

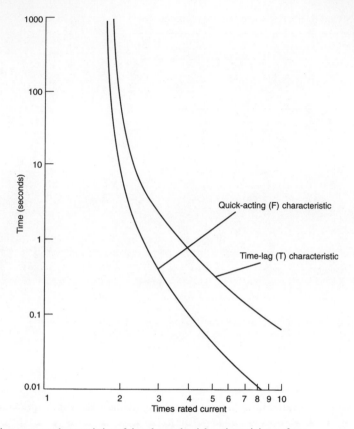

Fig. 8.10 Time–current characteristics of time-lag and quick-acting miniature fuses

automotive fuselink is shown in Fig. 8.8; the element in this fuse is visible through the plastic casing.

(b) *Low-voltage fuses*

A wide range of low-voltage fuses are available for industrial and domestic applications. These fuses have ratings appropriate for national or international single-phase or three-phase supplies, for example 220 V, 230 V, 240 V, 400 V, 415 V, 660 V and 690 V.

Widely differing systems for domestic protection are used in different countries and these cannot be described separately here. As an example, in the UK current-limiting cartridge fuses are used in plugs which supply appliances, the consumer unit supplying an entire property may have current-limiting cartridge fuses, semi-enclosed fuses or miniature circuit breakers and another fuse is installed by the supply authority on the incoming supply.

Industrial fuses may have general-purpose (type 'g') fuselinks which will operate correctly at any current between 1.6 times the rated current (the conventional fusing current) and the breaking capacity. Such fuses must not be replaced by type 'a' back-up fuselinks, which have a higher minimum breaking current and do not necessarily operate safely below this current; this type 'a' back-up fuselink is used to save space.

Table 8.3 Peak arc voltages for semiconductor fuselinks

Supply voltage (V)	Peak arc voltage (V)
230	700
400	900
690	1400

Type 'gG' fuses are used for general application. These have a full breaking capacity range and they provide protection for cables and transformers and back-up protection for circuit breakers. Specialized fuses are available for the protection of motors, semiconductors, street lighting, pole-mounted transformers and other purposes. Reference 8A provides further detail. Common applications are motor protection and semiconductor protection and these are described briefly.

Fuselinks for motor-starter protection must be able to withstand starting pulses without deterioration. Type 'gM' fuselinks are designed for this purpose and they have a dual rating. A designation 100M160, for example means that the fuselink has a continuous rating of 100 A and the general-purpose characteristics of a fuselink rated at 160 A.

Fuselinks for semiconductor protection are designed to operate with an arc voltage which does not damage the semiconductor device; this voltage is therefore lower than for other types of fuselink. Arc voltages at several supply voltages are shown in Table 8.3 for typical semiconductor fuselinks.

Semiconductor fuselinks also have lower let-through I^2t and cut-off current because semiconductors are susceptible to damage by heat and overcurrents. These fuses operate at higher temperatures than normal to achieve the necessary protection, and forced air-cooling may be used to increase their current rating.

(c) High-voltage fuses

High-voltage fuses can be of current-limiting or non-current-limiting type. The latter are expulsion fuses which do not contain the arc products when they operate; they can be very noisy and are therefore normally used outdoors.

Current-limiting high-voltage fuses are enclosed (as already shown in Fig. 8.5) and they may be used for the protection of motors, transformers and shunt power capacitors. The rated current of the fuselink is normally higher than the expected current. These fuses are normally used in three-phase systems and they are tested at 87 per cent of their rated voltage. In a three-phase earthed neutral system the voltage rating should be at least 100 per cent of the line-to-line voltage, and in a single-phase system it should be at least 115 per cent of the circuit voltage.

Further information can be found in references 8A, 8B and 8C.

8.2.4   Test methods

(a) Type tests

Before production of a type of fuselink commences, type tests are performed to ensure that pre-production fuselink samples comply with relevant national or international standards. Measurements of power dissipation, time–current characteristic, overload withstand capability, breaking capacity and resistance are included in these type tests.

(b) *Production tests*

Routine testing of many important fuse characteristics is not possible because tests, such as breaking capacity are destructive. Extensive testing in production would also be very costly. Fuse manufacturers therefore make production fuselinks as identical as possible to the samples used for type testing.

The quality of fuselinks depends upon the quality of the components supplied to the fuse manufacturer. Key items, such as barrels, filling material, element material and endcaps are therefore regularly inspected and tested when received.

The dimensions and straightness of barrels are checked and their ability to withstand mechanical and thermal shock and internal pressure is tested. Endcap dimensions are checked to ensure that they fit closely over the barrel. The moisture content, chemical composition and grain size of the filler are measured. The diameter or thickness of the element wire or tape are checked and its resistance per metre is measured. Where elements are produced from tape and notched, the dimensions and pitch of the notches are tightly controlled.

During assembly, checks are made to ensure that the fuselink is completely filled with sand and that the element resistance is correct. After assembly, the overall dimensions are checked and the resistance is once more measured. A visual check including the markings is then made.

Other tests are made in the case of specialized fuselinks. For example, the conditions of the elements in a high-voltage fuselink is examined using X-ray photography.

In addition to these routine tests, manufacturers may also occasionally take sample fuselinks from production and subject them to some or all of the type tests.

(c) *Site checks*

Before use, every fuselink should be checked visually for cracks and tightness of endcaps and the resistance should be checked. It should also be checked that the ratings, especially current, voltage, breaking capacity and time–current characteristic, are correct for the application. In the case of semi-enclosed, rewireable fuses, care should be taken to use the appropriate diameter of fuse wire. Fuse holders should be checked to ensure that the clips or means of connection are secure and correctly aligned.

If a fuselink has been dropped onto a hard surface or subjected to other mechanical stress it should not be used; damage may not be visible but it could cause the fuse to malfunction with potentially serious results.

If a fault occurs and the fuselink is overloaded, it should be replaced even if it has not been operated. This situation arises especially in three-phase systems where one or two of the three fuses may operate to clear the fault.

Fuselinks (as opposed to rewireable fuses) cannot be safely repaired; they must always be replaced.

8.2.5 Standards

Many national and international standards exist because of the number of different fuse types.

Tables 8.4, 8.5 and 8.6 summarize the position for miniature, low-voltage and high-voltage fuses, respectively. The IEC recommendations are listed, together with related EN and BS standards and North American standards covering the same field.

Table 8.4 Comparison of international, regional and national standards for miniature fuses

IEC/BS/EN	Subject	North American
60127	Miniature fuses	
60127-1	• definitions and general requirements	
60127-2	• cartridge fuselinks	
60127-3	• subminiature fuselinks	
60127-4	• universal modular fuses (UMFs)	
60127-5	• guidelines for quality assessment	
60127-6	• fuse holders	
60127-10	• user guide	
	Blade-type electric fuses	SAE J 1284
	Miniature blade-type electric fuses	SAE J 2077
	Blade fuses – 42 V system	SAE J 2576
	Fuses for supplementary overcurrent protection	UL 248-14
	Automotive glass tube fuse	UL 275

In order to comply with the EMC directive, the following statement has been added to most of the UK fuse standards:

> Fuses within the scope of this standard are not sensitive to normal electromagnetic disturbances, and therefore no immunity tests are required. Significant electromagnetic disturbance generated by a fuse is limited to the instant of its operation. Provided that the maximum arc voltages during operation in the type test comply with the requirements of the clause in the standard specifying maximum arc voltage, the requirements for electromagnetic compatibility are deemed to be satisfied.

8.3 Protection relays

8.3.1 Principles of design and operation

Systems incorporating protection relays can disconnect high currents in high-voltage circuits which are beyond the scope of fuse systems.

In general, relays operate in the event of a fault by closing a set of contacts or by triggering a thyristor. This results in the closure of a trip-coil circuit in the circuit breaker which then disconnects the fault. The presence of the fault is detected by current transformers, voltage transformers or bimetal strips.

Electromechanical and solid-state relays are both widely used, but the latter are becoming more widespread because of their bounce-free operation, long life, high switching speed and additional facilities that can be incorporated into the relay. Additional facilities can, for instance, include measurement of circuit conditions and transmission of the data to a central control system by a microprocessor relay. This type of relay can also monitor its own function and diagnose any problems that are found. Solid-state relays can perform any of the functions of an electromechanical relay whilst occupying less space, but electromechanical relays are less susceptible to interference and transients. Electromechanical relays also have the advantage of providing complete isolation and they are generally cheaper than solid-state devices.

Contacts in electromechanical relays may have to close in the event of a fault after years of inactivity and twin sets of contacts can be used to improve reliability. The contact material must be chosen to withstand corrosive effects of a local environment because a film of corrosion would prevent effective contact being made. Dust in the atmosphere

Table 8.5 Comparison of international, regional and national standards for low-voltage fuses

IEC	EN/HD	BS	Subject	North American
60269			Low voltage fuses	
60269-1	EN 60269-1	88-1 (related to 1361 and 1362)	• general applications	
60269-2	EN 60269-2	88-2.1	• supplementary requirements for fuses for use by authorized persons	
60269-2-1	HD 630.2.1	related to 88-2.2 88-5 88-6	• standardized examples – bolted – wedge tightening – offset blade	
60269-3	EN 60269-3		• supplementary requirements for fuses for use by unskilled persons	
60269-3-1	HD 630.3.1	related to 1361 1362	• standardized examples – house service and consumer unit – plug top	
60269-4	EN 60294-4	88-4	• supplementary requirements for the protection of semiconductor devices	
60294-4-1	EN 60294-4-1		• standardized examples	
TR 61459			– Co-ordination between fuses and contactors/motor starters	
TR 61818			Application guide	
			General requirements	UL 248-1
			Class C fuses	UL 248-2
			Class CA and CB fuses	UL 248-3
			Class CC fuses	UL 248-4
			Class G fuses	UL 248-5
			Class H non-renewable fuses	UL 248-6
			Class H renewable fuses	UL 248-7
			Class J fuses	UL 248-8
			Class K fuses	UL 248-9
			Class L fuses	UL 248-10
			Plug fuses	UL 248-11
			Class R fuses	UL 248-12
			Semiconductor fuses	UL 248-13
			Supplemental fuses	UL 248-14
			Class T fuses	UL 248-15
			Test limiters	UL 248-16
			Fuse holders	UL 512
			Application guide for low-voltage ac non-integrally fused power circuit breakers (using separately mounted current-limiting fuses)	IEEE C37.27

can also increase the contact resistance and result in failure. Both corrosion and dust contamination can be avoided by complete enclosure and sealing of the relay.

The contacts must also withstand arcing during bounce on closure and when opening, but this is usually less important than resisting corrosion. Because of the need for high reliability, the contacts are usually made of or plated with gold, platinum, rhodium, palladium, silver or various alloys of these metals.

Table 8.6 Comparison of international, regional and national standards for high-voltage fuses

IEC	EN	BS	Subject	North American ANSI
60282			High-voltage fuses	
60282-1	60282-1	EN 60282-1	• current-limiting fuses	
60282-2		2692-2	• expulsion and similar fuses	
60549		related to 5564	High-voltage fuses for the protection of shunt capacitors	
60644	60644	EN 60644	Specification for high-voltage fuselinks for motor circuits	
60787		6553	Guide for the selection of fuselinks for transformer circuit applications	
			Service conditions and definitions for high-voltage fuses, distribution enclosed single-pole air switches, fuse disconnecting switches, and accessories	C37.40
			Design tests for high-voltage fuses, distribution enclosed single-pole air switches, and accessories	C37.41
			Specifications for distribution cut-outs and fuselinks	C37.42
			Specifications for distribution oil cut-outs and fuselinks	C37.44
			Specifications for power fuses and fuse disconnecting switches	C37.46
			Specifications for distribution fuse disconnecting switches, fuse supports and current-limiting fuses	C37.47
			Guide for application, operation and maintenance of high-voltage distribution cut-outs and fuselinks, secondary fuses, distribution enclosed single-pole air switches, power fuses, fuse disconnecting switches and accessories	C37.48

Solid-state relays are not affected directly by corrosion or dust, but temperature and humidity may effect them if conditions are severe enough.

Electromechanical relays operate by induction, attraction or thermally, where a bimetallic strip is used to detect overcurrent. The first two types are most common and their principles are described, along with those of solid-state relays, in the following sections. Further information on protection relays can be found in references 8C, 8D and 8E.

Special arrangements are necessary for the protection of dc systems and these are discussed in section 7.6.5.

(a) Induction relays

An induction relay has two electromagnets, labelled E_1 and E_2 in Fig. 8.11. Winding A of electromagnet E_1 is fed by a current transformer which detects the current in the protected circuit. Winding B in electromagnet E_1 is a secondary, and it supplies the winding on E_2. The phases of the currents supplied to E_1 and E_2 differ and therefore the magnetic fluxes produced by the two electromagnets have different phases. This results in a torque on the disc mounted between the electromagnets, but the disc can only move when a certain torque level is reached because it is restricted by a hair spring or a stop. Normal currents in the protected circuit do not therefore cause movement of the disc.

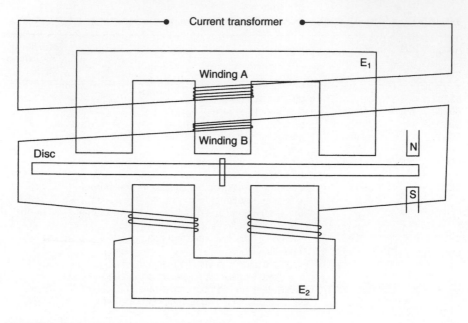

Fig. 8.11 Principle of operation of an induction relay

When the disc does turn, its speed depends upon the current supplied by the current transformer and the eddy current braking effected by a permanent magnet located near the edge of the disc. When the disc rotates through a certain angle, the relay contacts close and the time for this to occur can be adjusted by the position of the stop or the angle through which the disc has to rotate. This adjustment allows protection co-ordination to be achieved by means of 'time grading'. For example, in the radial feeder shown in Fig. 8.12, the minimum time taken for the protection relay to operate

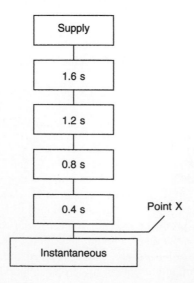

Fig. 8.12 Time grading

would be set higher at points closer to the supply. A fault at point X, a considerable distance from the supply would cause operation of the relay set at a minimum time of 0.4 s and the fault would be disconnected before it caused operation of relays nearer the supply, thus preventing the unnecessary tripping of healthy circuits.

Induction relays have Inverse Definite Minimum Time (IDMT) time–current characteristics in which the time varies inversely with current at lower fault currents, but attains a constant minimum value at higher currents. This constant minimum value depends upon the adjustments previously described.

Further adjustment is possible by means of tappings on the relay winding A in Fig. 8.11. For example, if a current transformer has a secondary winding rated at 1 A, tappings could be provided in the range 50 per cent to 200 per cent in 25 per cent steps, corresponding to currents of 0.5 A to 2 A in 0.25 A steps. If the circuit is uprated, it may then be possible to adjust the relay rather than replace it. For example, if the 100 per cent setting is used when the maximum current expected in the protected circuit is 400 A, the 150 per cent setting could be used if the maximum current is increased to 600 A.

(b) *Attracted-armature relays*

The basis of operation of an attracted-armature relay is shown in Fig. 8.13. The electromagnet pulls in the armature when the coil current exceeds a certain value and the armature is linked to the contacts and when it moves, it opens normally closed contacts and closes normally open contacts. The time required for operation is only a few seconds, and it depends upon the size of the current flowing in the coil.

These devices are called instantaneous relays. They have a range of current settings which are provided by changing the tapping of the coil or by varying the air gap between the electromagnet and the armature.

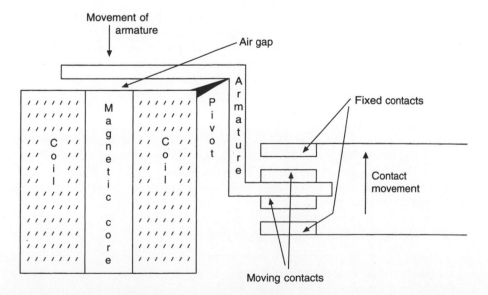

Fig. 8.13 Principle of operation of an attracted-armature relay

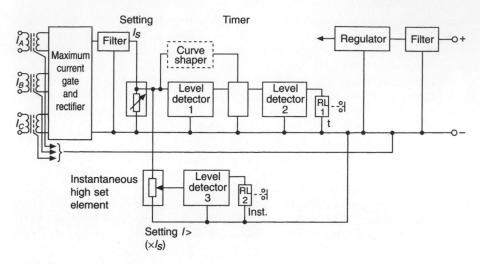

Fig. 8.14 Components of a solid-state time-delayed overcurrent relay

(c) Solid-state relays

The first solid-state relays were based on transistors and performed straightforward switching, but now they can often perform much more complicated functions by means of digital logic circuits, microprocessors and memories.

Currents and voltages are measured by sampling incoming analogue signals, and the results are stored in digital form. Logic operations, such as comparison are then performed on the data to determine whether the relay should operate to give an alarm or trip a circuit breaker. Digital data can also be stored on the computer for subsequent analysis, for instance, after a fault has occurred.

Solid-state relays may incorporate a variety of additional circuits. Figure 8.14 shows for example a circuit which imposes a time delay on the output signal from a relay. Such circuits may also be used to shape the time–current characteristic of a protection relay in various ways. A comparison of typical time–current characteristics from electromechanical and solid-state relays is shown in Fig. 8.15. There are many other possible functions, such as power supervision to minimize power use, an arc sensor to over-ride time delays and measurement of true rms values in the presence of harmonics.

Solid-state relays may require shielding against electromagnetic interference arising from electrostatic discharges or high-voltage switching. Optical transmission of signals is sometimes used to reduce the effects of this interference.

The electronics in solid-state relays can be damaged by moisture, and the relays are usually encapsulated to prevent this.

8.3.2 Rating principles and properties

All components of the protection system including current transformers, relays and circuit breakers must have the correct current, voltage and frequency rating, and the I^2t let-through and interrupting capacity of the entire system depends upon the circuit breaker. A protection relay must have a minimum operating current which is greater

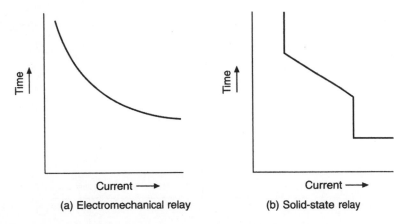

Fig. 8.15 Comparison of time–current characteristics of electromechanical and solid-state relays

than the rated current of the protected circuit, and other properties of the relay must be chosen correctly in order for it to operate the circuit breaker as required by the application.

Manufacturers publish information regarding the selection, installation and use of relays. The following points in particular will need to be considered.

Heat is generated within a relay in use and if several relays are grouped together in an enclosed space, provision should be made to ensure that temperature rises are not excessive.

Protection levels, time delays and other characteristics of both electromechanical and solid-state relays can be changed on site. For example, overvoltage protection may be set to operate at levels between 110 per cent and 130 per cent, and Fig. 8.16 shows the effect of adjusting the operating time of an overcurrent relay.

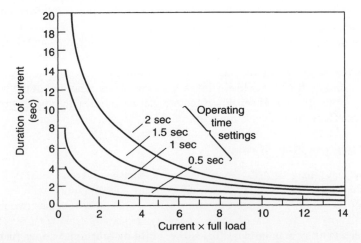

Fig. 8.16 Adjustment of time–current characteristics for an overcurrent relay

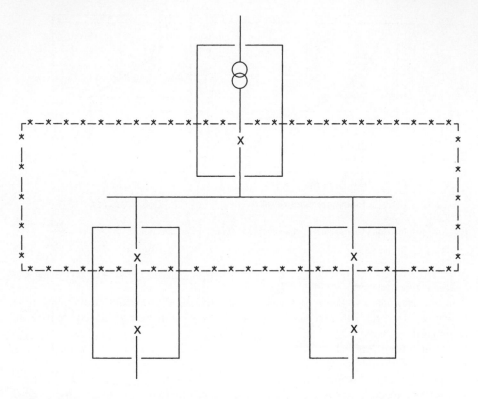

Fig. 8.17 Overlapping zones of protection. The protection zone marked —✗—✗—✗ overlaps the zones marked —. X marks the positions of circuit breakers with their associated protection relays

In general, electromechanical relays can be adjusted continuously and solid-state relays are adjusted in steps. Many other adjustments can be made. For example, DIP switches in solid-state relays can be used to set the rated voltage and frequency and to enable phase-sequence supervision.

The adjustment of operating times is used to provide time-grading and co-ordination, which has already been explained with reference to Fig. 8.12. An alternative system uses protection relays which operate only when a fault is in a clearly defined zone; this is illustrated in Fig. 8.17. To achieve this, a relay compares quantities at the boundaries of a zone. Protection based on this principle can be quicker than with time-graded systems because no time delays are required.

8.3.3 Main classes of relay

Protection relays may be 'all-or-nothing' types, such as overcurrent tripping relays, or they may be measuring types which compare one quantity with another. An example of the latter is in synchronization, when connecting together two sources of power.

A protection relay may be classified according to its function. Various functions are noted, and details for a wide range of functions are given in reference 8F.

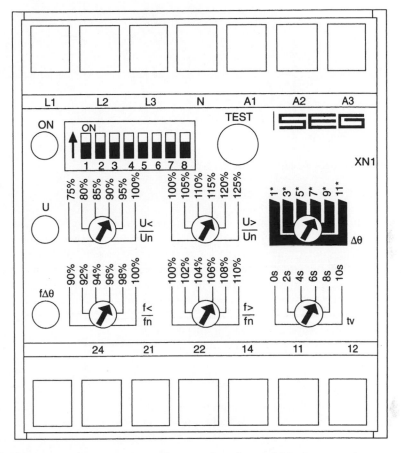

Fig. 8.18 Mains decoupling relay showing adjustments for voltage and frequency protection

Common applications are:

- undervoltage and overvoltage detection
- overcurrent detection
- overfrequency and underfrequency detection

These functions may be combined in a single relay. For example, the relay shown in Fig. 8.18 is used in power stations and provides overvoltage and undervoltage, overfrequency and underfrequency protection, and it rapidly disconnects the generator in the case of a failure in the connected power system.

Another application in power systems is the protection of transmission lines by distance relays. Current and voltage inputs to a distance relay allow detection of a fault within a predetermined distance from the relay and within a defined zone. The fault impedance is measured and if it is less than a particular value, then the fault is within a particular distance. This is illustrated in the $R–X$ diagrams shown in case (a) of Fig. 8.19; if the R and X values derived from the measured fault impedance $R + iX$ result in a point within the circle, then the relay operates. If directional fault detection is required, the area of operation is moved, as shown in case (b).

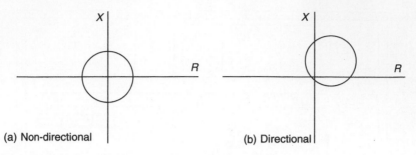

(a) Non-directional

(b) Directional

Fig. 8.19 Operating areas for distance protection relays

Another form of protection for lengths of conductor is the pilot wire system. Current transformers are placed at each end of a conductor and they are connected by pilot wires. Relays determine whether the currents at the two ends of the conductor are the same, and they operate if there is an excessive difference. In the balanced voltage system shown in Fig. 8.20(a), no current flows in the pilot wire unless there is a fault. In a balanced current system, current does flow in the pilot wire in normal conditions, and faults are detected by differences in voltage at the relays which are connected between the pilot wires; this is shown in Fig. 8.20(b).

Other applications involving relays include:

- the protection of motor starters against overload
- checking phase balance
- the protection of generators from loss of field
- the supervision of electrical conditions in circuits

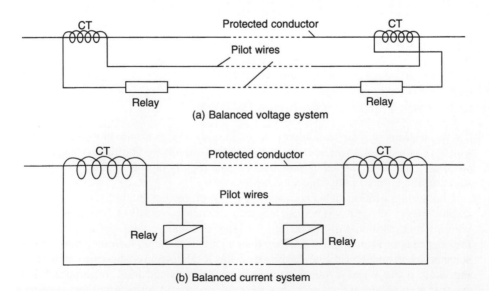

(a) Balanced voltage system

(b) Balanced current system

Fig. 8.20 Pilot wire protection systems

8.3.4 Test methods

(a) *Production tests*

Manufacturers of electromechanical relays often produce their own components and inspect them before assembly. Components for solid-state relays are generally bought in from specialist manufacturers and there is an incoming check to eliminate those which would be subject to early failure or excessive drift. Other characteristics, such as memory are checked as appropriate.

After assembly, manufacturers test all protection relays to ensure that they comply with relevant national and international standards. The calibration of adjustable settings is checked.

Fault conditions that the relays are designed to protect against can be simulated, and typical inputs to relays from current or voltage transformers can be duplicated by test sets; this can be used to check, for example, the correct functioning of relays for phase comparison and the proper disconnection of overcurrent and earth faults of various impedances. Such tests can be performed by setting up a test circuit or by means of a computer-based power system simulator which controls the inputs to the relays. The latter method allows the effect of high-frequency transients and generator faults to be investigated, and it is independent of an actual power supply; it may, for example, be used for the testing of selectable protection schemes in distance relays.

Other tests include the following:

- *environmental tests* are performed to check, for example, the effects on performance of temperature and humidity
- *impact, vibration and seismic tests* are performed on both solid-state and electromechanical relays, although the latter are more prone to damage from such effects
- *voltage transients* are potentially damaging to solid-state relays, and the relays are tested to ensure they can withstand a peak voltage of 5 kV with a rise time of 1.2 μs and a decay time of 50 μs

(b) *Site tests*

Primary injection tests are applied during initial commissioning and these should show up any malfunction associated with protection relays. Secondary injection tests should be performed if there is a maloperation which may be related to the protection relay. Details of these injections tests are given in reference 8E.

Periodic inspection and testing is necessary throughout the lifetime of a protection relay and computerized equipment is available for this purpose. Some relatively complicated protection schemes incorporate automatic checking systems which send test signals at regular intervals; the test signals can also be sent manually. Digital relays may include continuous self-checking facilities.

8.3.5 Standards

There are many standards covering various types of relay and aspects of their use. Some standards which apply to relays in general are also relevant to protection relays and some apply specifically to protection relays. The key IEC recommendations together with equivalent BS and EN standards and related North American standards are summarized in Table 8.7.

Table 8.7 Comparison of international, regional and national standards for protection relays

IEC	EN/CECC	BS	Subject	North American
60255			Electrical relays	
60255-23	EN 60255-23	EN 60255-23	• contact performance	
61810-1	EN 61810-1	EN 61810-1	• all-or-nothing electrical relays	
60255-3	EN 60255-3	EN 60255-3 142 secns 3.1	• single input energizing quantity measuring relays with dependent or independent time	
60255-5	EN 60255-5	EN 60255-5	• insulation co-ordination for measuring relays and protection equipment – requirements and tests	
60255-6	EN 60255-6	EN 60255-6	• measuring relays and protection equipment	
61810-7		QC 160000-1	• test and measurement procedures for all-or-nothing relays	
60255-8	EN 60255-8	EN 602558	• thermal electrical relays	
62246-1	EN 62246-1	EN 62246-1	• dry reed make contact units	
61811-1	EN 61811-1	EN 61811-1	• electromechanical non-specified time all-or-nothing relays of assessed quality. Generic specification	
60255-11			• interruptions to alternating component (ripple) in dc auxiliary energizing quantity of measuring relays	
60255-12			• directional relays and power relays with two input energizing quantities	
60255-13			• biased (percentage) differential relays	
60255-14		5992-4	• endurance test for electrical relay contacts – preferred values for contact loads	
60255-15		5992-5	• endurance test for electrical relay contacts – characteristics of test equipment	
60255-16			• impedance measuring relays	
61811-10	EN 61811-10	EN 61811-10	• electromechanical relays of assessed quality	
60255-21	EN 60255-21	EN 60255-21	• vibration, shock, bump and seismic tests on measuring relays and protection equipment	
60255-22	EN 60255-22	EN 60255-22	• electrical disturbance tests	
			Relays and relay systems associated with electric power apparatus	ANSI/IEEE C37.90
			Surge withstand capability tests for relays and relay systems associated with electrical power apparatus	ANSI/IEEE C37.90.1
			Guide for protective relay applications to power transformers	ANSI/IEEE C37.91
			Guide for power system protective relay applications for audio tones over telephone channels	ANSI/IEEE C37.93
			Guide for protective relaying of utility – consumer interconnections	ANSI/IEEE C37.95

Table 8.7 (*contd*)

IEC	EN/CECC	BS	Subject	North American
			Guide for protective relay applications to power system buses	ANSI/IEEE C37.97
			Seismic testing of relays	ANSI/IEEE C37.98
			Qualifying class 1E protective relays and auxiliaries for nuclear power generating stations	ANSI/IEEE C37.105
			Testing procedures for relays and electrical and electronic equipment	ANSI/EIA 407-A
			Solid state relays	ANSI/EIA 443

References

8A. Williams, D.J.A., Turner, H.W. and Turner, C. *User's Guide to Fuses* (2nd edn), ERA Technology Ltd, UK, 1993.

8B. Wright, A. and Newbery, P.G. *Electric Fuses* (2nd edn), Institution of Electrical Engineers, London, 1995.

8C. Wright, A. and Christopoulos, C. *Electric Power System Protection*, Chapman & Hall, London, 1993.

8D. GEC Measurements, *Protective relays-Applications guide* (3rd edn), The General Electric Co., UK, 1987.

8E. Jones, G.R., Laughton, M.A. and Say, M.G. *Electrical Engineer's Reference Book*, Butterworth-Heinemann, Oxford, UK, 1993.

8F. Recommended practice for protection and co-ordination of industrial and commercial power systems, *IEEE*, Wiley Interscience, 1986.

Chapter 9

Wires and cables

V.A.A. Banks and P.H. Fraser

With amendments by
D. Gracias
Pirelli Cables Ltd

A.J. Willis
Brand-Rex Ltd

9.1   Scope

Thousands of cable types are used throughout the world. They are found in applications ranging from fibre-optic links for data and telecommunication purposes through to EHV underground power transmission at 275 kV or higher. The scope here is limited to cover those types of cable which fit within the general subject matter of the pocket book.

This chapter therefore covers cables rated between 300/500 V and 19/33 kV for use in the public supply network, in general industrial systems and in domestic and commercial wiring. Optical communication cables are included in a special section. Overhead wires and cables, submarine cables and flexible appliance cords are not included.

Even within this relatively limited scope, it has been necessary to restrict the coverage of the major metallic cable and wire types to those used in the UK in order to give the cursory appreciation which is the main aim.

9.2   Principles of power cable design

9.2.1   Terminology

The voltage designation used by the cable industry does not always align with that adopted by users and other equipment manufacturers, so clarification may be helpful.

A cable is given a voltage rating which indicates the maximum circuit voltage for which it is designed, not necessarily the voltage at which it will be used. For example, a cable designated 0.6/1 kV is suitable for a circuit operating at 600 V phase-to-earth and 1000 V phase-to-phase. However it would be normal to use such a cable on distribution and industrial circuits operating at 230/400 V in order to provide improved safety and increased service life. For light industrial circuits operating at 230/400 V it would be normal to use cables rated at 450/750 V, and for domestic circuits operating at 230/400 V, cable rated at 300/500 V would often be used. Guidance on the cables that are suitable for use in different locations is given in BS 7540.

The terms LV (Low Voltage), MV (Medium Voltage) and HV (High Voltage) have different meanings in different sectors of the electrical industry. In the power cable industry, the following bands are generally accepted and these are used in this chapter:

LV – cable rated from 300/500 V to 1.9/3.3 kV

MV – cables rated from 3.8/6.6 kV to 19/33 kV

HV – cable rated at greater than 19/33 kV

Multi-core cable is used in this chapter to describe power cable having two to five cores. Control cable having seven to 48 cores is referred to as *multi-core control cable*.

Cable insulation and sheaths are variously described as *thermoplastic*, *thermosetting*, *vulcanized*, *cross-linked*, *polymeric* or *elastomeric*. All extruded plastic materials applied to cable are *polymeric*. Those which would re-melt if the temperature during use is sufficiently high are termed *thermoplastic*. Those which are modified chemically to prevent them from re-melting are termed thermosetting, cross-linked or vulcanized. Although these materials will not re-melt, they will soften and deform at elevated temperatures, if subjected to excessive pressure. The main materials within the two groups are as follows:

- *thermoplastic*
  - polyethylene (PE)
  - medium-density polyethylene (MDPE)
  - polyvinyl chloride (PVC)
- *thermosetting, cross-linked or vulcanized*
  - cross-linked polyethylene (XLPE)
  - ethylene-propylene rubber (EPR)

Elastomeric materials are polymeric. They are rubbery in nature, giving a flexible and resilient extrusion. Elastomers such as EPR are normally cross-linked.

9.2.2   General considerations

Certain design principles are common to power cables, whether they are used in the industrial sector or by the electricity supply industry.

For many cable types the conductors may be of copper or aluminium. The initial decision made by a purchaser will be based on price, weight, cable diameter, availability, the expertise of the jointers available, cable flexibility and the risk of theft. Once a decision has been made, however, that type of conductor will generally then be retained by that user, without being influenced by the regular changes in relative price which arise from the volatile metals market.

For most power cables the form of conductor will be solid aluminium, stranded aluminium, solid copper (for small wiring sizes) or stranded copper, although the choice may be limited in certain cable standards. Solid conductors provide for easier fitting of connectors and setting of the cores at joints and terminations. Cables with stranded conductors are easier to install because of their greater flexibility, and for some industrial applications a highly flexible conductor is necessary.

Where cable route lengths are relatively short, a multi-core cable is generally cheaper and more convenient to install than single-core cable. Single-core cables are sometimes used in circuits where high load currents require the use of large conductor sizes, between 500 mm^2 and 1200 mm^2. In these circumstances, the parallel connection of two or more multi-core cables would be necessary in order to achieve the required rating and this presents installation difficulties, especially at termination boxes.

Single-core cable might also be preferred where duct sizes are small, where longer cable runs are needed between joint bays or where jointing and termination requirements dictate their use. It is sometimes preferable to use 3-core cable in the main part of the route length, and to use single-core cable to enter the restricted space of a termination box. In this case, a transition from one cable type to the other is achieved using trifurcating joints which are positioned several metres from the termination box.

Armoured cables are available for applications where the rigours of installation are severe and where a high degree of external protection against impact during service is required. *Steel Wire Armour (SWA)* cables are commonly available although *Steel Tape Armour (STA)* cables are also available. Generally, SWA is preferred because it enables the cable to be drawn into an installation using a pulling stocking which grips the outside of the oversheath and transfers all the pulling tension to the SWA. This cannot normally be done with STA cables because of the risk of dislocating the armour tapes during the pull. Glanding arrangements for SWA are simpler and they allow full usage of its excellent earth fault capability. In STA, the earth fault capability is much reduced and the retention of this capability at glands is more difficult. The protection offered against a range of real-life impacts is similar for the two types.

9.2.3 Paper-insulated cables

Until the mid-1960s, paper-insulated cables were used worldwide for MV power circuits. There were at that time very few alternatives apart from the occasional trial installation or special application using PE or PVC insulation.

The position is now quite different. There has been a worldwide trend towards XLPE cable and the UK industrial sector has adopted XLPE-insulated or EPR-insulated cable for the majority of MV applications, paper-insulated cable now being restricted to minor uses, such as extensions to older circuits or in special industrial locations. The use of paper-insulated cables for LV has been superseded completely by polymeric cables in all sectors throughout the world.

The success of polymeric-insulated cables has been due to the much easier, cleaner and more reliable jointing and termination methods that they allow. However, because of the large amount of paper-insulated cable still in service and its continued specification in some sectors, such as the regional public supply networks for MV circuits, its coverage here is still appropriate.

Paper-insulated cables comprise copper or aluminium phase conductors which are insulated with lapped paper tapes, impregnated with insulating compound and sheathed with lead, lead alloy or corrugated aluminium. For mechanical protection, lead or lead alloy sheathed cables are finished off with an armouring of steel tapes or steel wire and a covering of either bitumenized hessian tapes or an extruded PVC or PE oversheath. Cables which are sheathed with corrugated aluminium need no further metallic protection, but they are finished off with a coating of bitumen and an extruded PVC oversheath. The purpose of the bitumen in this case is to provide additional corrosion protection should water penetrate the PVC sheath at joints, in damaged areas or by long-term permeation.

There are, therefore, several basic types of paper-insulated cable and these are specified according to existing custom and practice as much as to meet specific needs and budgets. Particular features of paper-insulated cables used in the electricity supply industry and in industrial applications are described in sections 9.3.1.1 and 9.3.2.1, respectively.

The common element is the paper insulation itself. This is made up of many layers of paper tape, each applied with a slight gap between the turns. The purity and grade of the paper is selected for best electrical properties and the thickness of the tape is chosen to provide the required electrical strength.

In order to achieve acceptable dielectric strength, all moisture and air is removed from the insulation and replaced by *Mineral Insulating Non-Draining (MIND) compound*. Its waxy nature prevents any significant migration of the compound during the lifetime of the cable, even at full operating temperature. This is in contrast to oil-filled HV cables, utilizing a lower viscosity impregnant which must be pressurized throughout the cable service life to keep the insulation fully impregnated. Precautions are taken at joints and terminations to ensure that there is no local displacement of MIND compound which might cause premature failure at these locations. The paper insulation is impregnated with MIND compound during the manufacture of the cable, immediately before the lead or aluminium sheath is applied.

A 3-core construction is preferred in most MV paper-insulated cables. The three cores are used for the three phases of the supply and no neutral conductor is included in the design. The parallel combination of lead or aluminium sheath and armour can be used as an earth continuity conductor, provided that circuit calculations prove its adequacy for this purpose. Conductors of 95 mm^2 cross section and greater are sector-shaped so that when insulated they can be laid up in a compact cable construction. Sector-shaped conductors are also used in lower cross sections, down to 35 mm^2, 50 mm^2 and 70 mm^2 for cables rated at 6 kV, 10 kV and 15 kV, respectively.

The 3-core 6.6 kV cables and most 3-core 11 kV cables are of *belted* design. The cores are insulated and laid up such that the insulation between conductors is adequate for the full line-to-line voltage (6.6 kV or 11 kV). The laid-up cores then have an additional layer of insulating paper, known as the *belt layer,* applied and the assembly is then lead sheathed. The combination of core insulation and belt insulation is sufficient for phase-to-earth voltage between core and sheath (3.8 kV or 6.35 kV).

A 15 kV, 22 kV and 33 kV 3-core cables and some 11 kV 3-core cables are of *screened* design. Here each core has a metallic screening tape and the core insulation is adequate for the full phase-to-earth voltage. The screened cores are laid up and the lead or aluminium sheath is then applied so that the screens make contact with each other and with the sheath.

The bitumenized hessian serving or PVC oversheath is primarily to protect the armour from corrosion in service and from dislocation during installation. The PVC oversheath is now preferred because of the facility to mark cable details, and its clean surface gives a better appearance when installed. It also provides a smooth firm surface for glanding and for sealing at joints.

9.2.4   Polymeric cables

PVC and PE cables were being used for LV circuits in the 1950s and they started to gain wider acceptance in the 1960s because they were cleaner, lighter, smaller and easier to install than paper-insulated types. During the 1970s the particular benefits of XLPE and HEPR insulations were being recognized for LV circuits and today it is these cross-linked insulations, mainly XLPE, which dominate the LV market with PVC usage in decline for power circuits, although still used widely for low voltage wiring circuits. The LV XLPE cables are more standardized than MV polymeric types, but even so there is a choice of copper or aluminium conductor (circular or shaped), single-core or multi-core, SWA or unarmoured, and PVC or *Low Smoke and Fume (LSF)* sheathed. A further option is available for LV in which the neutral and/or earth conductor is a layer of wires applied concentrically around the laid-up cores rather than as an insulated core within the cable. In this case, the concentric earth conductor can replace the armour layer as the protective metal layer for the cable. This concentric wire design is mainly used in the LV electricity distribution network, whereas the armoured design is primarily used in industrial applications.

Polyethylene and PVC were shown to be unacceptable for use as general MV cable insulation in the years following the 1960s because their thermoplastic nature resulted in significant temperature limitations. The XLPE and EPR were required in order to give the required properties. They allowed higher operating and short circuit temperatures within the cable, as well as the advantages of easier jointing and terminating than for paper-insulated cables. This meant that in some applications a smaller conductor size could be considered than had previously been possible in the paper-insulated case.

The MV polymeric cables comprise copper or aluminium conductors insulated with XLPE or EPR and covered with a thermoplastic sheath of MDPE, PVC or LSF material. Within this general construction there are options of single-core or 3-core types, individual or collective screens of different sizes and armoured or unarmoured construction. Single-core polymeric cables are more widely used than single-core paper-insulated cables, particularly for electricity supply industry circuits. Unlike paper-insulated cables, MV polymeric 3-core cables normally have circular-section cores. This is mainly because the increases in price and cable diameter are usually outweighed in the polymeric case by simplicity and flexibility of jointing and termination methods using circular cores.

Screening of the cores in MV polymeric cables is necessary for a number of reasons, which combine to result in a two-level screening arrangement. This comprises extruded semiconducting layers immediately under and outside the individual XLPE or EPR insulation layer, and a metallic layer in contact with the outer semiconducting layer. The semiconducting layers are polymeric materials containing a high proportion of carbon, giving an electrical conductivity well below that of a metallic conductor, but well above that required for an insulating material.

The two semiconducting layers must be in intimate contact in order to avoid partial discharge activity at the interfaces, where any minute air cavity in the insulation would cause a pulse of charge to transfer to and from the surface of the insulation in each half-cycle of applied voltage. These charge transfers result in erosion of the insulation surface and premature breakdown. In order to achieve intimate contact, the insulation and screens are extruded during manufacture as an integral triple layer and this is applied to the individual conductor in the same operation. The inner layer is known as the conductor screen and the outer layer is known as the core screen or dielectric screen.

When the cable is energized, the insulation acts as a capacitor and the core screen has to transfer the associated charging current to the insulation on every half-cycle of the voltage. It is therefore necessary to provide a metallic element in contact with the core screen so that this charging current can be delivered from the supply. Without this metallic element, the core screen at the supply end of the cable would have to carry a substantial longitudinal current to charge the capacitance which is distributed along the complete cable length, and the screen at the supply end would rapidly overheat as a result of excessive current density. However the core screen is able to carry the current densities relating to the charging of a cable length of say 200 mm, and this allows the use of a metallic element having an intermittent contact with the core screen, or applied as a collective element over three laid-up cores. A 0.08-mm thick copper tape is adequate for this purpose.

The normal form of *armouring* is a single layer of wire laid over an inner sheath of PVC or LSF material. The wire is of galvanized steel for 3-core cables and aluminium for single-core cables. Aluminium wire is necessary for single-core cables to avoid magnetizing or eddy-current losses within the armour layer. In unarmoured cable, the screen is required to carry the earth fault current resulting from the failure of any equipment being supplied by the cable or from failure of the cable itself. In this case, the copper tape referred to previously is replaced by a screen of copper wires of cross section between say 6 mm^2 and 95 mm^2, depending on the earth fault capacity of the system.

9.2.5   Low Smoke and Fume (LSF) and fire performance cables

Following a number of fire disasters in the 1980s, there has been a strong demand for cables which behave more safely in a fire. Cables have been developed to provide the following key areas of improvement:

- improved resistance to ignition
- reduced flame spread and fire propagation
- reduced smoke emission
- reduced acid gas or toxic fume emission

An optimized combination of these properties is achieved in LSF cables, which provide all of the above-mentioned characteristics.

The original concept of LSF cables was identified through the requirements of underground railways in the 1970s. At that time, the main concern was to maintain sufficient visibility such that orderly evacuation of passengers through a tunnel could be managed if the power to their train were interrupted by a fire involving the supply cables. This led to the development of a smoke test known as the '3-metre cube', this being based on the cross section of a London Underground tunnel. This test is now defined in BS EN 50268. The reduced emissions of the toxic fumes also ensured passengers escape was not impaired. PVC sheathed cables can, by suitable use of highly flame retarded PVC materials, be designed to provide good resistance to ignition and flame spread, but they produce significant volumes of smoke and toxic fumes when burning. On this basis the LSF materials become specified for underground applications. The tests for reduced flame propagation are defined in BS EN 50265 (single cable) and BS EN 50266 (grouped cables), with tests for acid gas emission being defined in BS EN 50267.

The demand for LSF performance has since spread to a wide range of products and applications and LSF now represents a generic family of cables. Each LSF cable will meet the 3-metre cube smoke emission, ignition resistance and acid gas emission tests, but fire propagation performance is specified as appropriate to a particular product and application. For instance, a power cable used in large arrays in a power station has very severe fire propagation requirements, while a cable used in individual short links to equipment would have only modest propagation requirements. Hence BS EN 50265 would be appropriate to assess the single cable, but BS EN 50266 comprises of a number of categories to cater for varying numbers of cables grouped together.

Additionally, there has been a significant growth in demand for cables that are expected to continue to function for a period in a fire situation, enabling essential services to continue operation during the evacuation of buildings or during the initial fire fighting stages. In these cables, in addition to these LSF properties, the insulation is expected to maintain its performance in a fire. This insulation may be achieved by use of a compacted mineral layer, mica/glass tapes or a ceramifiable silicone layer. Specific designs are described in section 9.3.3.

9.3   Main classes of cable

9.3.1   Cables for the electricity supply industry

9.3.1.1   MV paper-insulated cables

Until the late 1970s, the large quantities of paper-insulated lead covered (PILC) cables used in the UK electricity supply industry for MV distribution circuits were manufactured according to BS 6480. These cables also incorporated steel wire armour (SWA) and bitumenized textile beddings or servings. An example is illustrated in Fig. 9.1. The lead sheath provided an impermeable barrier to moisture and a return path for earth fault currents and the layer of SWA gave mechanical protection and an improved earth fault capacity. PILC cable continues to be specified by a few utilities; although conversion to XLPE designs are planned.

Following successful trials and extensive installation in the early 1970s, a new standard (ESI 09-12) was issued in 1979 for Paper-Insulated Corrugated Aluminium Sheathed (PICAS) cable. This enabled the electricity supply industry to replace expensive lead sheath and SWA with a corrugated aluminium sheath which offered a high degree of mechanical protection and earth fault capability, while retaining the proven reliability of paper insulation. The standard was limited to three conductor cross sections

Fig. 9.1 Lead-sheathed paper-insulated MV cable for the electricity supply industry (courtesy of Pirelli Cables)

Fig. 9.2 Paper-Insulated Corrugated Aluminium Sheathed (PICAS) cable for the electricity supply industry (courtesy of Pirelli Cables)

(95 mm^2, 185 mm^2 and 300 mm^2) using stranded aluminium conductors with belted paper insulation; although it later included designs with screened paper insulation. An example of PICAS cable is shown in Fig. 9.2. A PICAS cable was easier and lighter to install than its predecessor and it found almost universal acceptance in the UK electricity supply sector. It is still being specified by a few utilities, but as with PILC, a switch to XLPE designs is planned.

9.3.1.2 MV polymeric cables

High-quality XLPE cable has been manufactured for over 25 years. IEC 502 (revised in 1998 as IEC 60502) covered this type of cable and was first issued in 1975. A comparable UK standard BS 6622 was issued in 1986 and revised in 1999.

The following features are now available in MV XLPE cables and these are accepted by the majority of users in the electrical utility sector:

- copper or aluminium conductors
- semiconducting conductor screen and core screen (which may be fully bonded or easily strippable)
- individual copper tape or copper wire screens

- PVC, LSF or MDPE bedding
- copper wire collective screens
- steel wire or aluminium armour
- PVC, LSF or MDPE oversheaths

Early experience in North America in the 1960s resulted in a large number of premature failures, mainly because of poor cable construction and insufficient care in avoiding contamination of the insulation. The failures were due to *water treeing*, which is illustrated in Fig. 9.3. In the presence of water, ionic contaminants and oxidation products, electric stress gives rise to the formation of tree-like channels in the XLPE insulation. These channels start either from defects in the bulk insulation (forming *bow-tie trees*) or at the interfaces between the semiconducting screens and the insulation (causing *vented trees*). Both forms of trees cause a reduction in electrical strength of the insulation and can eventually lead to breakdown. Water treeing has

Fig. 9.3 Example of water treeing in a polymeric cable

largely been overcome by better materials in the semiconducting screen and by improvements in the quality of the insulating materials and manufacturing techniques, and reliable service performance has now been established.

The UK electricity supply industry gradually began to adopt XLPE-insulated or EPR-insulated cable for MV distribution circuits from the early 1990s in place of the PILC or PICAS cables. Each distribution company has specified the best construction for its particular needs. An example of the variation between companies is the difference in practice between solidly bonded systems and the use of earthing resistors to limit the earth fault currents. In the former case, the requirement might typically be to withstand an earth fault current of 13 kA for three seconds. In the latter case, only 1 kA for one second might be specified and the use of single-core cable with a copper wire screen in place of a 3-core cable with a collective copper wire screen or SWA is viable. Different cross-sectional areas of copper wire screen may be specified depending on the earth fault level in the intended installation network. The majority of specific designs being used by the UK electricity supply industry are now incorporated into BS 7870-4.10 (for single core) or BS 7870-4.20 (for 3-core). Examples of XLPE cable designs being used or considered by the UK distribution companies are shown in Figs 9.4, 9.5 and 9.6. The latter shows the most commonly adopted design for 11 kV networks, with a similar design used for 33 kV networks although with stranded copper conductors due to the higher load transfer requirements. The MDPE sheaths are specified for buried installations, with LSF used for tunnel applications.

Fig. 9.4 Example of lead-sheathed XLPE-insulated cable for use in the UK electricity supply industry (courtesy of Pirelli Cables)

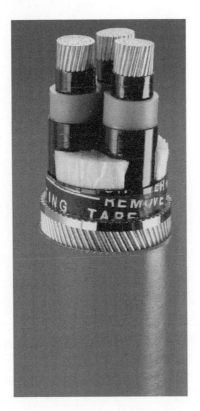

Fig. 9.5 Example of 3-core SWA XLPE-insulated cable for use in the UK electricity supply industry (courtesy of Pirelli Cables)

9.3.1.3   LV polymeric cables

Protective Multiple Earthed (PME) systems which use *Combined Neutral and Earth (CNE) cables* have become the preferred choice in the UK public supply network, both for new installations and for extensions to existing circuits. This is primarily because of the elimination of one conductor by the use of a common concentric neutral and earth, together with the introduction of new designs which use aluminium for all phase conductors.

Before CNE types became established, 4-core paper-insulated sheathed and armoured cable was commonly used. The four conductors were the three phases and neutral, and the lead sheath provided the path to the substation earth. The incentive for PME was the need to retain good earthing for the protection of consumers. With the paper cables, while the lead sheath itself could adequately carry prospective earth fault currents back to the supply transformer, the integrity of the circuit was often jeopardised by poor and vulnerable connections in joints and at terminations. By using the neutral conductor of the supply cable for this purpose the need for a separate earth conductor was avoided.

The adoption of 0.6/1 kV cables with extruded insulation for underground public supply in the UK awaited the development of cross-linked insulation systems with a performance similar to paper-insulated systems in overload conditions. An example of the

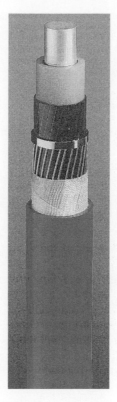

Fig. 9.6 Example of single-core copper wire screen XLPE-insulated cable for use in the UK electricity supply industry (courtesy of Pirelli Cables)

cables which have been developed is the *Waveform CNE type* which is XLPE-insulated and has the neutral/earth conductor applied concentrically in a sinusoidal form. Insulated solid aluminium phase conductors are laid up to form a three-phase cable and the CNE conductor consists of a concentric layer of either aluminium or copper wires.

If the wires in the CNE conductor are of aluminium, they are sandwiched between layers of unvulcanized synthetic rubber compound to give maximum protection against corrosion. This construction is known as *Waveconal* and is illustrated in Fig. 9.7. Where the CNE conductors are of copper, they are partially embedded in the rubber compound without a rubber layer over the wires. This is termed *Wavecon* and is illustrated in Fig. 9.8. Some electricity companies initially adopted Wavecon types because of concern over excessive corrosion in the aluminium CNE conductor, but through standardization all companies had moved to the use of copper wire design by 2001. Waveform cables are manufactured in accordance with BS 7870-3.4.

Both waveform types are compact, with cost benefits. The aluminium conductors and synthetic insulation result in a cable that is light and easy to handle. In addition, the waveform lay of the CNE conductors enables service joints to be readily made without cutting the neutral wires, as they can be formed into a bunch on each side of the phase conductors.

Fig. 9.7 Construction of a 'Waveform' XLPE-insulated CNE cable; 'Waveconal' (courtesy of Pirelli Cables)

9.3.2   Industrial cables

'Industrial cables' are defined as those power circuit cables which are installed on the customer side of the electricity supply point, but which do not fall into the category of 'wiring cables'.

Generally these cables are rated 0.6/1 kV or above. They are robust in construction and are available in a wide range of sizes. They can be used for distribution of power around a large industrial site or for final radial feeders to individual items of plant. Feeder cables might be fixed or in cases, such as coal-face cutting machines and mobile cranes they may be flexible trailing or reeling cables.

Many industrial cables are supplied to customers' individual specifications and since these are not of general interest they are not described here. The following sections focus on types which are manufactured to national standards and which are supplied through cable distributors and wholesalers for general use.

9.3.2.1   Paper-insulated cables

For ratings between 0.6/1 kV and 19/33 kV, paper-insulated cables for fixed installations were supplied in the UK to BS 480, and then to BS 6480 following metrication in 1969. These cables comprise copper or aluminium phase conductors insulated with lapped paper tapes, impregnated with MIND compound and sheathed with lead or lead alloy. For mechanical protection they were finished with an armouring of steel tapes or steel wire and a covering of bitumenized hessian tapes or an extruded PVC oversheath.

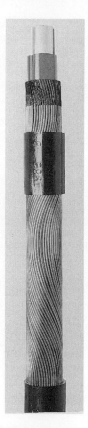

Fig. 9.8 Construction of a 'Waveform' XLPE-insulated CNE cable; 'Wavecon' (courtesy of Pirelli Cables)

The 3-core cables of this type with SWA have been preferred for most applications and these have become known as *Paper-Insulated Lead-Covered Steel Wire Armoured (PILCSWA) cables*. Single-core cables to BS 6480 do not have armouring; this is partly because the special installation conditions leading to the selection of single-core do not demand such protection and partly because a non-magnetic armouring, such as aluminium would be needed to avoid eddy current losses in the armour. These single-core cables are known as *Paper-Insulated Lead Covered (PILC)*.

It has already been observed that paper-insulated cables are now seldom specified for industrial use, but BS 6480 remains an active standard.

9.3.2.2 Polymeric cables for fixed installations

The XLPE-insulated cables manufactured to BS 5467 are generally specified for 230/400 V and 1.9/3.3 kV LV industrial distribution circuits. These cables have superseded the equivalent PVC-insulated cables to BS 6346 because of their higher current rating, higher short-circuit rating and better availability.

For MV applications in the range 3.8/6.6 kV to 19/33 kV, XLPE-insulated wire-armoured cables to BS 6622 are usually specified.

Multi-core LV and MV cables are normally steel wire armoured. This armouring not only provides protection against impact damage for these generally bulky and exposed cables, but it is also capable of carrying very large earth fault currents and provides a very effective earth connection.

Single-core cables are generally unarmoured, although aluminium wire armoured versions are available. Single-core cables are usually installed where high currents are present (for instance in power stations) and where special precautions will be taken to avoid impact damage. For LV circuits of this type, the most economic approach is to use unarmoured cable with a separate earth conductor, rather than to connect in parallel the aluminium wire armour of several single-core cables. For MV applications, each unarmoured cable has a screen of copper wires which would together provide an effective earth connection.

Even in the harsh environment of coal mines, XLPE-insulated types are now offered as an alternative to the traditional PVC- and EPR-insulated cables used at LV and MV, respectively. In this application the cables are always multi-core types having a single or double layer of SWA. The armour has to have a specified minimum conductance because of the special safety requirements associated with earth faults and this demands the substitution of some steel wires by copper wires for certain cable sizes.

Where LSF fire performance is needed, LV wire-armoured cables to BS 6724 are the established choice. These cables are identical in construction and properties to those made to BS 5467 except for the LSF grade of sheathing material and the associated fire performance. Cables meeting all the requirements of BS 6724 and, in addition, having a measure of fire resistance such that they continue to function in a fire are standardized in BS 7846, further details of which are given in section 9.3.3. Similarly, BS 7835 for MV wire-armoured cable, which is identical to BS 6622 apart from the LSF sheath and fire performance, was issued in 1996 and revised in 2000.

The only other type of standardized cable used for fixed industrial circuits is multi-core control cable, often referred to as auxiliary cable. Such cable is used to control industrial plant, including equipment in power stations. It is generally wire-armoured and rated 0.6/1 kV. Cables of this type are available with between 5 and 48 cores. The constructions are similar to 0.6/1 kV power cables and they are manufactured and supplied to the same standards (BS 5467, BS 6346 and BS 6724, as appropriate).

9.3.2.3   Polymeric cables for flexible connections

Flexible connections for both multi-core power cables and multi-core control cables are often required in industrial locations. The flexing duty varies substantially from application to application. At one extreme a cable may need to be only flexible enough to allow the connected equipment to be moved occasionally for maintenance. At the other extreme the cable may be needed to supply a mobile crane from a cable reel or a coal-face cutter from cable-handling gear.

Elastomeric-insulated and sheathed cable is used for all such applications. This may have flexible stranded conductors (known as 'class 5') or highly flexible stranded conductors (known as 'class 6'). Where metallic protection or screening is needed, this comprises a braid of fine steel or copper wires. For many flexible applications the cable is required to have a resistance to various chemicals and oils.

Although flexible cables will normally be operated on a 230/400 V supply, it is normal to use 450/750 V rated cables for maximum safety and integrity.

A number of cable types have been standardized in order to meet the range of performance requirements and the specification for these is incorporated into BS 6500. Guidance on the use of the cables is provided in this standard and further information is available in BS 7540.

9.3.3   Wiring cables

The standard cable used in domestic and commercial wiring in the UK since the 1960s is a flat PVC twin-and-earth type, alternatively known as 6242Y cable. This comprises a flat formation of PVC-insulated live and neutral cores separated by a bare earth conductor, the whole assembly being PVC-sheathed to produce a flat cable rated at 300/500 V. Cable is also available with three insulated cores and a bare earth, for use on double-switched lighting circuits. These forms of cable are ideal for installation under cladding in standard-depth plaster. They are defined in BS 6004, which covers a large size range, only the smaller sizes of which are used in domestic and commercial circuits.

There are other cable types included in BS 6004 which have more relevance to non-domestic installations. These include cables in both flat and circular form, similar to the 6242Y type but with an insulated earth conductor. Circular cables designated 6183Y are widely used in commercial or light industrial areas, especially where many circuits are mounted together on cable trays. Also included in BS 6004 are insulated conductors designated 6491X which are pulled into conduit or trunking in circuits where mechanical protection or the facility to re-wire are the key factors.

Of recent years, LSF versions of the twin-and-earth cables and the conduit wires have become available and are being used in installations where particular emphasis on fire performance is required. Such cables are included in BS 7211. These are commonly known as 6242B (for twin flat) or 6491B (for single-core conduit wire).

A significant change occurred in 2004 for all fixed wiring in electrical installations in the UK. An amendment was published to BS 7671 (the IEE Wiring Regulations) which specified new cable core colours to bring the UK more closely into line with practice in mainland Europe; the term *harmonized core colours* is often used. For single-phase fixed installations, the red phase and black neutral are replaced by brown phase and blue neutral, as used for many years in flexible cables for appliances. For three phase cables, the new phase colours are brown, black and grey instead of red, yellow and blue, with the neutral now blue instead of black. In both cases, the protective conductor is still identified by a green–yellow combination. An alternative for three-phase cables is for all phase cores to be brown and marking of L1, L2 and L3 to be carried out at terminations. The neutral will be blue again in this case. Electrical installations commenced after 31 March 2004 may use either the new harmonized core colours or the pre-existing colours, but not both. New installations after 31 March 2006 must only use the harmonized core colours.

An alternative type of cable with outstanding impact and crush strength is *mineral-insulated cable (MICC)* manufactured to BS EN 60702-1. This is often known by its trade name, *Pyrotenax*. In a MICC cable, the copper line and neutral conductors are positioned inside a copper sheath, the spaces between the copper components being filled with heavily compacted mineral powder of insulating grade. Pressure or impact applied to the cable merely compresses the powder in such a way that the insulation integrity is maintained. The copper sheath often acts as the circuit earth conductor. An oversheath is not necessary but is often provided for reasons of appearance or for external marking. An MICC cable has a relatively small cross section and is easy to install.

In shopping and office complexes or in blocks of flats there may be a need for a distribution sub-main to feed individual supply points or meters. If this sub-main is to be installed and operated by the owner of the premises, then a 0.6/1 kV split-concentric service cable to BS 4553 may be used. This comprises a phase conductor insulated in PVC or XLPE, around which is a layer of copper wires and an oversheath. Some of the copper wires are bare and these are used as the earth conductor. The remainder are polymer-covered and they make up the neutral conductor. For larger installations, 3-core versions of this cable are available to manufacturers' specification.

In circuits supplying equipment for fire detection and alarm, emergency lighting and emergency supplies, regulations dictate that the cables will continue to operate during a fire. This continued operation could be ensured by measures, such as embedding the cable in masonry, but may be achieved by cables which are fire-resistant in themselves. BS 5839-1:2002, the code of practice for fire detection and fire alarm systems for buildings, recommends the use of fire resisting cables for mains power supply circuits and all critical signal paths in such systems. Fire-resistance tests for cables are set down in BS 6387, BS 8434-1, BS 8434-2 and EN 50200. The latter three tests are called up in BS 5839-1 although two levels of survival time are specified, 30 minutes for 'standard' and 120 minutes for 'enhanced'. The three types of cable are recognized in the code of practice are to BS EN 60702-1 (as described earlier), BS 7846 and BS 7629 (both as described below).

The MICC cables to BS EN 60702-1 should comply with the 'enhanced' performance, since the mineral insulation is unaffected by fire. An MICC cable will only fail when the copper conductor or sheath melts and where such severe fires might occur the cable can be sheathed in LSF material to assist in delaying the onset of melting. MICC cable is also categorized CWZ in BS 6387.

A number of alternatives to MICC cables for fire resistance have been developed and standardized. Some rely on a filled silicone rubber insulation which degrades to an insulating char, which continues to provide separation between the conductors so that circuit integrity is maintained during a fire. Other types supplement standard insulation with layers of mica tape so that even if the primary insulation burns completely the mica tape provides essential insulation to maintain supplies during the fire. Both cable types are standardized in BS 7629, and in addition to complying with performance levels up to CWZ of BS 6387, designs also may comply with either the 'standard' or 'enhanced' performance required by BS 5839-1.

Some circuits requiring an equivalent level of fire resistance need to be designed with larger cables than are found in BS 7629. Such circuits might be for the main emergency supply, fire-fighting lifts, sprinkler systems and water pumps, smoke extraction fans, fire shutters or smoke dampers. These larger cables are standardized in BS 7846, which includes the size range and LSF performance of BS 6724, but through the use of layers of mica tape to supplement the insulation these cables can be supplied to the CWZ performance level in BS 6387, and additionally to the 'standard' or 'enhanced' performance levels specified in BS 5839-1. An additional fire test category in BS 7846, called F3, may be considered to be more appropriate for applications where the cable might be subject to fire, impact and water spray in combination during the fire.

9.4   Parameters and test methods

There are a large number of cable and material properties which are controlled by the manufacturer in order to ensure fitness for purpose and reliable long-term service performance. However, it is the operating parameters of the finished installed cable which are of most importance to the user in cable selection. The major parameters of interest are as follows:

- current rating
- capacitance
- inductance
- voltage drop
- earth loop impedance
- symmetrical fault capacity
- earth fault capacity

These are dealt with in turn in the following sections.

9.4.1   Current rating

The current rating of each individual type of cable could be measured by subjecting a sample to a controlled environment and by increasing the load current passing through the cable until the steady-state temperature of the limiting cable component reached its maximum permissible continuous level. This would be a very costly way of establishing current ratings for all types of cable in all sizes, in all environments and in all ambient temperatures. Current ratings are therefore obtained using an internationally-accepted calculation method, published in IEC 60287. The formulae and reference material properties presented in IEC 60287 have been validated by correlation with data produced from laboratory experiments.

Current ratings are quoted in manufacturers' literature and they are listed in IEE Wiring Regulations (BS 7671) for some industrial, commercial and domestic cables. The ratings are quoted for each cable type and size in air, in masonry, direct-in-ground and in underground ducts. Derating factors are given so that these quoted ratings can be adjusted for different environmental conditions, such as ambient temperature, soil resistivity or depth of burial.

Information is given in BS 7671 and in the IEE Guidance Notes on the selection of the appropriate fuse or mcb to protect the cable from overload and fault conditions, and general background is given in sections 8.2 and 8.3.

9.4.2   Capacitance

The capacitance data in manufacturers' literature is calculated from the cable dimensions and the permittivity of the insulation.

For example, the star capacitance of a 3-core belted armoured cable to BS 6346 is the effective capacitance between a phase conductor and the neutral star point. It is calculated using the following formula:

$$C = \frac{\varepsilon_o}{18\ln[(d+t_1+t_2)/d]}(\mu F / km) \tag{9.1}$$

where ε_o = relative permittivity of the cable insulation (8.0 for PVC)
  d = diameter of the conductor (mm)
  t_1 = thickness of insulation between the conductors (mm)
  t_2 = thickness of insulation between conductor and armour (mm)

Equation 9.1 assumes that the conductors are circular in section. For those cables having shaped conductors, the value of capacitance is obtained by multiplying the figure obtained using eqn 9.1 by an empirical factor of 1.08.

The calculated capacitance tends to be conservative, that is the actual capacitance will always be lower than the calculated value. However, if an unusual situation arises in which the cable capacitance is critical, then the manufacturer is able to make a measurement using a capacitance bridge.

If the measured capacitance between cores and between core and armour is quoted, then the star capacitance can be calculated using eqn 9.2:

$$C = \frac{9C_x - C_y}{6} \ (\mu F/km) \tag{9.2}$$

where C_x = measured capacitance between one conductor and the other
             two connected together to the armour
       C_y = measured capacitance between three conductors connected
             together and the armour

9.4.3   Inductance

The calculation of cable inductance L for the same example of a 3-core armoured cable to BS 6346 is given by eqn 9.3 as follows:

$$L = 1.02 \times \{0.2 \times \ln[2Y/d] + k\} \ (mH/km) \tag{9.3}$$

where d = diameter of the conductor (mm)
       Y = axial spacing between conductors (mm)
       k = a factor which depends on the conductor make-up
          (k = 0.064 for 7-wire stranded
             0.055 for 19-wire stranded
             0.053 for 37-wire stranded
             0.050 for solid)

The same value of cable inductance L is used for cables with circular- or sector-shaped conductors.

9.4.4   Voltage drop

BS 7671 specifies that within customer premises the voltage drop in cables is to be a maximum value of 4 per cent. It is therefore necessary to calculate the voltage drop along a cable.

The cable manufacturer calculates voltage drop assuming that the cable will be loaded with the maximum allowable current which results in the maximum allowable operating temperature of the conductor. The cable impedance used for calculating the voltage drop is given by eqn 9.4.

$$Z = \{R^2 + (2\pi f L - 1/2\pi f C)^2\}^{1/2} \ (\Omega/m) \tag{9.4}$$

where R = ac resistance of the conductor at maximum conductor
             temperature (Ω/m)
       L = inductance (H/m)
       C = capacitance (F/m)
       f = supply frequency (Hz)

The voltage drop is then given by eqns 9.5 and 9.6:
For single-phase circuits:

$$\text{voltage drop} = 2Z \ (\text{V/A/m}) \tag{9.5}$$

and for three-phase circuits:

$$\text{voltage drop} = (3)^{1/2}Z \ (\text{V/A/m}) \tag{9.6}$$

9.4.5 Symmetrical and earth fault capacity

It is necessary that cables used for power circuits are capable of carrying any fault currents that may flow, without damage to the cable; the requirements are specified in BS 7671. This assessment demands a knowledge of the maximum prospective fault currents on the circuit, the clearance characteristics of the protective device (as explained in Chapter 8) and the fault capacity of the relevant elements in the cable. For most installations it is necessary to establish the let-through energy of the protective device and to compare this with the adiabatic heating capacity of the conductor (in the case of symmetrical and earth faults) or of the steel armour (in the case of earth faults).

The maximum let-through energy (I^2t) of the protective device is explained in Chapter 8. It can be obtained from the protective device manufacturer's data. In practice the value will be less than that shown by the manufacturer's information because of the reduction in current during the fault which results from the significant rise in temperature and resistance of the cable conductors.

The fault capacity of the cable conductor and armour can be obtained from information given in BS 7671 and the appropriate BS cable standard, as follows:

$$k^2S^2 = \text{adiabatic fault capacity of the cable element} \tag{9.7}$$

where S = the nominal cross section of the conductor *or*
         the nominal cross section of, say, the armour (mm^2)
    k = a factor reflecting the resistivity, temperature coefficient, allowable
         temperature rise and specific heat of the metallic cable element
         (k = 115 for a PVC-insulated copper conductor within the cable
           = 176 for an XLPE-insulated copper earth conductor external to
             the cable
           = 46 for the steel armour of an XLPE-insulated cable)

In practice it will be found that provided the cable rating is at least equal to the nominal rating of the protective device and the maximum fault duration is less than 5 seconds, the conductors and armour of the cables to BS will easily accommodate the let-through energy of the protective device.

It is also important that the impedance of the supply cable is not so high that the protective device takes too long to operate during a zero-sequence earth fault on connected equipment. This is important because of the need to protect any person in contact with the equipment, by limiting the time that the earthed casing of the equipment, say, can become energized during an earth fault.

This requirement, which is stated in BS 7671, places restrictions on the length of the cable that can be used on the load side of a protective device, and it therefore demands knowledge of the earth fault loop impedance of the cable. Some cable manufacturers have calculated the earth fault impedance for certain cable types and the data are presented in specialized literature. These calculations take account of the average temperature of each conductor and the reactance of the cable during the fault. The values are supported by independent experimental results. BS 7671 allows the use of such manufacturer's data or direct measurement of earth fault impedance on a completed installation.

9.5 Optical communication cables

The concept of using light to convey information is not new. There is a historical evidence that Aztecs used flashing mirrors to communicate and in 1880 Alexander Graham Bell first demonstrated his photophone, in which a mirror mounted on the end of a megaphone was vibrated by the voice to modulate a beam of sunlight, thereby transmitting speech over distances up to 200 m.

Solid-state photodiode technology has its roots in the discovery of the light-sensitive properties of selenium in 1873, used as the detector in Bell's photophone. The Light Emitting Diode (LED) stems from the discovery in 1907 of the electroluminescent properties of silicon, and when the laser was developed in 1959, the components of an optical communication system were in place, with the exception of a suitable transmission medium.

The fundamental components of a fibre optic system are shown in Fig. 9.9. This system can be used for either analogue or digital transmissions, with a transmitter which converts electrical signals into optical signals. The optical signals are launched through a joint into an optical fibre, usually incorporated into a cable. Light emitting from the fibre is converted back into its original electrical signal by the receiver.

9.5.1 Optical fibres

An optical fibre is a dielectric waveguide for the transmission of light, in the form of a thin filament of very transparent silica glass. As shown in Fig. 9.10, a typical fibre comprises a core, the cladding, a primary coating and sometimes a secondary coating or buffer. Within this basic construction, fibres are further categorized as multi-mode or single-mode fibres with a step or graded index.

The core is the part of the fibre which transmits light, and it is surrounded by a glass cladding of lower refractive index. In early fibres, the homogeneous core had a constant refractive index across its diameter, and with the refractive index of the cladding also constant (at a lower value) the profile across the whole fibre diameter (as shown in Fig. 9.11(a)) became known as a *step index*. In this type of fibre, the light rays can be envisaged as travelling along a zigzag path of straight lines, kept within the core by total reflection at the inner surface of the cladding. Depending on the angle of the rays to the fibre axis, the path length will differ so that a narrow pulse of light entering the fibre will become broader as it travels. This sets a limit to the rate at which pulses can be transmitted without overlapping and hence a limit to the operating bandwidth.

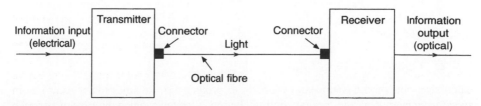

Fig. 9.9 Basic fibre optic system

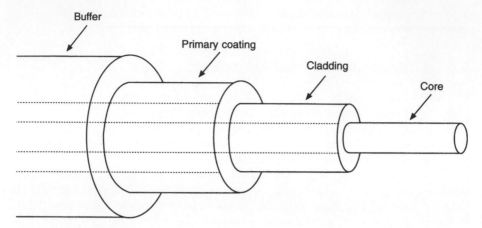

Fig. 9.10 Basic optical fibre

To minimize this effect, which is known as *mode dispersion*, fibres have been developed in which the homogeneous core is replaced by one in which the refractive index varies progressively from a maximum at the centre to a lower value at the interface with the cladding. Figure 9.11(b) shows such a *graded index fibre,* in which the rays no longer follow straight lines. When they approach the outer parts of the core, travelling temporarily faster, they are bent back towards the centre where they travel more slowly. Thus the more oblique rays travel faster and keep pace with the slower rays travelling nearer the fibre centre. This significantly reduces the pulse broadening effect of step index fibres.

The mode dispersion of step index fibres has also been minimized by the development of *single-mode fibres*. As shown in Fig. 9.11(c), although it is a step index fibre, the core is so small (of the order of 8 μm in diameter) that only one mode can propagate.

Fibre manufacture involves drawing down a preform into a long thin filament. The preform comprises both core and cladding, and for graded index fibres, the core contains many layers with dopants being used to achieve the varying refractive index. Although the virgin fibre has a tensile strength comparable to that of steel, its strength is determined by its surface quality. Microcracks develop on the surface of a virgin fibre in the atmosphere, and the lightest touch or scratch makes the fibre impractically fragile. Thus it must be protected, in line with the glass drawing before it touches any solid object, such as pulleys or drums, by a protective coating of resin, acetate or plastic material, known as the *primary coating*.

Typically the primary coating has a thickness of about 60 μm, and in some cases a further layer of material called the *buffer* is added to increase the mechanical protection.

Another type of optical fibre has a plastic construction, with either step or graded index cores. Although larger in size (up to 1.0 mm cladding diameter) and with higher transmission losses than glass fibres, plastic optical fibres have economic and handling advantages for short distance, low data rate communication systems.

9.5.2 Optical cable design

The basic aim of a transmission cable is to protect the transmission medium from its environment and the rigours of installation. Conventional cables with metallic conductors are designed to function effectively in a wide range of environments, as

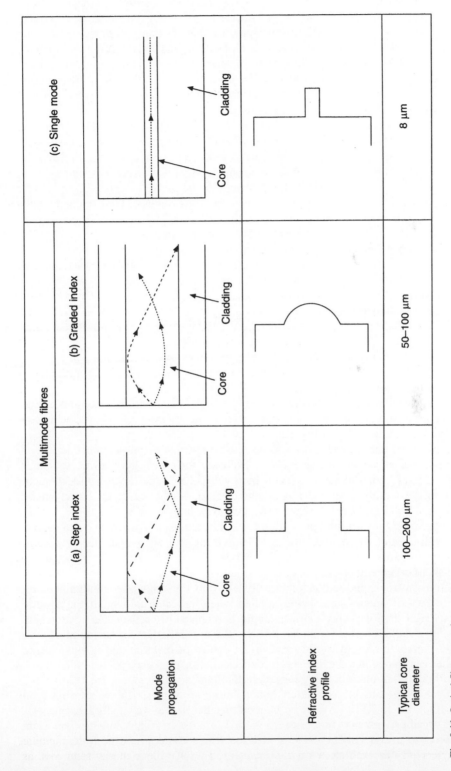

Fig. 9.11 Optical fibre categories

mentioned in sections 9.2 and 9.3. However, optical fibres differ significantly from copper wires to an extent that has a considerable bearing on cable designs and manufacturing techniques. The transmission characteristics and lifetime of fibres are adversely affected by quite low levels of elongation, and lateral compressions can produce small kinks or sharp bends which create an increase in attenuation loss known as *microbending loss*. This means that cables must protect the fibre from strain during installation and service, and they must cater for longitudinal compression that occurs, for example with a change in cable temperature.

Fibre life in service is influenced by the presence of moisture as well as stress. The minute cracks which cover the surface of all fibres can grow if the fibre is stressed in the presence of water, so that the fibre could break after a number of years in service. Cables must be able to provide a long service life in such environments as tightly packed ducts which are filled with water.

The initial application of optical cables was the trunk routes of large telecommunications networks, where cables were directly buried or laid in ducts in very long lengths, and successful cable designs evolved to take into account the constraints referred to earlier. The advantages of fibre optics soon led to interest in other applications, such as computer and data systems, premises cabling, military systems and industrial control. This meant that cable designs had to cater for tortuous routes of installation in buildings, the flexibility of patch cords and the arduous environments of military and industrial applications. Further opportunities for optical cables are presented by installations in existing rights-of-way, such as sewers, gas pipes and water lines without the need for costly civil engineering works.

Nevertheless, many of the conventional approaches to cable design can be used for optical cables, with modification to take into account the optical and mechanical characteristics of fibres and their fracture mechanics.

Cables generally comprise several elements or individual transmission components, such as copper pairs, or one or more optical fibres. The different types of element used in optical cables are shown in Fig. 9.12.

The primary-coated fibre can be protected by a *buffer* of one or more layers of plastic material as shown in Fig. 9.12(a). Typically for a two-layer buffer, the inner layer is of a soft material acting as a cushion with a hard outer layer for mechanical protection, the overall diameter being around 850 µm. In other cases, the buffer can be applied with a sliding fit to allow easy stripping over long lengths.

In *ruggedized fibres* further protection for a buffered fibre is provided by surrounding it with a layer of non-metallic synthetic yarns and an overall plastic sheath. This type of arrangement is shown in Fig. 9.12(b).

When one or more fibres are run loosely inside a plastic tube, as shown in Fig. 9.12(c), they can move freely and will automatically adjust to a position of minimum bending strain to prevent undue stress being applied when the cable is bent. If the fibre is slightly longer than the tube, a strain margin is achieved when the cable is stretched, say during installation, and for underground and duct cables the tube can be filled with a gel to prevent ingress of moisture. Correct choice of material and manufacturing technique can ensure that the tube has a coefficient of thermal expansion similar to that of the fibre, so that microbending losses are minimized with temperature excursions.

Optical fibres can be assembled into a linear array as a *ribbon*, as shown in Fig. 9.12(d). Up to 12 fibres may be bonded together in this way or further encapsulated if added protection is required.

In order to prevent undue cable elongation which could stress the fibres, optical cables generally incorporate a *strength member*. This may be a central steel wire or

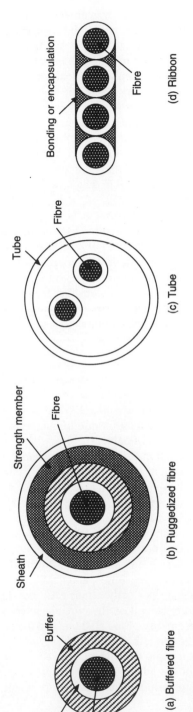

Fig. 9.12 Optical cable elements

strand, or non-metallic fibreglass rods or synthetic yarns. The strength member should be strong, light and usually flexible, although in some cases a stiff strength member can be used to prevent cable buckling which would induce microbending losses in the fibres. Strength members are shown in the cable layouts in Fig. 9.13(b) and Fig. 9.13(c).

The strength member can be incorporated in a *structural member* which is used as a foundation for accommodating the cable elements. An example is shown in Fig. 9.13(c), where a plastic section with slots is extruded over the strength member with ribbons inserted into the slots to provide high fibre count cables.

A *moisture barrier* can be provided either by a continuous metal sheath or by a metallic tape with a longitudinal overlap, bonded to the sheath. Moisture barriers can be of aluminium, copper or steel and they may be flat or corrugated. In addition, other cable interstices may be filled with gel or water-swellable filaments to prevent the longitudinal ingress of moisture.

Where protection from external damage is required, or where additional tensile strength is necessary, *armouring* can be provided; this may be metallic or non-metallic. For outdoor cables, an *overall sheath* of polyethylene is applied. For indoor cables the sheath is often of low-smoke zero-halogen materials for added safety in the event of fire.

Although the same basic principles of cable construction are used, the wide range of applications result in a variety of cable designs, from simplex indoor patch cords to cables containing several thousand fibres for arduous environments, to suboceanic cables. Figure 9.13 shows just a few examples.

9.5.3   Interconnections

The satisfactory operation of a fibre optic system requires effective jointing and termination of the transmission medium in the form of fibre-to-fibre splices and fibre connections to repeaters and end equipment. This is particularly important because with very low loss fibres the attenuation due to interconnections can be greater than that due to a considerable length of cable.

For all types of interconnection there is an *insertion loss*, which is caused by *Fresnel reflection* and by misalignment of the fibres. Fresnel reflection is caused by the changes in refractive index at the fibre–air–fibre interface, but it can be minimized by inserting into the air gap an index-matching fluid with the same refractive index as the core.

Misalignment losses arise from three main sources as shown in Fig. 9.14. Interconnection designs aim to minimize these losses. *End-face separation* (Fig. 9.14(a)) allows light from the launch fibre to spread so that only a fraction is captured by the receive fibre; this should therefore be minimized. Normally the fibre cladding is used as the reference surface for aligning fibres, and the fibre geometry is therefore important, even when claddings are perfectly aligned. Losses due to *lateral misalignment* (Fig. 9.14(b)) will therefore depend on the core diameter, non-circularity of the core, cladding diameter, non-circularity of the cladding and the concentricity of the core and cladding in the fibres to be jointed. *Angular misalignment* can result in light entering the receive fibre at such an angle that it cannot be accepted. It follows that very close tolerances are required for the geometry of the joint components and the fibres to be jointed, especially with single-mode fibres with core diameters of 8 μm and cladding diameters of 125 μm.

The main types of interconnections are fibre splices and demountable connectors.

Fibre splices are permanent joints made between fibres or between fibres and device pigtails. They are made by *fusion splicing* or mechanical alignment. In fusion splicing, prepared fibres are brought together, aligned and welded by local heating

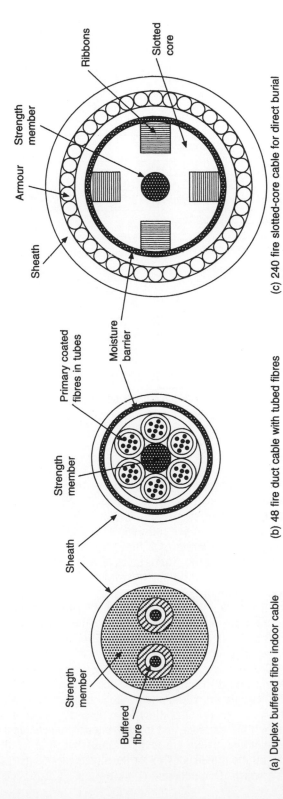

(a) Duplex buffered fibre indoor cable

(b) 48 fire duct cable with tubed fibres

(c) 240 fire slotted-core cable for direct burial

Fig. 9.13 Examples of optical cables

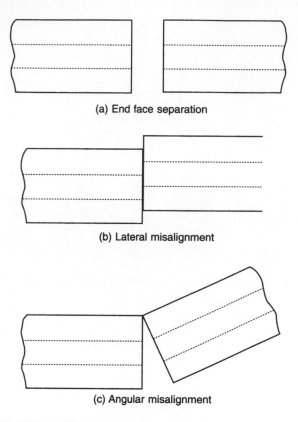

(a) End face separation

(b) Lateral misalignment

(c) Angular misalignment

Fig. 9.14 Sources of misalignment loss

combined with axial pressure. Sophisticated portable equipment is used for fusion splicing in the field. This accurately aligns the fibres by local light injection and it carries out the electric arc welding process automatically. Nevertheless, a level of skill is required in the preparation of the fibres, stripping the buffers and coatings and cleaving the fibres to achieve a proper end face. There are a number of *mechanical techniques* for splicing fibres which involve fibre alignment by close tolerance tubes, ferrules and v-grooves, and fixing by crimps, glues or resins. Both fusion and mechanical splicing techniques have been developed to allow simultaneous splicing of fibres which are particularly suitable for fibre ribbons. For a complete joint, the splices must be incorporated into an enclosure which is suitable for a variety of environments, such as underground chambers or pole tops. The enclosure must also terminate the cables and organize the fibres and splices, and cassettes are often used where several hundred splices are to be accommodated.

Demountable connectors provide system flexibility, particularly at and within the transmission equipment and distribution panels, and they are widely used on patch cords in certain data systems. As with splices, the connector must minimize Fresnel and misalignment loss, but it must also allow for repeated connection and disconnection, it must protect the fibre end face and it must cater for mechanical stress, such as tension, torsion and bending. There are many designs but in general the tolerances which are achievable on the dimensions of the various components result in a higher optical loss than in a splice. Demountable connectors have also been developed for multiple-fibre simultaneous connection, with array designs being particularly suitable for fibre ribbons. For connector-intensive systems, such as office data systems, use is made of factory-predetermined cables and patch cords to reduce the need for on-site termination.

9.5.4   Installation

Optical fibre cables are designed so that normal installation practices and equipment can be used wherever possible, but as they generally have a lower strain limit than metallic cables special care may be needed in certain circumstances and manufacturer's recommendations regarding tensile loads and bending radii should be followed.

Special care may be required in the following circumstances:

- because of their light weight, optical cables can be installed in greater lengths than metallic cables. For *long underground ducts* access may be needed at intermediate points for additional winching effort and space should be allowed for larger 'figure 8' cable deployment.
- *mechanical fuses* and *controlled winching* may be necessary to ensure that the rated tensile load is not exceeded
- *guiding equipment* may be necessary to avoid subjecting optical cables to unacceptable bending stresses, particularly when the cable is also under tension
- when installing cables in trenches the footing should be free from stones. These could cause microbending losses.
- in buildings, and particularly in risers, cleats and fixings should not be overtightened, or appropriate designs should be used to prevent compression and the resulting microbending losses
- indoor cable routes should provide turning points if a large number of bends is involved. Routes should be as straight as possible.
- excess lengths for jointing and testing of optical cables are normally greater that those required for metallic cables
- where non-metallic optical cables are buried, consideration of the subsequent location may be necessary. Marker posts and the incorporation of a location wire may be advisable.

Blown fibre systems have been developed as a means of avoiding fibre overstrain for complex route installation and of allowing easy system upgrading and future proofing. It results in low initial capital costs and provides for the distribution of subsequent costs. Initially developed by British Telecom, the network infrastructure is created by the most appropriate cabling method, being one or a group of empty plastic tubes. As and when circuit provision is required, one or more fibres can be blown by compressed air into the tubes. Individual tubes can, by means of connectors, be extended within buildings up to the fibre terminating equipment. The efficient installation of fibres into the tube network often requires the use of specially-designed fibres and equipment, such as air supply modules, fibre insertion tools and fibre pay-offs. For installation it is necessary to follow the instructions provided by the supplier, taking into account the requirements for the use of portable electrical equipment and compressed air, and the handling, cutting and disposal of optical fibres. A novel variation of this system is a data cable used for structural wiring systems. In a 'figure-8' configuration one unit comprises a 4-pair data cable and the other an empty tube, so that when an upgrade is required to an optical system the appropriate fibre can be blown in without the need for recabling.

9.6   Standards

9.6.1   Metallic wires and cables

Most generally available cables are manufactured to recognized standards which may
be national, European or international. Each defines the construction, the type and
quality of constituent materials, the performance requirements and the test methods for
the completed cable.

The IEC standards cover those cables which need to be standardized to facilitate
world trade, but this often requires a compromise by the parties involved in the
preparation and acceptance of a standard. Where cables are to be used in a particular
country, the practices and regulations in that country tend to encourage the more
specific cable types defined in the national standards for that country. BS remains the
most appropriate for use in the UK, and for the main cable types described in this
chapter reference has therefore been made mainly to the relevant BS.

Some cables rated at 450/750 V or less have through trade become standard
throughout the EU, and these have been incorporated into Harmonization Documents
(HDs). Each EU country must then publish these requirements within a national
standard. A harmonized cable type in the UK for instance would still be specified to
the relevant BS and the cable would, if appropriate, bear the <HAR> mark.

The key standards for metallic wires and cables which have been referred to in the
chapter are listed in Table 9.1.

Table 9.1 International and national standards for metallic wires and cables

IEC	HD	BS	Subject
	603	4553	600/1000 V PVC-insulated single-phase split concentric cables with copper conductors
60502-1		5467	Cables with thermosetting insulation up to 600/1000 V and up to 1900/3300 V
60227	21	6004	Non-armoured PVC-insulated cables rated up to 450/750 V
60245	22	6007	Non-armoured rubber-insulated cables rated up to 450/750 V
		6346	PVC-insulated cables
		6387	Performance requirements for cables required to maintain integrity under fire conditions
60055	621	6480	Impregnated paper-insulated lead sheathed cables up to 33 000 V
	621	(EA 09-12)	Paper-insulated corrugated aluminium sheathed 6350/11000 V cable
60227 and 60245	21 and 22	6500	Insulated flexible cords and cables rated up to 450/750 V
60502-2	620	6622	Cables with XLPE or EPR insulation from 3800/6600 V up to 19 000/33 000 V
		6724	600/1000 V and 1900/3300 V armoured cables having thermosetting insulation with low emission of smoke and corrosive gases in fire
	22	7211	Non-armoured cables having thermosetting insulation rated up to 450/750 V with low emission of smoke and corrosive gases in fire
		7629	Fire resistant thermosetting insulated cables rated at 300/500 V with limited circuit integrity in fire
60364		7671	Requirements for electrical installations: IEE Wiring Regulations (16th edition)

Table 9.1 (*contd*)

IEC	HD	BS	Subject
60287		7769	Electric cables – calculation of current rating
		7835	Cables with XLPE or EPR insulation from 3800/6600 V up to 19 000/33 000 V with low emission of smoke and corrosive gases in fire
	603	7870-3.40	Polymeric insulated cables for distribution rated at 600/1000 V
	620	7870-4.10	Polymeric insulated cables for distribution rated from 3800/6600V up to 19000/33000 V: single-core cable with copper wire screens
	620	7870-4.20	Polymeric insulated cables for distribution rated from 3800/6600 V up to 19 000/33 000 V: 3 core cable with collective copper wire screens
	626	7870-5	Polymeric insulated aerial bundled cables rated 600/1000 V for overhead distribution
	604	7870-6	Polymeric insulated cables for generation rated at 600/1000 V and 1900/3300 V
	622	7870-7	Polymeric insulated cables for generation rated from 3800/6600 V up to 19 000/33 000 V

9.6.2   Optical communication cables

For communication systems and their evolution to be effective, standardization must be at an international level. Optical fibre and cable standardization in IEC started in 1979. The ENs which have been published generally use IEC standards as a starting point but they incorporate any special requirements for sale within the EU where European Directives may apply.

Table 9.2 summarizes the main standards in the areas of optical fibres, optical cables, connectors, connector interfaces and test and measurement procedures for interconnecting devices.

Table 9.2 International and national standards for optical fibres, optical cables, connectors, connector interfaces and test and measurement procedures for interconnecting devices

IEC	EN	BS	Subject
	50174	EN 50174	Information technology – Cabling installation
60793-1	60793-1	EN 60793-1	Optical fibres: Measurement and test procedures
60793-2	60793-2	EN 60793-2	Optical fibres: Product specifications
60794-1-1	60794-1-1	EN 60794-1-1	Optical fibre cables: Generic specification
60794-1-2	60794-1-2	EN 60794-1-2	Optical fibre cables: Basic test procedures
60794-2	60794-2	EN 60794-2	Indoor optical fibre cables
60794-3	60794-3	EN 60794-3	Outdoor optical fibre cables
60794-4	60794-4	EN 60794-4	Aerial optical cables along overhead lines
60869	60869	EN 60869	Fibre optic attenuators
60874	60874	EN 60874	Connectors for optical fibres and cables
60875	60875	EN 60875	Fibre optic branching devices
60876	60876	EN 60876	Fibre optic spatial switches
61202	61202	EN 61202	Fibre optic isolators
61274	61274	EN 61274	Fibre optic adaptors
61300	61300	EN 61300	Fibre optic interconnecting devices and passive components test and measurement procedures

(*contd*)

Table 9.2 (*contd*)

IEC	EN	BS	Subject
61314	61314	EN 61314	Fibre optic fanouts
61753	61753	EN 61753	Fibre optic interconnecting devices and passive components performance standards
61754	61754	EN 61754	Fibre optic connector interfaces
61977	61977	EN 61977	Fibre optic filters
61978	61978	EN 61978	Fibre optic compensators
62005	62005	EN 62005	Reliability of fibre optic interconnecting devices and passive components
62077	62077	EN 62077	Fibre optic circulators
62099	62099	EN 62099	Fibre optic wavelength switches
62134	62134	EN 2134	Fibre optic enclosures
	181000 to 181104	EN 181000 to 181104	Fibre optic branching devices
	186000 to 186310	EN 186000 to 186310	Connector sets for optical fibres and cables
	187103	EN 187103	Optical fibre cables for indoor applications
	187105	EN 187105	Single-mode optical cables for duct or buried installation

References

9A.   Moore, G.F. *Electric Cables Handbook* (3rd edn), Blackwell Scientific Publications Ltd, 1997.
9B.   Heinhold, L. *Power Cables and their Application* (3rd revision), Siemens AG, 1990.

Chapter 10

Motors, motor control and drives

Professor W. Drury
Control Techniques, Emerson Industrial Automation

10.1   Introduction

Electric motors can be found in applications from computer disk drives, domestic appliances, automobiles to industrial process lines, liquid pumping, conveyors, weaving machines and many more. Modern building services are heavily reliant upon motors and drives, which can be found at the heart of air handling and elevator systems. Even in the world of theatre and film, the electric motor is at the centre of the action, moving scenery and allowing actors to perform death-defying feats in complete safety. If it moves, it is reasonable to expect to find that an electric motor is somehow responsible.

The flexibility of power transmission that was introduced by the electric motor has been harnessed and controlled by the application of torque, speed and position controllers, which all fall under the generic term of drives or *Variable Speed Drives (VSD)*. This precision of control has been central to all aspects of industrial automation and has opened up new and demanding applications such as automatic production and sectional process lines, machine tool axis control, glass engraving, embroidery machines and precision polishing machines. It has also facilitated considerable reductions in energy consumption by regulation of flow through speed control in fan and pump type loads, where power consumption is proportional to the cube of the speed. This ability to reduce energy consumption continues to be a major stimulus to growth in the variable speed drives market.

Many ac motors operate at fixed speed, being connected directly to the fixed-frequency supply system, but of the 70 per cent of all electrical power which flows through electric motors, a significant proportion passes through semiconductor conversion in the form of drives. The importance of this technology is self-evident, though the selection of appropriate equipment is often less clear. It would be a difficult, if not impossible, task to detail every motor type and associated power conversion circuit available, and the focus here will be on those of greatest practical importance in the broad base of industries.

All electric motors have a stationary component, the stator, and a moving component, the rotor. In conventional motors, the rotor turns within the stator, but special motors are available for applications such as material handling conveyors where the rotor rotates outside the stator. In linear motors, the rotor moves along the path of the stator. In all cases, the rotor and stator are separated by an air-gap. The stator and rotor usually have a laminated steel core to reduce the losses arising from time-varying magnetic fields (as explained in section 3.2).

Almost all electric motors adhere to common fundamental principles of operation. This so-called unified theory of electric machines is important in the design and analysis of motor performance, but is not always helpful in understanding the principles of operation, and is therefore not included here. Only the characteristics relevant to operation and control of the practically important motors are presented.

10.2   The direct current (DC) motor

History will recognize the vital role played by dc motors in the development of industrial power transmission systems, the dc machine being the first practical device to convert electrical power into mechanical power. Inherently straightforward operating characteristics, flexible performance and high efficiency have encouraged the widespread use of dc motors in many types of industrial drive applications. The basic construction of a dc motor is shown in Fig. 10.1.

Standard dc motors are readily available in one of two main forms:

- *wound-field*, where the magnetic flux in the motor is controlled by the current flowing in a field or excitation winding, usually located on the stator
- *permanent magnet*, where the magnetic flux in the motor is created by permanent magnets which have a curved face to create a constant air-gap to the conventional armature, located on the rotor. These are commonly used at powers up to approximately 3 kW.

Torque in a dc motor is produced by the product of the magnetic field created by the field winding or magnets and the current flowing in the armature winding. The action of a mechanical commutator switches the armature current from one winding to another to maintain the relative position of the current to the field, thereby producing torque independent of rotor position.

The circuit of a shunt-wound dc motor (Fig. 10.2) shows the armature M, the armature resistance R_a and the field winding. The armature supply voltage V_a is supplied typically from a controlled thyristor system and the field voltage V_f from a separate bridge rectifier.

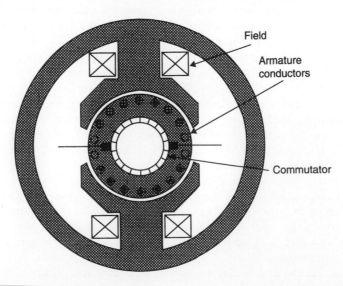

Fig. 10.1 DC motor in schematic form

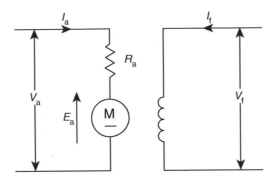

Fig. 10.2 Shunt wound dc motor

As the armature rotates, an *electromotive force* (emf) E_a is induced in the armature circuit and is called the *back-emf* since it opposes the applied voltage V_a (according to Lenz's Law, see section 2.2.4). The E_a is related to armature speed and main field flux by

$$E_a = k_1 n \phi \tag{10.1}$$

where n is the speed of rotation, ϕ is the field flux and k_1 is a motor constant. From Fig. 10.1 it is seen that the terminal armature voltage V_a is given by

$$V_a = E_a + I_a R_a \tag{10.2}$$

Multiplying each side of eqn 10.2 by I_a gives:

$$V_a I_a = E_a I_a + I_a^2 R_a \tag{10.3}$$

(or total power supplied = power output + armature losses). The interaction of the field flux and armature flux produces an armature torque as given in eqn 10.4.

$$\text{Torque } M = k_2 I_f I_a \tag{10.4}$$

where k_2 is a motor constant and I_f is the field current. This confirms the straightforward and linear characteristic of the dc motor and consideration of these simple equations will show its controllability and inherent stability. The speed characteristic of a motor is generally represented by curves of speed against input current or torque and its shape can be derived from eqns 10.1 and 10.2:

$$k_1 n \phi = V_a - (I_a R_a) \tag{10.5}$$

If the flux is held constant by holding the field current constant in a properly compensated motor then

$$n = k_2 [V_a - (I_a R_a)] \tag{10.6}$$

From eqns 10.4 and 10.6, it follows that full control of the dc motor can be achieved through control of the field current and the armature current. In the dc shunt wound motor shown in Fig. 10.2, these currents can be controlled independently. Most industrial dc motor controllers or drives are voltage fed; that is to say that a voltage is applied, and the current is controlled by measuring the current and adjusting the voltage to give the desired current. This basic arrangement is shown in Fig. 10.3.

DC motors exist in other formats. The series dc motor shown in Fig. 10.4 has the field and armature windings connected in series. In this case the field current and

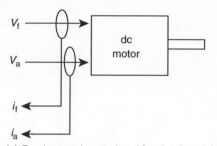

(a) Fundamental control and feedback variables

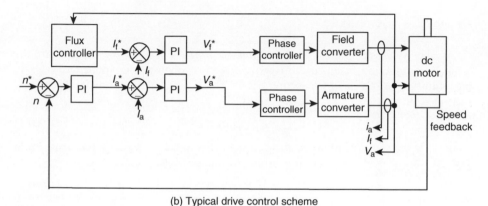

(b) Typical drive control scheme

Fig. 10.3 Control structure for a shunt wound dc motor

armature current are equal and show characteristically different performance results, though still defined by eqns 10.4 and 10.6.

In the shunt motor the field flux ϕ is only slightly affected by armature current, and the value of I_aR_a at full load rarely exceeds 5 per cent of V_a, giving a torque–speed curve shown typically as a in Fig. 10.6, where speed remains sensibly constant over a wide range of load torque.

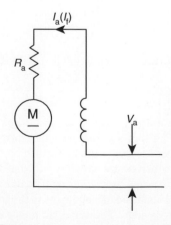

Fig. 10.4 Schematic of series dc motor

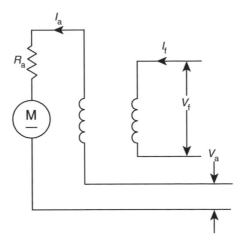

Fig. 10.5 Compound dc motor

The compound-wound dc motor shown in Fig 10.5 combines both shunt and series characteristics.

The shape of the torque–speed characteristic is determined by the resistance values of the shunt and series fields. The slightly drooping characteristic (curve **b** in Fig. 10.6) has the advantage in many applications of reducing the mechanical effects of shock loading.

The series dc motor curve (**c** in Fig. 10.6) shows that the initial flux increases in proportion to current, falling away due to magnetic saturation. In addition the armature circuit includes the resistance of the field winding and the speed becomes roughly inversely proportional to the current. If the load falls to a low value the speed increases dramatically, which may be hazardous, so, the series motor should not normally be

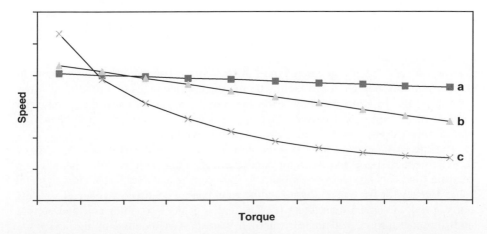

Fig. 10.6 Torque–speed characteristic (**a** - shunt wound dc motor, **b** - compound dc motor, **c** - series dc motor)

used where there is a possibility of load loss. But because it produces high values of torque at low speed and its characteristic is falling speed with load increase, it is useful in applications such as traction and hoisting, and some mixing duties where initial stiction is dominant.

Under semiconductor converter control with speed feedback from a tachogenerator, the shape of the speed–load curve is largely determined within the controller. It has become standard to use a shunt dc motor with converter control even though the speed–load curve, when under open-loop control is often slightly drooping.

The power-speed limit for the dc motor is approximately 3×10^6 kW rev/min, due to restrictions imposed by the commutator.

10.3 The cage induction motor

This simplest form of ac induction motor or asynchronous motor is the basic, universal workhorse of industry. Its general construction is shown in Fig. 10.7. It is usually designed for fixed-speed operation, larger ratings having such features as deep rotor bars to limit *Direct on Line (DOL)* starting currents. Electronic variable speed drive technology is able to provide the necessary variable voltage, current and frequency that the induction motor requires for efficient, dynamic and stable variable-speed control.

Modern electronic control technology is able not only to render the ac induction motor satisfactory for many modern drive applications but also to extend greatly its application and enable users to take advantage of its low capital and maintenance costs. More striking still, microelectronic developments have made possible the highly dynamic operation of induction motors by the application of flux vector control. The practical effect is that it is now possible to drive an ac induction motor in such a way as to obtain a dynamic performance in all respects better than could be obtained with a phase-controlled dc drive combination.

The stator winding of the standard industrial induction motor in the integral kilowatt range is three-phase and is sinusoidally distributed. With a symmetrical three-phase supply connected to these windings, the resulting currents set up, in the air-gap between the stator and the rotor, a travelling wave magnetic field of constant magnitude and moving at synchronous speed. The rotational speed of this field is f/p revolutions per second, where f is the supply frequency (hertz) and p is the number of pole pairs (a four-pole motor, for instance, having two pole pairs). It is more usual to express speed in revolutions per minute, as $60 f/p$ (rpm).

The emf generated in a rotor conductor is at a maximum in the region of maximum flux density and the emf generated in each single rotor conductor produces a current, the consequence being a force exerted on the rotor which tends to turn it in the direction of the flux rotation. The higher the speed of the rotor, the lower the speed of the rotating stator flux field relative to the rotor winding, and therefore the smaller is the emf and the current generated in the rotor cage or winding.

The speed when the rotor turns at the same rate as that of the rotating field is known as synchronous speed and the rotor conductors are then stationary in relation to the rotating flux. This produces no emf and no rotor current and therefore no torque on the rotor. Because of friction and windage the rotor cannot continue to rotate at synchronous speed; the speed must therefore fall and as it does so, rotor emf and current, and therefore torque, will increase until it matches that required by the losses and by any load on the motor shaft. The difference in rotor speed relative to that of the

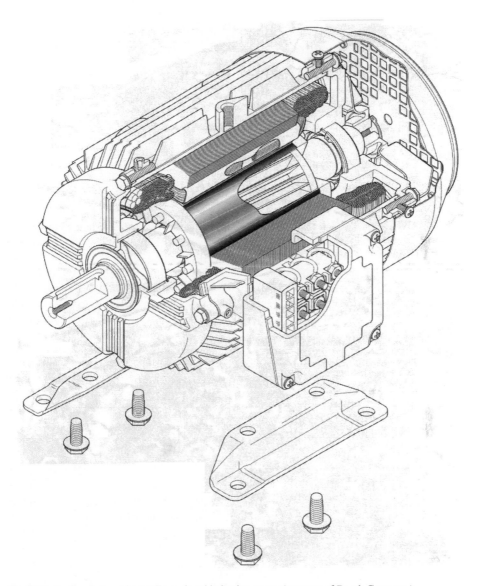

Fig. 10.7 Sectional view of a totally enclosed induction motor (courtesy of Brook Crompton)

rotating stator flux is known as the *slip*. It is usual to express slip as a percentage of the synchronous speed. Slip is closely proportional to torque from zero to full load.

The most popular squirrel cage induction motor is of a 4-pole design. Its synchronous speed with a 50 Hz supply is therefore $60\,f/p$, or 1500 rpm. For a full-load operating slip of 3 per cent, the speed will then be $(1-s)60\,f/p$, or 1455 rpm.

10.3.1 Torque characteristics

A disadvantage of the squirrel cage machine is its fixed rotor characteristic. The starting torque is directly related to the rotor circuit impedance, as is the percentage slip

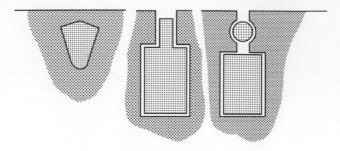

Fig. 10.8 Typical rotor bar profiles

when running at load and speed. Ideally, a relatively high rotor impedance is required for good starting performance (torque against current) and a low rotor impedance provides low full-load speed slip and high efficiency.

This problem can be overcome to a useful extent for DOL application by designing the rotor bars with special cross sections as shown in Fig. 10.8 so that rotor eddy currents increase the impedance at starting when the rotor flux (slip) frequency is high. Alternatively, for special high starting torque motors, two or even three concentric sets of rotor bars are used. Relatively costly in construction but capable of a substantial improvement in starting performance, this form of design produces an increase in full load slip. Since machine losses are closely proportional to working speed slip, increased losses may require such a high starting torque machine to be derated.

The curves in Fig. 10.9 indicate squirrel cage motor characteristics. In the general case, the higher the starting torque the greater the full load slip. This is one of the important parameters of squirrel cage design as it influences the operating efficiency.

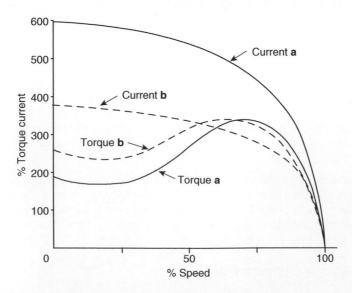

Fig. 10.9 Typical torque–speed and current–speed curves (**a** - standard motor, **b** - high torque motor (6 per cent slip))

10.3.2   Voltage–frequency relationship

To convert a constant speed motor operating direct-on-line to a variable speed drive using an inverter it is necessary to consider the effect of frequency on flux and torque. The magnitude of the field created by the stator winding is controlled broadly by the voltage impressed upon the winding by the supply. This is because the resistance of the winding results in only a small voltage drop, even at full load current, and therefore in the steady state the supply voltage must be balanced by the emf induced by the rotating field. This induced emf depends on the product of three factors:

- the total flux per pole, which is usually determined by the machine designer
- the total number of turns per phase in the stator winding
- the rate of field rotation or frequency

For inverter operation the speed of field rotation for which maximum voltage is appropriate is known as the *base speed*. The consequence of reducing the supply frequency can readily be deduced from the aforementioned relationship. For the same flux the induced emf in the stator winding will be proportional to frequency, hence the voltage supplied to the machine windings must be correspondingly reduced in order to avoid heavy saturation of the core. This is valid for changes in frequency over a wide range. The voltage–frequency relationship should therefore be linear if a constant flux is to be maintained within the machine, as the motor designer intended. If flux is constant, so is the motor torque for a given stator current, and the drive therefore has a constant-torque characteristic.

Although constant *v/f* control is an important underlying principle, it is appropriate to point out departures from it which are essential if a wide speed range is to be covered. First, operation above the base speed is easily achieved by increasing the output frequency of the inverter above the normal mains frequency; two or three times the base speed is easily obtained. The output voltage of an inverter cannot usually be made higher than its input voltage therefore the *v/f* characteristic is typically like that shown in Fig. 10.10(a). Since *v* is constant above base speed, the flux will fall as the frequency is increased after the output voltage limit is reached. In Fig. 10.10(b), the machine flux falls in direct proportion to the *v/f* ratio. Although this greatly reduces the core losses, the ability of the machine to produce torque is impaired and less mechanical load is needed to draw full load current from the inverter. The drive is said to have a constant-power characteristic above base speed. Many applications not requiring full torque at high speeds can make use of this extended speed range. Secondly, departure from constant *v/f* is beneficial at very low speeds, where the voltage drop arising from the stator resistance becomes significantly large. This voltage drop is at the expense of flux, as shown in Fig. 10.10(b). To maintain a truly constant flux within the machine the terminal voltage must be increased above the constant *v/f* value to compensate for the stator resistance effect. Indeed, as output frequency approaches zero, the optimum voltage becomes the voltage equal to the stator *IR* drop. Compensation for stator resistance is normally referred to as *voltage boost* and almost all inverters offer some form of adjustment so that the degree of voltage boost can be matched to the actual winding resistance. It is normal for the boost to be gradually tapered to zero as the frequency progresses towards base speed. Figure 10.10(c) shows a typical scheme for *tapered boost*. It is important to appreciate that the level of voltage boost should increase if a high starting torque is required, since in this case the *IR* drop will be greater by virtue of the increased stator current. In this case automatic load-dependent boost control is

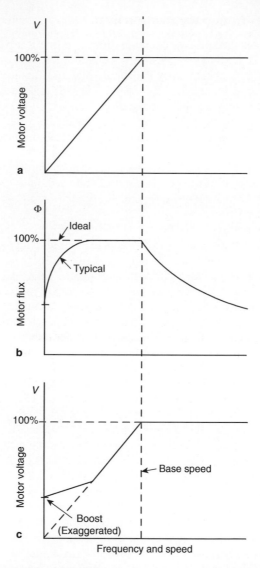

Fig. 10.10 Voltage–frequency characteristics (**a** - linear v/f below base speed, **b** - typical motor flux with linear v/f (showing fall in flux at low frequency as well as above base speed), **c** - modified v/f characteristic with low frequency boost (to compensate for stator resistance effects in steady state))

useful in obtaining the desired low speed characteristics. Such a strategy is referred to as *constant v/f control* and is a feature of most commercially available ac drives, though more advanced open-loop strategies are now becoming more available.

So far the techniques described have been based on achieving constant flux within the air-gap of the machine or, if that is not possible, then the maximum flux. Constant flux is the ideal condition if the highest torque is required because the load cannot be predicted with certainty, or if the most rapid possible acceleration time is desired.

A large number of inverters are used however for variable air volume applications where control of airflow is obtained by variable speed fans. The torque required by a fan follows a square law characteristic with respect to speed and reducing the speed of a fan by 50 per cent will reduce its torque requirement to 25 per cent. As the load is entirely predictable there is no need for full torque capability and hence flux to be maintained, and higher motor efficiency can be obtained by operating at a reduced flux level. A further benefit is that acoustic noise, a major concern in air conditioning equipment, is significantly reduced. It is therefore common for inverters to have an alternative square law *v/f* characteristic or, ideally, a self-optimizing economy feature so that rapid acceleration to meet a new speed demand is followed by settling to a highly efficient operating point. A loss of stability may result from such underfluxing with some control strategies.

10.3.3   Flux vector control

High-performance operation of ac motors can be achieved by the use of *flux vector control*. A detailed explanation is beyond the scope of this book, but because this subject is prone to considerable misrepresentation, a few words of explanation may be helpful.

The purpose of flux vector control is to decouple the flux-producing current from the torque-producing current, giving a control structure equivalent to that of the dc motor shown in Fig. 10.3(a). This is shown in Fig. 10.11.

Here, v_{sx} is the demand to the voltage source for flux-producing current and v_{sy} is the demand to the voltage source for torque-producing current. These demands are fed through polar to rectangular (or quadrature to magnitude and phase) conversion using a calculated phase angle, θ_{ref}, of the rotor position with respect to the field. Unlike the dc motor controller it is not possible to measure directly the flux-producing and torque-producing currents and in flux vector control, these quantities have to be calculated from the measured motor currents, known applied voltages and motor parameters. The performance and quality of this form of control system is critically dependent upon the accuracy of the calculation of the phase angle, θ_{ref}. If the rotor position is measured, as is the case in *closed-loop flux vector* systems, then the phase

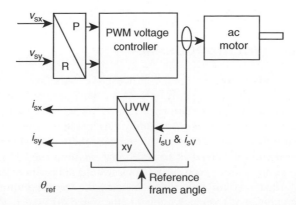

Fig. 10.11 Control structure for an ac motor

angle can be calculated very precisely, and excellent flux and torque control is possible at frequencies including 0 Hz. Key performance measures include dynamics, where the performance of dc motors is easily exceeded, and torque linearity, which can be controlled within 2 per cent over the entire operating range in motoring and braking operation. In open-loop control systems, where there is no rotor speed or position measurement, the quality of the control system is critically dependent upon the performance of the phase angle estimator. Modern industrial systems have to balance performance with robustness in operation, and robustness when a standard controller is applied to different industrial motors.

10.4   The slipring induction motor

The wound rotor or slipring ac machine addresses some of the disadvantages of the cage induction motor, but with the handicap of extra cost and the complexity of brushes and insulated rotor windings.

With the correct value of (usually) resistance inserted in the rotor circuit, a near-unity relationship between torque and supply current at starting can be achieved, such as 100 per cent full load torque (FLT), with 100 per cent full load current (FLC) and 200 per cent FLT with 200 per cent FLC. This is comparable with the starting capability of the dc machine. Not only high starting efficiency but also smoothly controlled acceleration historically gave the slipring motor a great popularity for lift, hoist and crane applications. It has had a similar popularity with fan engineers, providing a limited range of air volume control (either 2:1 or 3:1 reduction) at constant load by the use of speed-regulating variable resistances in the rotor circuit. Although a fan presents a square-law torque–speed characteristic, so that motor currents fall considerably with speed, losses in the rotor regulator at lower motor speeds are still relatively high, severely limiting the useful speed range.

Efficient variable-speed control of slipring motors can be achieved by slip energy recovery in which the the slip frequency on the rotor is converted to supply frequency. It is possible to retrofit variable frequency inverters to existing slipring motors simply by short-circuiting the slipring terminations (ideally on the rotor thereby eliminating the brushes) and treating the motor as a cage machine.

Variable voltage control of slipring motors has been used extensively, notably in crane and lift applications, though these applications are now largely being met by flux vector drives.

10.5   The ac synchronous motor

In a synchronous motor, torque can be produced at synchronous speed. This is achieved by a field winding, generally wound on the rotor, and dc excited so that it produces a rotor flux which is stationary relative to the rotor. Torque is produced when the rotating three-phase field produced by currents in the stator winding and the rotor field are stationary relative to each other, hence there must be physical rotation of the rotor at *synchronous speed* n_s in order that its field travels in step with the stator field axis. At any other speed a rotor pole would approach alternately a stator 'north' pole field, then a 'south' pole field, changing the resulting torque from a positive to a negative value at a frequency related to the speed difference, the mean torque being zero.

A typical inverter for variable speed control automatically regulates the main stator voltage to be in proportion to motor frequency. It is possible to arrange an excitation control loop which monitors the main stator voltage and increases the excitation field voltage proportionately.

The ac synchronous motor has attractive features for inverter variable speed drive applications, particularly at ratings of 40 kW and above. Not least is overall cost when compared with a cage induction motor and inverter, or a dc shunt wound motor with converter alternatives. In applications requiring a synchronous speed relationship between multiple drives, or precise speed control of single large drives the ac synchronous motor with inverter control system appears attractive, freedom from brushgear maintenance, good working efficiency and power factor being the main considerations.

10.6   The brushless servomotor

A synchronous machine with permanent magnets on the rotor is the heart of the modern brushless servomotor drive. The motor stays in synchronism with the frequency of supply, though there is a limit to the maximum torque which can be developed before the rotor is forced out of synchronism, *pull-out torque* being typically between 1.5 and 4 times the continuously rated torque. The torque–speed curve is therefore simply a vertical line.

The industrial application of brushless servomotors has grown significantly for the following reasons:

- reduction of price of power conversion products
- establishment of advanced control of PWM inverters
- development of new, more powerful and easier to use permanent magnet materials
- the developing need for highly accurate position control
- the manufacture of all these components in a very compact form

They are, in principle, easy to control because the torque is generated in proportion to the current. In addition, they have high efficiency, and high dynamic responses can be achieved.

Brushless servomotors are often called brushless dc servomotors because their structure is different from that of dc servomotors. They rectify current by means of transistor switching within the associated drive or amplifier, instead of a commutator as used in dc servomotors. Confusingly, they are also called ac servomotors because brushless servomotors of the synchronous type (with a permanent magnet rotor) detect the position of the rotational magnetic field to control the three-phase current of the armature. It is now widely recognized that *brushless ac* refers to a motor with a sinusoidal stator winding distribution which is designed for use on a sinusoidal or PWM inverter supply voltage. *Brushless dc* refers to a motor with a trapezoidal stator winding distribution which is designed for use on a square wave or block commutation inverter supply voltage.

The brushless servomotor lacks the commutator of the dc motor, and has a device (the drive, sometimes referred to as the amplifier) for making the current flow according to the rotor position. In the dc motor, increasing the number of commutator segments reduces torque variation. In the brushless motor, torque variation is reduced by making the coil three-phase and, in the steady state, by controlling the current of each phase into a sine wave.

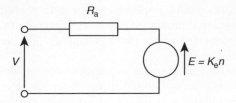

Fig. 10.12 Simplified equivalent circuit

10.6.1   Stationary torque characteristics

The simple equivalent circuit of Fig. 10.12 represents a motor which uses permanent magnets to supply the field flux. This is a series circuit comprising the armature resistance R_a, and back emf E. If the voltage drop across the transistors is ignored, the equation for the voltage is

$$V = R_a I_a + K_e n \tag{10.7}$$

K_e is known as the *back emf constant* of the motor, and n is the speed. Therefore the torque T is given by

$$T = K_t I_a = (K_t / R_a)(V - K_e n) \tag{10.8}$$

K_t is known as the *torque constant* of the motor.

Figure 10.13 shows the relation between T and n at two different voltages. The torque decreases linearly as the speed increases and the slope is $K_t K_e / R_a$ which is independent of the terminal voltage and the speed. Such characteristics make speed or position control relatively easy.

The starting torque and the speed are given by

$$T_s = K_t V / R_a \tag{10.9}$$

$$n_o = V / K_e \tag{10.10}$$

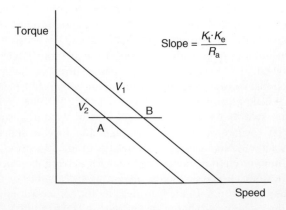

Fig. 10.13 Torque–speed characteristics

In modern drive systems, a flux vector controller is applied to many brushless servomotors. The control is almost exactly as for the induction motor except zero flux-producing current is demanded for operation in the normal frequency range.

10.7 The reluctance motor

The reluctance motor is arguably the simplest synchronous motor of all, the rotor consisting of a set of iron laminations shaped so that it tends to align itself with the field produced by the stator.

The stator winding is identical to that of a three-phase induction motor. The rotor is different, containing saliency which provides a preferred path for the flux. This is the feature which tends to align the rotor with the rotating magnetic field, making it a form of synchronous machine. In order to start the motor a form of cage needs to be incorporated into the rotor design, and the motor can then start as an induction motor. Once a higher speed is reached, the reluctance torque 'pulls in' the rotor to run synchronously in much the same way as a permanent magnet rotor.

Reluctance motors may be used on both fixed frequency supplies and inverter supplies. These motors tend to be one frame size larger than a similarly rated induction motor and have a low power factor (perhaps as low as 0.4) and poor pull-in performance. As a result, their industrial use has not been widespread except for special applications such as textile machines where large numbers of reluctance motors may be connected to a single 'bulk' inverter and may maintain synchronism. Even in this application, as the cost of inverters has reduced, bulk inverters are infrequently used and the reluctance motor is now rarely seen.

10.8 The switched reluctance motor

The Switched Reluctance (SR) motor is very different from the other polyphase machines described because both the stator and the rotor have salient poles. The motor can only be used in conjunction with its specific power converter and control, and consequently only overall characteristics are relevant.

The SR motor produces torque through the magnetic attraction which occurs between stator electromagnets and a corresponding set of salient poles formed on a simple rotor made only of ferromagnetic material. The intuitively straightforward principle of torque production is easily visualized in the very simple reluctance motor illustrated, in cross section, in Fig. 10.14.

If current is applied to the winding the rotor will turn until it reaches a position where it is aligned with the coils, at which point the inductance of the magnetic circuit is a minimum. The characteristic variation of inductance of a SR motor is shown in Fig. 10.15.

If the machine is lightly magnetically loaded and a modest level of torque is produced, then the steel from which the rotor and stator are made of will behave magnetically in an approximately linear fashion. That is, for a given number of turns on the windings, the magnetic flux of a phase varies approximately in proportion to the phase current. If linearity is assumed, then it can be shown that the torque produced as a function of angle θ is

$$T = [i^2(dL/d\theta)]/2 \qquad (10.11)$$

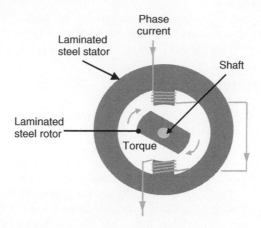

Fig. 10.14 Simple reluctance motor

Equation 10.11 shows that the torque is not dependent upon the direction of the current, but depends upon whether the current is applied when the inductance L is rising or falling with angular position.

The phase currents are always switched synchronously with the mechanical position of the rotor. At low speeds, the phases are energized over the entire region of rising inductance, and active current limiting is required from the controller. Torque is controlled by adjusting the magnitude of the phase current. As speed increases, the rise and fall times (especially the latter) of the phase current occupy significant rotor angle, and it is usual to advance the turn-on and turn-off angles with respect to rotor position. The torque is now controlled by both the current limit level and by the switching angles, though current is usually used as the primary control variable. At high speeds, the rise and fall times occupy still greater rotor angles. The current naturally self-limits and it is usual to control the torque using only the switching angles. The shape of the current waveform is greatly influenced by the high rate of change of inductance with respect to time.

By choosing appropriate switching angles and current levels, together with an appropriate electromagnetic design, the torque–speed characteristic of the switched

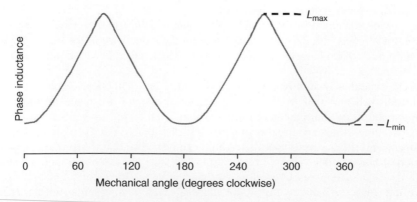

Fig. 10.15 Variation of inductance with rotor angle

Fig. 10.16 Cross section of three-phase 6–4 SR motor

reluctance drive can be tailored to suit the application. Furthermore, simply by changing the control parameter selection with torque and speed, a given machine design can be made to offer a choice of different characteristics.

The simple single-phase machine in Fig. 10.14 is capable of producing torque over only half of its electrical cycle. More demanding applications use higher pole numbers on the rotor and stator, with the stator poles wound and connected into multiple identical phases. Figure 10.16 illustrates the cross section of a three-phase 6–4 machine, diametrically opposed coils being connected together to form three-phase circuits denoted A, B and C.

The excitation of the phases is interleaved equally throughout the electrical period of the machine. This means that torque of the desired polarity can be produced continuously. The number of phases can in theory be increased without limit, but one to four phases are most common for commercial and industrial applications. Many different combinations of pole count are possible. It is sometimes beneficial to use more than one stator pole pair per phase, so that, for example, the 12–8 pole structure is commonly used for three-phase applications. Each phase circuit then comprises four stator coils connected and energized together. Increasing the phase number brings the advantage of smoother torque. Self-starting in either direction requires at least three phases.

These SR drives are finding application in high-volume appliances and some industrial applications which can take good advantage of their characteristics, notably high starting torque and where less importance is placed on the smoothness of rotation. Considerable advances have been made in improving the noise characteristics of this drive, but this can still be a limiting factor where a broad operating speed range is required.

10.9   Mechanical and duty cycle considerations

10.9.1   Mounting of the motor

Internationally agreed coding applies to a range of standard mountings for electric motors, dc and ac, which covers all the commonly required commercial arrangements. Within IEC 60034-7 (EN 60034-7) there are examples of all practical methods of mounting motors. The NEMA publishes alternative standards within NEMA Standards publication No. MG1: Motors and Generators.

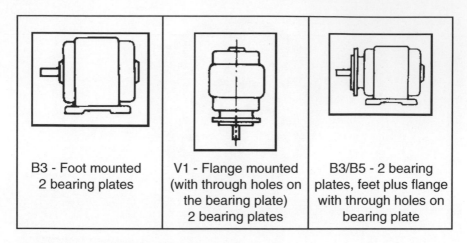

| B3 - Foot mounted 2 bearing plates | V1 - Flange mounted (with through holes on the bearing plate) 2 bearing plates | B3/B5 - 2 bearing plates, feet plus flange with through holes on bearing plate |

Fig. 10.17 Common mounting arrangements for motors

There are many standard mounting arrangements described in IEC 60034-7. The most usual types of construction for small and medium-sized motors are shown in Fig. 10.17.

10.9.2 Degree of protection

All types of electric motors are classified in accordance with a standard coding to indicate the degree of protection afforded by any design against mechanical contact and against various degrees of ambient contamination.

The designation defined in IEC 60034-5, (EN 60034-5) consists of the alphabets IP followed by two numerals signifying conformance with specific conditions. Additional information may be included by a supplementary alphabet following the second numeral. This system is contained within NEMA MG 1 but is not universally adopted by the industry in the United States. The first characteristic numeral indicates the degree of protection provided by the enclosure with respect to persons and also to the parts of the machine inside the enclosure. The commonly used numbers are given in Table 10.1.

The second characteristic numeral indicates the degree of protection provided by the enclosure with respect to harmful effects due to the ingress of water. The commonly used numbers are given in Table 10.2.

Motors for hazardous areas have to meet very special requirements and flameproof motors in particular have special enclosures. More general background on the IP coding system is given in section 16.2.2.1.

10.9.3 Duty cycles

The capacity of an electrical machine is often temperature-dependent, and the duty cycle of the application may significantly affect the rating. IEC 60034-1 defines a number of specific duty cycles with the designation S1 through to S10. For example, S3 describes the duty of an application with an intermittent duty shown in Fig. 10.18.

This form of duty rating refers to a sequence of identical duty cycles, each cycle consisting of an on-load and off-load period, the motor coming to rest during the latter. Starting and braking are not taken into account, it being assumed that the times taken up by these are short in comparison with the on-load period, and do not appreciably affect the heating of the motor.

Table 10.1 IP designation: first characteristic numeral

First characteristic numeral	Brief description	Definition
0	Non-protected machine	No special protection
2	Machine protected against solid objects greater than 12 mm in diameter	No contact by fingers or similar objects not exceeding 80 mm in length with or approaching live or moving parts inside the enclosure. Ingress of solid objects exceeding 12 mm in diameter.
4	Machine protected against solid objects greater than 1 mm in diameter	No contact with or approaching live or moving parts inside the enclosure by wires or strips of thickness greater than 1 mm in diameter.
5	Dust protected machine	No contact with or approaching live or moving parts within the machine. Ingress of dust is not totally prevented but dust does not enter in sufficient quantity to interfere with the satisfactory operation of the machine.
6	Dust-tight machine	No contact with or approach to live or moving parts inside the enclosure. No ingress of dust.

Table 10.2 IP designation: second characteristic numeral

Second characteristic numeral	Brief description	Definition
0	Non-protected machine	No special protection
1	Machine protected against dripping water	Dripping water (vertically falling drops) shall have no harmful effects
3	Machine protected against spraying water	Water falling as a spray at an angle up to 60° from the vertical shall have no harmful effect
4	Machine protected against splashing water	Water splashing against the machine from any direction shall have no harmful effect
5	Machine protected against water jets	Water projected by a nozzle against the machine from any direction shall have no harmful effect
6	Machine protected against heavy seas	Water from heavy seas or water projected in powerful jets shall not enter the machine in harmful quantities

When stating the motor power for this form of duty, it is necessary to state the *cyclic duration factor*, which is the on-time as a percentage of the cycle time. The duration for one cycle must be shorter than 10 minutes and preferred cyclic duration factors (15%, 25%, 40% and 60%) are specified.

10.10 Drive power circuits

The control of a dc motor is based upon the precise control of the voltage or current supplied to the armature and field windings. AC motor control is based upon the

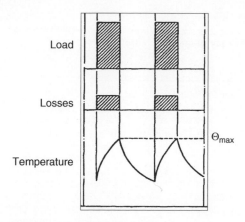

Fig. 10.18 Load, losses and temperature for duty type S3

precise control of the supply in terms of frequency, the magnitude of the voltage and, in some cases, the phase angle of the supply in relation to rotor position.

A large number of power semiconductor-based converter circuits exist to form voltage or current sources. The principles of many of these circuits are reviewed in Chapter 11 and only the most important practical types of circuit for motor drives are discussed in the following sections.

10.10.1   DC motor drive systems

In section 10.2 it is shown that complete control of a dc machine can be achieved by controlling the armature voltage V_a and the field current I_f. Two power converters are employed for this purpose in most variable speed drives which employ the separately excited dc machine. (In referring to the number of converters in a drive, it is common to ignore the field converter; this nomenclature will be adopted here). It is relatively common in simple drives for the field converter to be a single-phase uncontrolled bridge thereby applying fixed field voltage.

Where the variation in motor resistance with temperature, or a poorly regulated supply, results in unacceptable variations in field current, a controlled power converter is employed with current control. Such field controllers are further discussed later as applied to field-weakening control.

10.10.1.1   AC to dc power conversion

The three-phase fully controlled converter dominates all but the lowest powers where single-phase converters are used and Fig. 10.19 shows the power circuit together with associated ac/dc relationships. Figure 10.20 shows how the dc voltage can be varied by adjusting the firing delay angle α. The *pulse number p*, of this bridge is 6 and energy flow can be from ac to dc or dc to ac.

The salient 'ideal' characteristics of the three-phase fully controlled bridge are shown in Table 10.3.

The above characteristics are based upon idealized conditions of negligible ac inductance and constant dc current, but these are not often found in practice. It is not possible to consider all practical effects here, but the effect of dc link current ripple on

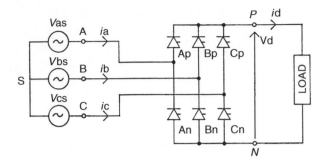

Fig. 10.19 Three-phase fully controlled bridge

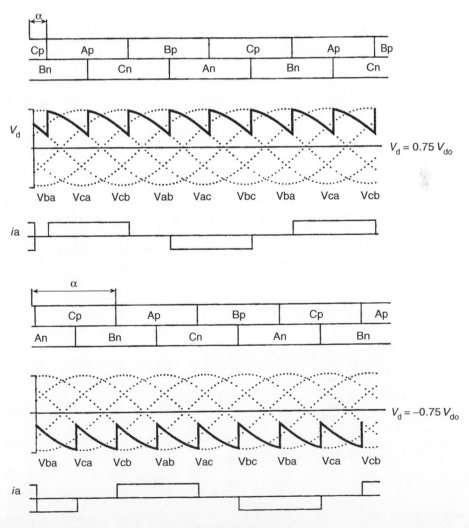

Fig. 10.20 Three-phase fully controlled bridge: dc voltage control

Table 10.3 Ideal characteristics of a three-phase fully controlled bridge

Firing angle	α
V_{do}	$\dfrac{3\sqrt{2}}{\pi} V_s$
V_d / V_{do}	$\cos \alpha$
I_s / I_d	$\sqrt{(3/2)}$
Overall Power Factor	$(3/\pi) \cdot \cos \alpha$
Supply Current nth harmonic/I_d	0 for $n = 3,6,9...$
	0 for n even
	$\sqrt{6}/n\pi$ for n odd
Phase of supply current harmonics	$n\alpha$

ac supply harmonics is of great practical industrial importance mainly in relation to three-phase bridges. Practical experience has led to the adoption by many of $I_5 = 0.25I_1$ (the ideal being $0.2I_1$), $I_7 = 0.13I_1$ (the ideal being $0.14I_1$), $I_{11} = 0.09I_1$ (the ideal being $0.11I_1$) and $I_{13} = 0.07I_1$ (the ideal being $0.08I_1$). In general, the amplitudes of higher harmonics are rarely of significance with regard to supply distortion. Under conditions of very high dc current ripple, the 5th harmonic can assume a considerably higher value than that quoted here. A practical example would be an application with a very capacitive dc load such as a voltage source inverter; in such a case where no smoothing choke is used, I_5 could be as high as $0.5I_1$.

10.10.1.2 Single-converter drives

Figure 10.21 shows a single-converter dc drive. In its most basic form the motor will drive the load in one direction only without braking or reverse running. It is said to be a *single-quadrant drive*, only operating in one quadrant of the torque–speed characteristic. Such drives have wide application from simple machine tools to fans, pumps, extruders, agitators and printing machines.

For full *four-quadrant* operation, a drive is required to operate in both the forward and reverse directions and to provide regenerative braking. This is illustrated in Fig. 10.22. A single fully controlled converter can still be used but some means of reversing, either the field or armature connections, as shown in Fig. 10.23, must be added.

Reversal of the armature current can involve bulky (high current) reversing switches, but due to the low inductance of the armature circuit it can be completed in

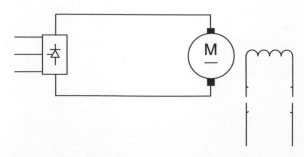

Fig. 10.21 Single-phase converter dc drive

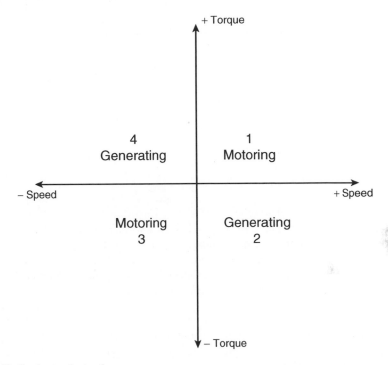

Fig. 10.22 Quadrants of operation

typically 0.2 seconds. Field current reversal takes longer, typically in the order of 1 second, but lower cost reversing switches may be used. The field reversal time can be reduced by using higher voltage field converters to force the current. Forcing voltages up to 4 per unit are used but care must be taken not to over-stress the machine. This increased voltage cannot be applied continuously and either a switched ac supply or a controlled field converter is required. Armature and field reversal techniques are used where torque reversals are infrequent such as hoists, presses, lathes and centrifuges.

10.10.1.3 Dual-converter drives

When a four-quadrant drive is required to change the direction of torque rapidly, the delays associated with reversing switches described earlier may be unacceptable and a dual converter comprising two fully controlled power converters connected in

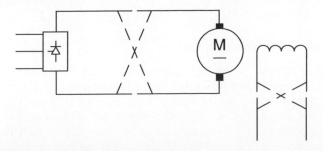

Fig. 10.23 Single-phase converter reversing/regenerative dc drive

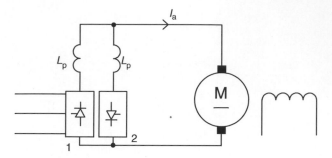

Fig. 10.24 Single-phase dual-converter dc drive

inverse-parallel can be used as shown in Fig. 10.24. Bridge 1 conducts when the armature current I_a is required to be positive and bridge 2 when it is required to be negative.

There are two common forms of dual converter. In the first, both bridges are
controlled simultaneously to give the same mean output voltage. The instantaneous
voltages from the rectifying and inverting bridges cannot be identical, and reactors L_p
are included to limit the current circulating between them. The principal advantage of
this system is that when the motor torque, and hence current, is required to change
direction, there need be no delay between the conduction of one bridge and the other.
This is the *dual converter bridge with circulating current*.

In the second, the *circulating current-free dual converter*, only one bridge at a time
is allowed to conduct. The cost and losses associated with the L_p reactors can then be
eliminated, and economies can also be made in the drive control circuits. The penalty
is a time delay of typically 10 ms as the current passes through zero, while it is ensured
that the thyristors in one bridge have safely turned off before those in the second are
fired. This circulating current-free dual converter is the most common industrial
four-quadrant drive and is used in many demanding applications. Paper, plastics and
textile machines where rapid control of tension is required are good examples.

10.10.1.4 Field control

By weakening the field as speed increases, a constant power characteristic can be
achieved. The field can be controlled by a three-phase (or often a single-phase) fully
controlled bridge. By including this converter in a speed-control loop, it can be
arranged that as speed increases, the armature voltage rises to the point where it
matches the pre-set reference in the field controller. Above that speed, an error signal is
produced by the voltage loop, which causes the field controller to weaken the motor
field current and thereby restore armature voltage to the set-point level. The resulting
characteristics are shown in Fig. 10.25.

10.10.1.5 DC to dc power conversion

The principles of dc to dc power converters are explained in section 11.4.1. They provide the means to change one dc voltage to another; step-up (or boost) converters are
available and have potential for the future, but step-down (or buck) converters are of
greatest commercial interest and will be concentrated upon here. These converters
are usually supplied from an uncontrolled ac to dc converter or a battery supply;
their output can be used to control a dc machine as in the case of the controlled ac to
dc converters.

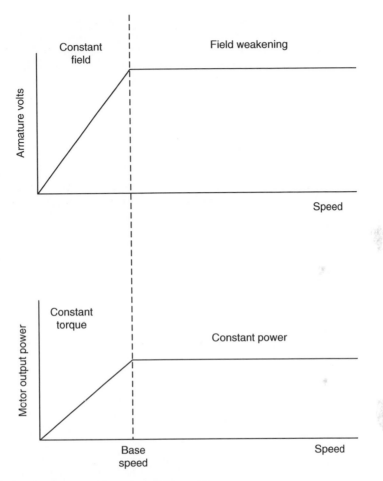

Fig. 10.25 Constant power operation using a field controller

Several important limitations of ac to dc converters are overcome by the dc–dc converter, as follows:

- the dc ripple frequency is determined by the ac and is, for a 50 Hz supply frequency, 100 Hz for single-phase and 300 Hz for three-phase fully controlled bridges. This means that additional smoothing components are often required when using high speed machines, permanent magnet motors or other special motors with low armature inductance.
- as a result of the delay inherent in thyristor switching (3.3 ms in a 50 Hz three-phase converter) the current control loop band width of the converter is limited to approximately 100 Hz, which is too low for many servo drive applications.
- thyristor controlled ac to dc converters have an inherently poor input power factor at low output voltages. The near-unity power factor can be achieved using an uncontrolled rectifier feeding a dc to dc converter.
- electronic short-circuit protection is not economically possible with thyristor converters. Protection is normally accomplished by fuses.

DC to dc converters are more complex and less efficient than ac–dc converters, but they find application mainly in dc servo drives, rail traction drives and small fractional drives employing permanent magnet motors. In the examples which follow, the circuits are illustrated showing bipolar transistors, but MOSFETs, IGBTs and, at higher powers, GTOs are widely used.

(a) *Single-quadrant step-down dc to dc converter*

The most basic dc to dc converter is shown in Fig. 10.26. The output voltage is changed by *pulse-width modulation (PWM)*, that is, by varying the time for which the transistor T is turned on and the voltage applied to the motor is therefore in the form of a square wave of varying period. The principle of PWM is explained in section 11.4.4. Because the motor is inductive the current waveform is smoothed, the flywheel diode D carrying the current whilst the transistor is turned off. The basic formulae relating the variables in Fig. 10.26 are as given in eqns 10.12 and 10.13.

$$V_a = V_{dc} t f \tag{10.12}$$

$$\Delta I_a = V_{dc}/4L_a f \tag{10.13}$$

where f is the frequency of transistor 'on pulse' (Hz), ΔI_a is the maximum deviation of armature current and t is the on-pulse duration.

Applications for this circuit are normally limited to drives below 5 kW and simple variable speed applications.

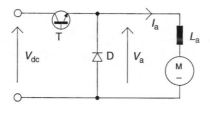

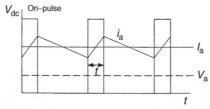

Fig. 10.26 Single quadrant dc to dc converter

(b) *Four-quadrant dc to dc converter (the H-bridge)*

Figure 10.27 shows a basic four-quadrant converter. During motoring, positive output transistors T1 and T4 are switched on during the on-period, whilst diodes D2 and D4 conduct during the off-period. When D2 and D4 conduct, the motor supply is reversed and consequently the voltage is reduced to zero at 50 per cent duty cycle. Any reduction of duty cycle below 50 per cent will cause the output voltage to reverse but with current in the same direction; hence the speed is reversed and the drive is regenerating.

With transistors T2 and T3 conducting, the current is reversed and hence the full four-quadrant operation is obtained. These converters are widely used in high-performance dc drives such as servos.

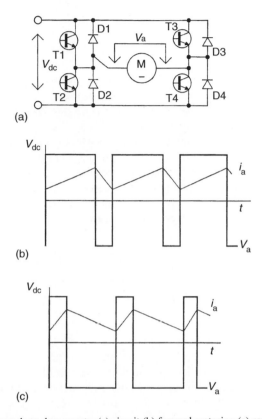

(a)

(b)

(c)

Fig. 10.27 Four-quadrant dc to dc converter (a) circuit (b) forward motoring (c) reverse braking

10.10.2   AC motor drive systems

10.10.2.1   AC to ac power converters with intermediate dc link

This category of ac drive commonly termed the *variable frequency inverter* is by far the most important in respect of the majority of industrial applications. It is considered here as a complete converter, although the input stages have been considered earlier in isolation, and their individual characteristics are, of course, applicable. Alternative input stages to some of the drives are also applicable.

The concept of these inverter drives is well understood, being rectification of fixed frequency, smoothing and then inverting to give variable frequency, variable voltage to supply an ac machine. Within this broad concept two major categories exist. First, in the *Voltage Source Inverter*, the converter impresses a voltage on the motor, and the impedance of the machine determines the current. Secondly, in the *Current Source Inverter*, the converter impresses a current on the motor, and the impedance of the machine determines the voltage. For most industrial applications the PWM voltage source inverter is applied, and only this converter will be considered here.

10.10.2.2   *General characteristics of a voltage source inverter*

Voltage source inverters can be considered as a voltage source behind an impedance, and consequently they are very flexible in their application. Major inherent features include the following:

- multi-motor loads can be supplied. This can be very economical in applications such as roller table drives and spinning machines.
- inverter operation is not dependent upon the motor characteristics. Various machines (induction, synchronous or even reluctance) can be used, provided the current drawn is within the current rating of the inverter although care should be taken where a low power factor motor is used to ensure that the inverter can provide the required reactive power.
- open-circuit protection is inherent. This is useful in applications where the cables between the inverter and the motor are insecure or subject to damage etc.
- the facility to ride through mains dips can easily be provided by buffering the dc voltage link with capacitance or, where necessary, a battery.
- motoring operation only in both directions is possible without the addition of resistive dumps for braking energy or expensive regenerative converters to feed energy back to the supply.

In the PWM inverter drive shown in Fig. 10.28, the dc link voltage is uncontrolled and derived from a simple diode bridge. The output voltage can be controlled electronically within the inverter by use of the PWM techniques explained in section 11.4.4. In this method, the transistors are switched on and off many times within a half cycle to generate a variable voltage output which is normally low in harmonic content. A PWM waveform is illustrated in Fig. 10.29.

A large number of PWM techniques exist, each having different performance notably in respect to the stability and audible noise of the driven motor. The use of PWM virtually eliminates low-speed torque pulsations since negligible low-order harmonics are present so this is an ideal solution where a drive system is to be used across a wide speed range.

Since voltage and frequency are both controlled with the PWM, quick response to changes in demand can be achieved. Furthermore, with a diode rectifier as the input circuit a high power factor, approaching unity, is offered to the incoming ac supply over the entire speed and load range.

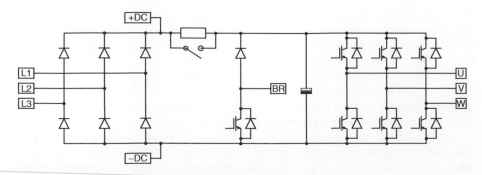

Fig. 10.28 Typical power circuit of a PWM voltage-fed inverter

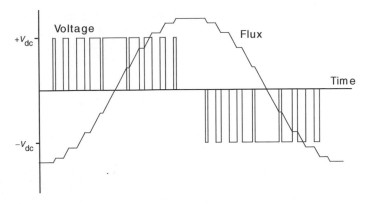

Fig. 10.29 PWM inverter output voltage and flux waveforms

The PWM inverter drive efficiency typically approaches 98 per cent but this figure is heavily affected by the choice of switching frequency, the losses being greater where a higher switching frequency is used. In practice the maximum fundamental output frequency is usually restricted to about 1 kHz for a transistor-based system. The upper frequency limit may be improved by making a transition to a less sophisticated PWM waveform with a lower switching frequency and ultimately to a square wave if the application demands it. However, with the introduction of faster switching power semiconductors, these restrictions to switching frequency and minimum pulse-width have been eased.

10.11   Effects of semiconductor power converters

The control of electric motors by power electronic converters has a number of significant effects. These are primarily due to the introduction of harmonic components into the voltage and current waveforms applied to the motor. In the case of ac machines which are normally considered to be fixed speed there are additional implications of variable speed operation, including mechanical speed limits and the possible presence of critical speeds within the operating speed range.

10.11.1   Effects upon dc machines

The effects due to deviation from a smooth dc supply are, in general, well understood by drive and motor manufacturers. The impact of ripple in the dc current increases the rms current, which leads to increased losses and hence reduced torque capacity. The harmonics associated with the current ripple lead to the now universal practice of using laminated magnetic circuits, which are designed to minimize eddy currents. With the chopper converters used in servo amplifiers and traction drives, frequencies in excess of 2 kHz can be impressed on the motor and special care is needed to select a motor with sufficiently thin laminations.

The ripple content of the dc currents significantly affects commutation within a dc machine. The provision of a smoothing choke can be extremely important, and a recommendation should be made by the motor manufacturer depending upon the supply converter used.

Besides the thermal and commutation impacts, the ripple current also results in pulsating torque, which can cause resonance in the drive train. Laminating the stator field poles not only improves the thermal characteristic of the motor but also its dynamic behaviour by decreasing the motor time constant.

10.11.2   Effects upon ac machines

It is often stated that standard ac motors can be used without problem on modern PWM inverters and while such claims may be largely justified, switching converters do have an impact and limitations do exist. The NEMA MG1:1987, Part 17A gives guidance on the operation of constant-speed cage induction motors on 'a sinusoidal bus with harmonic content' and general purpose motors used with variable-voltage or variable-frequency controls or both.

10.11.2.1   Machine rating: thermal effects

The operation of ac machines on a non-sinusoidal supply inevitably results in additional losses in the machine, which fall into three main categories:

- *Stator copper loss*. This is proportional to the square of the rms current, but additional losses in the winding conductors due to skin effect must also be considered,
- *Rotor 'copper' loss*. The rotor resistance is different for each harmonic current in the rotor. This is due to skin effect and is pronounced in deep bar rotors. The rotor 'copper' loss must be calculated independently for each harmonic and the increase caused by harmonic currents can be a significant component of the total losses, particularly with PWM inverters having higher harmonics for which slip and rotor resistance are high.
- *Iron loss*. This is increased by the harmonic components in the supply voltage. The increase in iron loss due to the main field is usually negligible, but there is a significant increase in loss due to end winding leakage and slew leakage fluxes at the harmonic frequencies.

The total increase in losses does result in increased temperatures within the motor, but these cannot be readily represented by a simple de-rating factor since the harmonic losses are not evenly distributed throughout the machine and the distribution will vary according to the design of the motor. This has special implications for machines operating in a hazardous atmosphere, and this is covered further in section 15.7.5.4.

Many fixed-speed motors have shaft-mounted cooling fans and operation away from the rated speed of the machine results in reduced or increased cooling. This needs to be taken into account when specifying a motor for variable-speed duty.

10.11.2.2   Machine insulation

The fast-rising voltage created by a PWM drive can result in a transiently uneven voltage distribution through a winding. For supply voltages up to 500 V, the voltage imposed by a correctly designed inverter is well within the capability of a standard motor of reputable manufacture, but for higher supply voltages an improved winding insulation system is generally required to ensure that the intended working life of the motor is achieved.

There can also be short-duration voltage over-shoots because of reflection effects in the motor cable, which is a system effect caused by the combined behaviour of the drive, cable and motor. The length of the motor cable can increase the peak motor voltage, but in applications with cables of 10 m or less, no special considerations are generally required. Output inductors (chokes) or output filters are sometimes used with drives for reasons such as long-cable driving capability or radio frequency suppression. In such cases no further precautions are required because these devices also reduce the peak motor voltage and increase its rise-time.

The IEC 60034-17 gives a profile for the withstand capability of a minimum standard motor, in the form of a graph of peak terminal voltage against voltage rise-time. The standard is based on research on the behaviour of motors constructed with the minimum acceptable level of insulation within the IEC motor standard family. Tests show that standard PWM drives with cable lengths of 20 m or more produce voltages outside the IEC 60034-17 profile. However most motor manufacturers produce, as a standard, machines with a capability substantially exceeding the requirements of IEC 60034-17.

10.11.2.3   Bearing currents

The sum of the three stator currents in an ac motor is ideally zero and there is no further path of current flow outside the motor, but in practice there are conditions which result in currents flowing through the bearings. These conditions include:

- *Magnetic asymmetry*. An asymmetric flux distribution within an electrical machine can result in an induced voltage from one end of the rotor shaft to the other. If the bearing breakover voltage is exceeded, a current flows through both the bearings. In some large machines, it is a common practice to fit an insulated bearing, usually at the non-drive end, to stop such currents.
- *Supply asymmetry*. With PWM inverter supplies, it is impossible to achieve perfect balance between the phases instantaneously, when pulses of different widths are produced. The resulting neutral voltage is not zero with respect to earth, and its presence equates to that of a common mode voltage source. This is sometimes referred to as a *zero sequence voltage*. It is proportional in magnitude to the dc link voltage in the inverter (itself proportional to the supply voltage), and has a frequency equal to the switching frequency of the inverter.

The risk of bearing currents can be minimized by adopting a grounding strategy which keeps all system components grounded at the same potential. This needs to be achieved for all frequencies and high inductance paths must be avoided, for instance keeping cable runs as short as possible. In addition, a low impedance path should be defined for the common-mode currents to return to the inverter. As the common-mode current flows through the three phase conductors in the supply cable, the best return path is through a shield around that cable. This could be in the form of a screen. Such measures are well defined by most reputable manufacturers in their EMC guidance and general guidelines are provided in Chapter 14.

10.12   The commercial drive

The block schematic shown in Fig.10.30 illustrates the normal arrangement of an inner torque or current loop with an outer speed loop. Alternatively, where torque control is

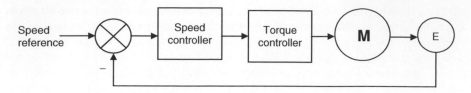

Fig. 10.30 Basic schematic diagram of a variable speed drive

required in, for example, a tensioning application, then the outer speed loop can be removed, or made subservient to a torque loop with an external torque demand.

However, the modern industrial drive comprises much more than a speed and torque controller. Recent reviews of industrial drive specifications and marketing literature from a broad range of suppliers confirm that the ability to turn the shaft of a motor could be considered a very small part of the feature set of a modern drive. Figure 10.31 is an illustration of the additional functionality frequently found; this generalization changes from supplier to supplier and by the sector of the drive market being considered.

It is rare that an industrial drive stands alone in an application. In the majority of cases, drives are part of a system and it is necessary for the parts of the system to communicate with one another, transmitting commands and data. This communication can be in many forms from traditional analogue signals through to wireless communication systems. The drives industry has been working to produce lower cost, higher performance drives, with good flexible and dynamic interfaces to other industrial products such as PLCs and HMIs. Other suppliers have taken a more holistic view of the needs of their customers, moving from a component supply situation to a solution provider. The ability to interface, efficiently and with appropriate dynamics, with other areas of a machine or factory automation system is of increasing importance in the design of industrial drives. As machine and factory control moves towards more

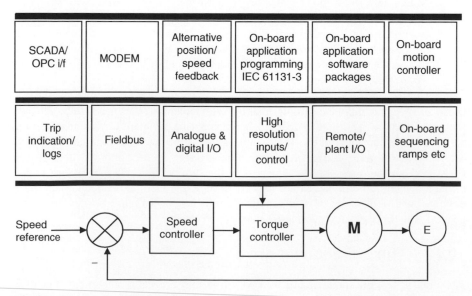

Fig. 10.31 Elements of a modern drive

distributed control structures, the drive will grow in importance as the hub a system controller. To fulfil this role, more flexibility and configurability in drives will result.

10.13   Standards

Key IEC and North American standards relating to electrical machines and drives are given in Table 10.4.

Table 10.4 National and international standards relating to motors and drives

IEC	Subject of standard	UL/NEMA
60034-1	Rotating electrical machines – Pt 1: Rating and performance	
60034-2	Rotating electrical machines – Pt 2: Methods for determining losses and efficiency of rotating electrical machinery from tests	
60034-5	Rotating electrical machines – Pt 5: Classification of degrees of protection provided by enclosures of electrical machines (IP code)	
60035-6	Rotating electrical machines – Pt 6: Methods of cooling (IC code)	
60035-7	Rotating electrical machines – Pt 7: Classification of types of construction, mounting arrangements and terminal box position (IM code)	
60034-8	Rotating electrical machines – Pt 8: Terminal markings and direction of rotation	
60034-9	Rotating electrical machines – Pt 9: Noise limits	
60034-11	Rotating electrical machines – Pt 11: Thermal protection	
60034-12	Rotating electrical machines – Pt 12: Starting performance of single-speed three-phase cage induction motors	
60034-14	Rotating electrical machines – Pt 14: Mechanical vibration of certain machines with shaft heights 56 mm and higher	
60034-15	Rotating electrical machines – Pt 15: Impulse voltage withstand levels of rotating ac machines with form-wound stator coils	
60034-17	Rotating electrical machines – Pt 17: Cage induction motors when fed from converters	
60034-18	Rotating electrical machines – Pt 18: Functional evaluation of insulation systems	
60034-19	Rotating electrical machines – Pt 19: Specific test methods for dc machines on conventional and rectifier-fed supplies	
60034-20-1	Rotating electrical machines – Pt 20-1: Control motors – stepping motors	
60034-25	Rotating electrical machines – Pt 25: Guide for the design and performance of cage induction motors specifically designed for converter supply	
60146-1	Semiconductor converters – General requirements and line commutated converters	
60146-2	Semiconductor converters – Pt 2: Self-commutating semiconductor converters including direct dc converters	
60146-6	Semiconductor converters – Pt 6: Application guide for the protection of Semiconductor converters against overcurrent by fuses	
61800-1	Adjustable speed electrical power drive systems – Pt 1: Rating specifications for low voltage adjustable speed dc power drive systems	

(contd)

Table 10.4 (*contd*)

IEC	Subject of standard	UL/NEMA
61800-2	Adjustable speed electrical power drive systems – Pt 2: Rating specifications for low voltage adjustable speed ac power drive systems	
61800-3	Adjustable speed electrical power drive systems – Pt 3: EMC requirements and specific test methods	
61800-4	Adjustable speed electrical power drive systems – Pt 4: Rating specifications for ac power drive systems above 1000 V ac and not exceeding 35 kV	
61800-5	Adjustable speed electrical power drive systems – Pt 5: Safety requirements	
61800-6	Adjustable speed electrical power drive systems – Pt 6: Guide for determination of types of load duty and corresponding current ratings	
61800-7	Adjustable speed electrical power drive systems – Pt 7: Generic interface and use of profiles for power drive systems	
	Standard for power conversion equipment	UL 508C
	Motors and generators	NEMA MG 1

References

10A. Drury, W.(ed), *The Control Techniques Drives and Controls Handbook,* The Institution of Electrical Engineers, 2001, ISBN 0 85296 793 4.
(A handbook detailing all important practical aspects of industrial variable speed drives and their successful application)

10B. Chalmers, B.J. Electric Motor Handbook, Butterworths, 1988, ISBN 0-408-00707-9.
(A practical reference book covering many aspects of characteristics, specification, design, selection, commissioning and maintenance)

10C. Vas, P. Sensorless Vector and Direct Torque Control, Oxford University Press, 1998, ISBN 0198564651.
(General background to the theory of vector control of motors)

Chapter 11

Power electronic circuits and devices

Professor A.J. Forsyth
The University of Manchester

11.1   Introduction

The use of solid-state techniques for the control and conversion of electrical power is now commonplace, with applications ranging from small dc power supplies for electronic devices, through actuator and motor drive systems, to large active filters and static power compensators for power transmission and distribution systems.

The aim here is to review the main classes of circuits and devices that are currently used, emphasizing the basic operating principles and characteristics. The opening sections cover the traditional diode-based ac–dc converters, starting with the three-pulse circuit and leading to the twelve-pulse configuration, and including the supply current characteristics. Thyristor phase control techniques are also briefly described. The following sections introduce the principal active devices that are in common use, namely the MOSFET and the IGBT (insulated gate bipolar transistor), then, starting with simple dc–dc converter switching cells, switching and inverter circuits are explained, leading to sinusoidal PWM methods. Finally, there is a short review of high-frequency power supplies.

11.2   Diode converters

11.2.1   Three-pulse rectifier

The three-pulse rectifier in Fig. 11.1 supplies a resistive load in series with a filter inductor, the inductor being large enough to ensure that the load current is continuous and ripple-free. The circuit is a half-wave rectifier, with each supply line, A, B and C, being connected through a single diode to the top of the load, the neutral wire forming the return path. The circuit waveforms follow directly from the assumption of a continuous and smooth load current:

- One of the three diodes must always be in conduction to provide a path for the load current.
- Only one diode may conduct at once, since otherwise two of the supply lines would have the same voltage, and this only occurs momentarily as the voltages cross each other.
- The diode that conducts at any instant is determined by the supply line with the largest positive voltage, since this voltage will forward bias the diode in that line and reverse bias the other two diodes.

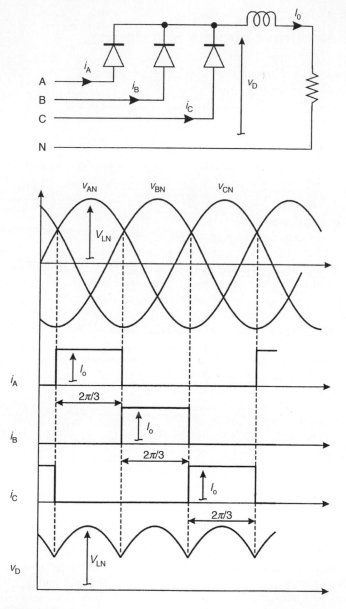

Fig. 11.1 Three-pulse rectifier

Therefore the diodes conduct in sequence for $2\pi/3$ radians or 120°, and the rectifier output voltage v_D is the maximum value of the three line-to-neutral voltages. Since the v_D waveform repeats every $2\pi/3$ radians, or three times each utility cycle, it is known as a three-pulse rectifier, and the output voltage ripple frequency is three times the supply frequency.

As the repetition period of v_D is $2\pi/3$ radians, the average value of the waveform may be calculated by integrating over this interval:

$$\text{average } v_D = \frac{1}{2\pi/3} \int_{\pi/6}^{5\pi/6} V_{LN} \sin\theta \, d\theta = \frac{3\sqrt{3}V_{LN}}{2\pi} = \frac{3V_{LL}}{2\pi} \tag{11.1}$$

where V_{LL} is the amplitude or peak value of the line-to-line supply voltage.

Since the circuit only draws a unidirectional current from each supply line, and also uses the neutral wire for the return current, it is rarely used in practice, but it is simple to understand, and is helpful in analyzing more complex circuits.

11.2.2 Six-pulse rectifier

Connecting two three-pulse circuits, one operating on the positive half cycle and the other on the negative half cycle, in a back-to-back configuration forms the six-pulse rectifier, as shown in Fig. 11.2. The upper three-pulse rectifier here is identical to that described in section 11.2.1, whilst the lower circuit is essentially formed from the first by reversing the diodes, enabling it to operate on the negative parts of the supply voltage waveform.

Since the average values of v_{D1} and v_{D2} are the same, the currents in the two load resistors will be the same, provided that these have equal values R. Under these conditions, the line currents will consist of 120° pulses of $+I_o$ as DA1, DB1 and DC1 conduct, and 120° pulses of $-I_o$ as DA2, DB2 and DC2 conduct. The negative pulses will flow between the positive pulses, making the line currents quasi-squarewaves.

Furthermore, since the two resistors carry equal currents, the neutral current will be zero, and the neutral connection may be removed without affecting the circuit operation. This leaves the standard three-wire, six-pulse rectifier of Fig. 11.3, into which the load is simply drawn as a constant current element, representing its highly inductive nature.

Figure 11.4 shows the detail of the two three-pulse output voltages v_{D1} and v_{D2} and the total output voltage $v_{DD} = v_{D1} + v_{D2}$. Since the ripple components in v_{D1} and v_{D2} are phase-shifted by $\pi/3$ radians, the resultant ripple in v_{DD} is reduced in amplitude and has a repetition interval of $\pi/3$ radians, giving six pulses per period of the input waveform. The average of v_{DD} is simply twice the average value of the three-pulse waveforms and from eqn 11.1:

$$\text{average } V_{DD} = 2\frac{3V_{LL}}{2\pi} = \frac{3V_{LL}}{\pi} \tag{11.2}$$

The six-pulse rectifier is widely used for the conversion of ac to dc, for example as the input stage of a variable speed induction motor drive system, described in Chapter 10.

11.2.3 Characteristics of the six-pulse input current

Fourier analysis of the quasi-squarewave currents drawn by the six-pulse rectifier shows that the harmonic amplitudes of the currents are given by eqn 11.3:

$$\text{harmonic amplitudes} = \frac{4I_o}{n\pi} \sin\left[\frac{n\pi}{3}\right] \text{ for odd values of } n \tag{11.3}$$

where n is the harmonic number. The fundamental component of each line current is in phase with the respective line-to-neutral voltage, as is evident from Fig. 11.2.

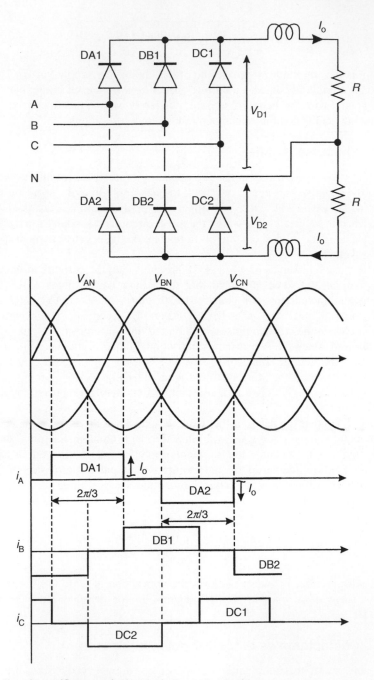

Fig. 11.2 Six-pulse rectifier as two back-to-back three-pulse circuits

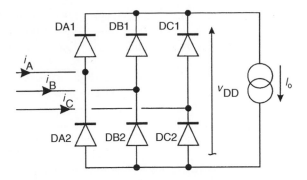

Fig. 11.3 Six-pulse rectifier

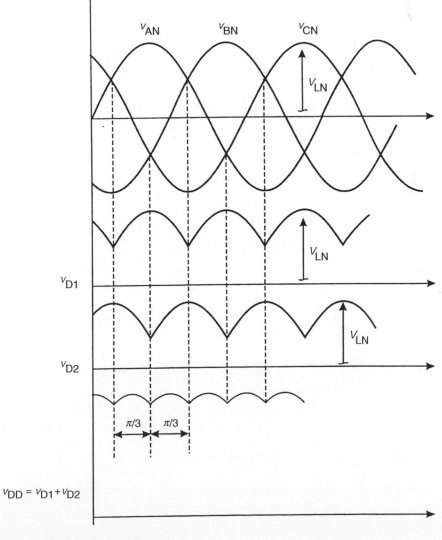

Fig. 11.4 Output voltage of six-pulse rectifier

The following conclusions may be drawn from eqn 11.3:

- the fundamental current has an amplitude of $2\sqrt{3}\,I_o/\pi$
- there are no even harmonics
- the sine function will be zero for $n = 3, 9, 15, 21, \ldots$, therefore, the harmonics that are odd multiples of three, known as triplens, will also be zero

Figure 11.5 shows the harmonic amplitudes of the quasi-squarewave current, normalized to the fundamental amplitude. The non-zero harmonics are of the order $6k \pm 1$, where $k = 1, 2, 3, \ldots$.

The rms value, I_{RMS} of the quasi-squarewave rectifier input current in Fig. 11.2 is:

$$I_{rms} = \sqrt{\frac{1}{2\pi}\int_0^{2\pi}[i(\theta)]^2\,d\theta}$$

$$= \sqrt{\frac{\int_0^{2\pi/3}I_o^2\,d\theta + \int_\pi^{5\pi/3}(-I_o)^2\,d\theta}{2\pi}} = \sqrt{2/3}\,I_o \qquad (11.4)$$

The active power transfer to loads that draw non-sinusoidal currents from a sinusoidal supply is solely due to the fundamental component of the current, and the six-pulse rectifier therefore has a power factor of less than unity even though the fundamental component of current is in phase with the voltage. The power factor of the six-pulse rectifier is given by eqns 11.5 and 11.6.

$$\text{Power Factor } \lambda = \frac{\text{Active Power}}{\text{Apparent Power}}$$

$$= \frac{\sqrt{3}V_{LL}I_{L(1)rms}\cos\phi_1}{\sqrt{3}V_{LL}I_{Lrms}} = \frac{I_{L(1)rms}}{I_{Lrms}}\cos\phi_1 \qquad (11.5)$$

where $I_{L(1)rms}$ is the rms fundamental component of line current, I_{Lrms} is the rms line current and ϕ_1 is the phase angle of the fundamental current with respect to the voltage. Since $\phi_1 = 0$, the power factor is:

$$\text{Power Factor } \lambda = \frac{I_{L(1)rms}}{I_{Lrms}} = \frac{(2\sqrt{3}I_o/\pi)/\sqrt{2}}{\sqrt{2/3}\,I_o} = \frac{3}{\pi} = 0.955 \qquad (11.6)$$

The practical operation of the rectifier differs slightly from this idealized representation because of the presence of source inductance, which slows down the edges of the quasi-squarewave input currents and gives rise to an output voltage regulation effect. Ripple current in the dc inductor also modifies the shape of the input currents.

11.2.4   Twelve-pulse rectifier with delta–star transformer

To enhance the characteristics of the six-pulse rectifier, reducing the input current harmonics and the output voltage ripple, multiple rectifiers may be combined with a phase-shifting device, a transformer or an autotransformer. As an example, Fig. 11.6

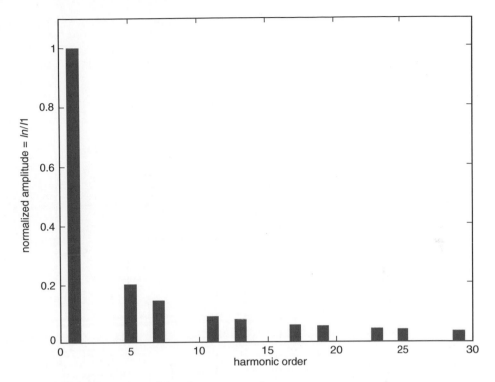

Fig. 11.5 Harmonic spectrum of six-pulse current waveform

shows one of the simplest ways of combining two six-pulse rectifiers to produce a twelve-pulse circuit.

The first rectifier in Fig. 11.6 is supplied through a delta–star transformer, whilst the second operates through a delta–delta transformer. The turns ratio of the delta–star transformer is $1:1/\sqrt{3}$, so the secondary line-to-line voltages have the same amplitude as the primary voltages, but are advanced in phase by 30°. The delta–delta transformer has a 1:1 turns ratio and its primary and secondary voltages and currents are identical, neglecting magnetizing current; it does not provide any phase shift. The two sets of transformer primaries are connected in parallel, whilst the rectifier outputs are connected in series across a common load. This results in the two six-pulse rectifiers delivering the same load current. The two transformers could be replaced by a single device having two sets of secondaries; one connected in star, the other in delta.

Figure 11.7 shows the current waveforms for the first supply line in the twelve-pulse rectifier. The diagram shows that in the two six-pulse rectifiers, each draw quasi-squarewave currents of value $\pm I_o$, i_{X1} and i_{X2} being the currents in the first input line to each rectifier. A phase shift of $\pi/6$ radians or 30° is shown between the two currents, i_{X1} is leading i_{X2} due to the 30° phase advance in the input voltage waveforms to the first rectifier, which is provided by the delta–star transformer.

The first transformer input current, i_{A1}, is formed by the combination of two reflected quasi-squarewave secondary currents, $(i_{X1} - i_{Z1})/\sqrt{3}$. The second transformer input current, i_{A2}, is the same as the secondary current i_{X2}, since this transformer is a simple delta–delta configuration. The resultant current drawn from the utility is $i_A = i_{A1} + i_{A2}$, and is seen to have a multi-level stepped shape somewhat closer to the ideal of a sinewave than either i_{A1} or i_{A2}.

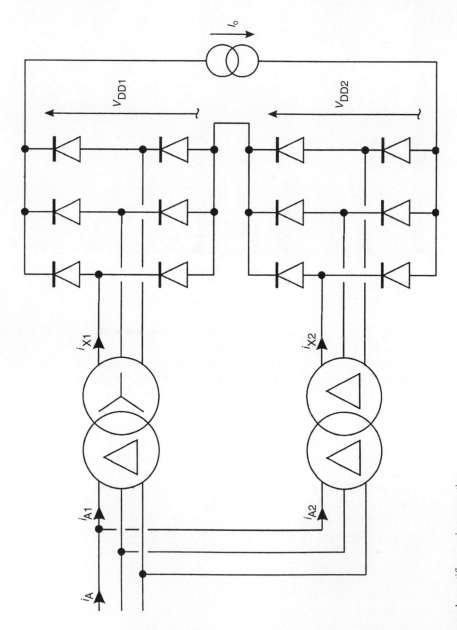

Fig. 11.6 Twelve-pulse rectifier – series connection

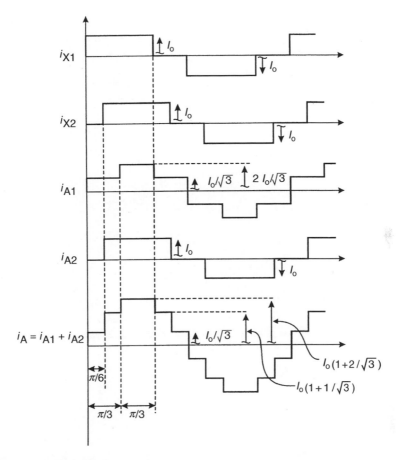

Fig. 11.7 Twelve-pulse rectifier waveforms

Due to the 30° phase shift between the two rectifiers, their six-pulse ripples will be out of phase, resulting in an overall output voltage which has a smaller ripple, the ripple frequency being twelve times the supply frequency. Analysis of the twelve-pulse current waveform reveals that the non-zero harmonics are now 1, 11, 13, 23, 25, ..., that is of order $12k \pm 1$ where $k = 1, 2, 3, ...$. The power factor of the twelve-pulse rectifier is increased to 0.9886 due to the reduced harmonic content of the input currents.

For lower output voltage and higher output current applications, the two rectifier outputs may be connected in parallel, however, an inter-phase reactor would normally then be used to prevent the circulation of currents between the two rectifiers. Twelve-pulse rectifiers would normally be used in higher power applications, or in environments having very stringent power quality specifications, such as an aircraft.

11.2.5 Controlled rectifiers

By replacing the diodes in the circuits described in the previous sections by thyristor devices, the output voltage of the rectifiers may be controlled, and a reversal of power flow is also possible if a source of energy is present in the dc circuit. The *thyristor* is a comparatively old power device that has limited control characteristics. A gate pulse must be applied to the control terminal to switch the device into its conducting state, however, once the device is in conduction it cannot be turned off and will only return to the off state when the circuit current naturally falls to zero.

The operating principle of *thyristor-controlled rectifiers* is illustrated using the three-pulse circuit in Fig. 11.1. The waveforms with thyristor control are shown in Fig. 11.8. As before, each device conducts the dc load current in sequence for one-third of the supply cycle, however, the conduction of each device is delayed by the firing delay angle α. This results in the output voltage v_D being modified. The average value of v_D is given by eqn 11.7:

$$\text{average } v_D = \frac{3V_{LL}}{2\pi}\cos\alpha \qquad (11.7)$$

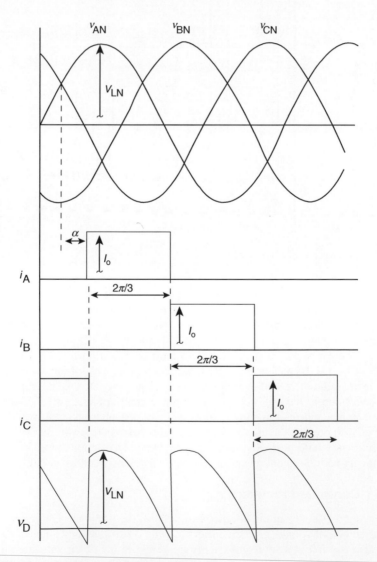

Fig. 11.8 Waveforms for phase-controlled three-pulse rectifier

The average output voltage falls to zero as α is increased to 90°. If a source of emf is present in the dc circuit, orientated to maintain the current flow, then α may be increased beyond 90° and the circuit enters the inversion mode of operation. The average value of V_D becomes negative, but the direction of current in the dc circuit is unchanged, and power flows from the energy source in the dc circuit into the ac system. To reverse the dc circuit current and obtain four-quadrant operation of the output, two rectifier circuits must be used in a back-to-back configuration.

The output voltage of the six- and twelve-pulse rectifiers may be controlled in a similar manner. However, a disadvantage of this mode of operation is that it results in a phase shift of the input currents, reducing the power factor. Thyristor-controlled rectifiers are widely used for the control of dc motor drives, as described in section 10.10.1.

11.3 Active devices

The *MOSFET* and *IGBT (insulated gate bipolar transistor)* are the predominant active power devices in use today, covering virtually all mainstream applications, the principal exception being the *GTO (gate turn off thyristor)*, which is found in specialized high power systems. The device symbols are shown in Fig. 11.9.

The MOSFET and IGBT both have an insulated gate control terminal and enhancement type control characteristics; they are normally off and the gate voltage must be increased beyond the threshold voltage (typically around 4 V) to bring the devices into conduction. Drive voltages of 12 or 15 V are normally used to ensure that the devices are fully switched on. Since the gate drive circuit must only charge and discharge the input capacitance of the device at the switching instants, the power consumption of the drive circuit is low, but pulse currents of several amperes are required to ensure rapid switching.

The MOSFET is a majority carrier device and is characterized by a constant on-state resistance, so the rms current must be used to estimate conduction losses. The on-state resistance has a positive temperature coefficient and typically rises by a factor of 1.5–2.0 for a 100°C temperature rise. The switching speed of the MOSFET is very high, current rise and fall times of tens of nanoseconds being achievable. The MOSFETs rated at a few hundred volts are available with current carrying capabilities of up to a hundred amperes, whilst devices with voltage ratings approaching 1000 V

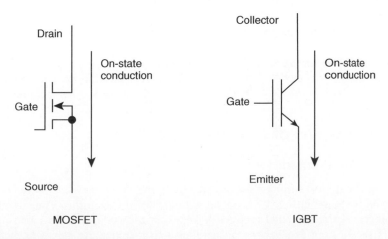

Fig. 11.9 MOSFET and IGBT symbols

tend to have current ratings of just a few amperes, the values of on-state resistance being correspondingly higher. Very few MOSFETs are available with voltage ratings in excess of 1000 V. The MOSFET is therefore used in lower power applications with switching frequencies of up to a few hundred kilohertz. The applications include high-frequency power supplies, dc–dc converters and small servo drives.

The IGBT is a minority carrier device, and through the use of conductivity modulation it is able to operate with much higher current densities than the MOSFET. The device is characterized by a constant on-state voltage, typically in the region of 2.0–3.0 V, requiring the use of the average forward current in estimates of on-state losses. The on-state voltage of the IGBT usually has a positive temperature coefficient, rising by approximately 20 per cent for a 100°C temperature rise. Due to the recombination time of the stored carriers within the device, the IGBT exhibits a tail current characteristic at turn off; the current rapidly falls to around 10 per cent of its on-state level then decays down to zero comparatively slowly, the overall switching time being a significant fraction of a microsecond. This effect limits the maximum operating frequency of the device to a few tens of kilohertz. The IGBTs are available with current ratings of up to several hundred amperes and with off-state voltages of up to several kilovolts. The devices are widely used in three-phase inverters and converters and have been used in small high-voltage dc power transmission systems.

11.3.1   Inductive switching waveforms

In the majority of power electronic circuits the operation of an active device results in the commutation of an inductive current to or from the device and a freewheel diode path. The simple equivalent circuit in Fig. 11.10 illustrates this basic switching process. The inductive current path is represented by a constant current element.

Assuming that the voltages and currents change linearly at the switching instants, then the turn-on and turn-off waveforms are as shown in Fig. 11.10. At the turn-on instant, the transistor current must rise beyond the full load current level by an amount I_{rr}, the *peak reverse recovery current* of the diode, before the diode can support reverse voltage and the transistor voltage can collapse to its on-state level. During the switching transient, the transistor experiences high instantaneous power dissipation and a significant energy loss. A similar effect is seen to occur at turn-off, where the transistor voltage must rise to the off-state level, forward biasing the diode, before the transistor current can fall to zero.

The average power loss in a transistor due to the inductive switching waveforms increases with operating frequency and this is a limit to the maximum operating frequency of a device. *Snubber circuits* have been used to shape the switching waveforms and limit the power losses in the devices but since the MOSFET and IGBT are much more robust than the bipolar power transistors that they replaced, these are now less common.

11.4   Principles of switching circuits

11.4.1   DC–DC converters

The simplest and most common dc–dc converter is the *step-down chopper* or *buck converter*, and is shown in Fig. 11.11 along with idealized waveforms. The circuit operates from a dc source, V_{in}, and supplies power at a lower voltage, V_o, to a load element, which is shown here as a resistor in parallel with a smoothing capacitor, but could

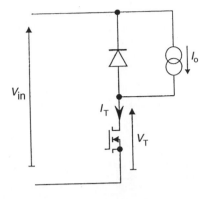

Equivalent circuit

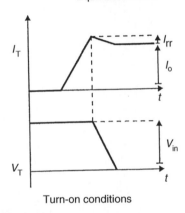

Turn-on conditions

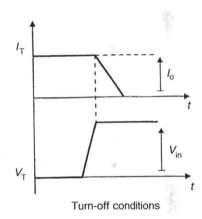

Turn-off conditions

Fig. 11.10 Inductive switching waveforms

instead be the armature of a dc motor, as described in section 10.10.1. The inductor acts as an energy storage buffer between the input and the output.

When the transistor is switched on the voltage V_D is equal to V_{in}, the freewheel diode is reverse biased, and a voltage of $V_{in} - V_o$ drives a linearly increasing current, I_L, through the inductor. When the transistor turns off, the inductor current diverts to the freewheel diode, voltage V_D is zero and there is a reverse voltage of V_o across the inductor resulting in a linear fall in current. The load voltage V_o is equal to the average of the V_D waveform since no average or dc voltage is dropped across an ideal inductor, and is given by eqn 11.8:

$$V_o = \text{average of } V_D = DV_{in} \qquad (11.8)$$

where D is the *duty ratio* or *on time-to-period ratio* of the transistor, and $0 \le D \le 1$.

The *step-up chopper* or *boost converter* (Fig. 11.12) is essentially a circuit dual of the buck converter. The transistor and diode again conduct the inductor current alternately. The inductor current rises linearly when the transistor is turned on, the inductor voltage being equal to V_{in}, and the current falls linearly when the transistor is turned off since the inductor voltage is then reversed with a value of $V_o - V_{in}$.

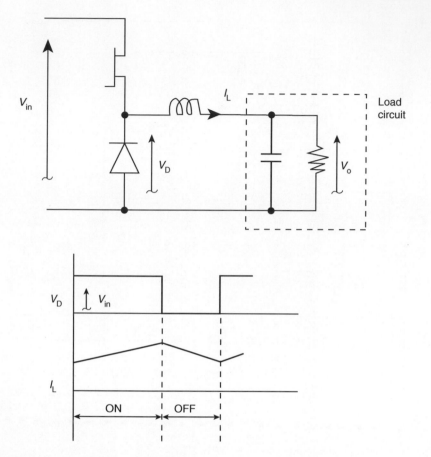

Fig. 11.11 Step-down chopper or buck converter

The average of the V_D waveform is now equal to the input voltage, and since the average of V_D is $V_o(1-D)$, the output voltage may be expressed in eqn 11.9:

$$V_o = \frac{V_{in}}{1-D} \qquad (11.9)$$

which confirms the voltage step-up operating characteristics of the circuit.

11.4.2 Two- and four-quadrant converters

By re-drawing the boost converter circuit with the input on the right, the buck and boost circuits may be combined to form the bidirectional converter shown in Fig. 11.3. To ensure proper circuit operation the transistors must be switched in anti-phase. With a dc source connected to the left hand terminals of the bidirectional converter and a load element on the right, the circuit operates as a buck converter and the inductor current flows to the right. If the dc source is connected to the right hand terminals and the load connected on the left, the inductor current will be reversed and the circuit will operate as a boost converter.

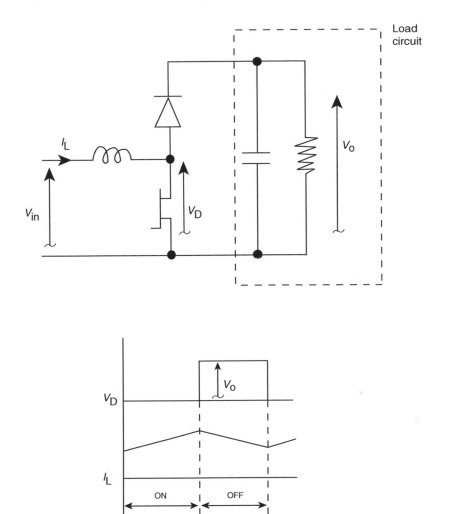

Fig. 11.12 Step-up chopper or boost converter

The circuit therefore provides two-quadrant operation with unidirectional voltage at both terminals, but bidirectional currents. The circuit could be used to control the charging and discharging of a battery, or to provide two-quadrant operation of a dc machine, with unidirectional voltage and therefore speed, but with bidirectional current and torque. The bidirectional converter in Fig. 11.13 is commonly referred to as an *inverter leg* since it forms the building block of dc–ac inverter circuits.

To achieve a four-quadrant output, two of the bidirectional circuits operating from a common dc source may be combined as shown in Fig. 11.14. The inductor is no longer shown but it is implied and it might typically be formed from the load impedance, such as the inductance of a motor.

The four-quadrant output is formed by the difference of the two bidirectional converter outputs. The resultant circuit is sometimes known as an *H-bridge* or a *single-phase inverter*, and it has a wide variety of applications, requiring different operating patterns for the transistors. The transistors in each leg of the circuit would normally be

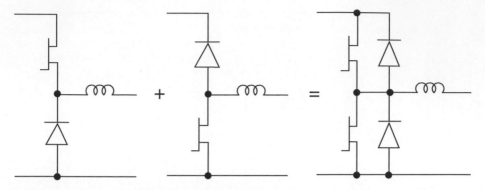

Fig. 11.13 Combination of the buck and boost converters

operated in anti-phase to ensure that the leg output voltages are always defined by the state of the transistors. Typical applications include:

- *four-quadrant drive for a dc machine*, enabling motoring and generating operation with both forward and reverse rotation, as described in section 10.10.1. Diagonally opposite transistors could be operated in synchronism with variable duty ratio. With a duty ratio of 0.5 the average machine voltage would be zero, and increasing or decreasing the duty ratio could produce positive or negative average voltages.
- *high frequency ac output* to drive a high-frequency transformer in a switched-mode power supply. Here the two legs could each be operated with 0.5 duty ratios, and a variable phase shift introduced between the legs to produce a controllable quasi-squarewave output voltage.
- *synthesizing an ac output from a dc source*, for example to supply ac equipment in the absence of a utility connection, or to interface a dc energy source such as a photovoltaic array with utility. In this case, diagonally opposite devices could be operated in synchronism with a 0.5 duty ratio, producing a squarewave ac output voltage, however, a filter circuit may be required to remove the unwanted harmonics from the squarewave. Alternatively, the devices could be operated at

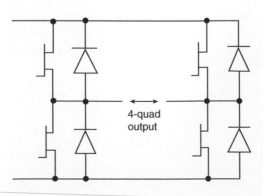

Fig. 11.14 Single-phase inverter

a much higher frequency than the utility using sinusoidal pulse width modulation techniques, discussed in section 11.4.4.

Since the single-phase inverter produces a four-quadrant output, it can not only be used to synthesize an ac output from a dc power source, but can also be used to control the flow of power from an ac source to a dc load. If sinusoidal PWM techniques are used, then it is known as a PWM rectifier.

11.4.3 Three-phase inverter

To synthesize a set of three-phase voltages from a dc source, three inverter legs are connected together as shown in Fig. 11.15. The circuit has two operating modes, either the transistors operate at the same frequency as the ac output waveforms, known as quasi-squarewave or six-step operation, or alternatively the devices operate at a much higher frequency than the ac output using a form of sinusoidal pulse width modulation.

Figure 11.16 illustrates the simpler *quasi-squarewave* operation of the three-phase inverter. The transistors in each leg operate in anti-phase with duty ratios of 0.5, the leg output voltages, measured with respect to a notional ground at the mid-point of the dc input, are therefore symmetrical squarewaves of $\pm V_{in}/2$. By displacing the switching actions in the three legs by $2\pi/3$ radians or $120°$ as shown, the resultant line-to-line output voltages form a set of mutually displaced quasi-squarewaves. The V_{AB} line-to-line voltage is shown as an example and consists of $120°$ intervals of positive and negative voltage separated by $60°$ intervals of zero voltage. Since the line-to-line output voltages are identical in form to the ideal input currents drawn by the six-pulse rectifier (Fig. 11.2) the waveforms will also have an identical frequency spectrum (Fig. 11.5), the non-zero harmonics being of order $6k \pm 1$. From eqn 11.3 the amplitude of the fundamental component of the line-to-line waveforms is $2\sqrt{3}\, V_{in}/\pi$.

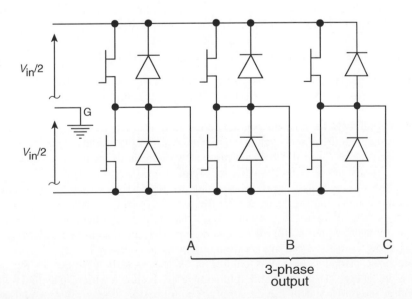

Fig. 11.15 Three-phase inverter

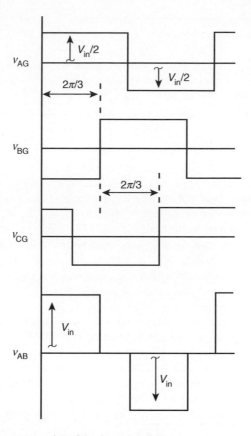

Fig. 11.16 Quasi-squarewave operation of the three-phase inverter

One of the largest application areas for the three-phase inverter is in variable speed drive systems for ac motors, as described in section 10.10.2.

11.4.4 Sinusoidal Pulse Width Modulation (PWM)

The most common operating mode for inverter circuits is with the transistors switching at a much higher frequency than the ac output waveform that is being synthesized, typically twenty times greater or more, which implies switching frequencies in the region of 1–20 kHz for a mains frequency inverter. The duty ratio of the transistors is varied throughout the mains cycle to shape the required sinusoidal output voltage. The main advantage of *Pulse Width Modulation (PWM)* methods for inverter control is the harmonic purity of the output waveform. Apart from the required fundamental component, the frequency spectrum of the output contains only switching frequency-related harmonics, and these are at high frequency, typically clustered around integer multiples of the switching frequency. These frequency components may be removed by a small high frequency filter if necessary, however, the load impedance itself often acts as a low pass filter, making additional filtering unnecessary.

The basic principle of PWM control is explained with reference to the single inverter leg in Fig. 11.17. The leg output voltage v_{AG} is shown with respect to a notional ground at the mid-point of the dc supply. The transistors operate in anti-phase with switching period T, the upper device having a duty ratio or on time-to-period ratio D,

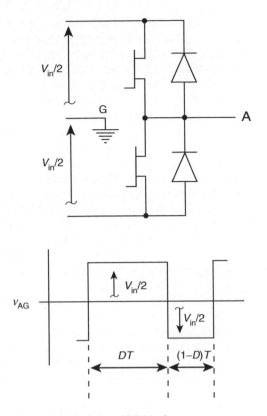

Fig. 11.17 Inverter leg and output voltage for one PWM cycle

and the lower device therefore having a duty ratio of $(1-D)$. The local average, $\bar{v}_{AG}$, of the v_{AG} waveform, taken across the arbitrary switching cycle in Fig. 11.17 is given by eqn 11.10:

$$\bar{v}_{AG} = \frac{V_{in}}{2}(2D-1) \tag{11.10}$$

By varying D from one cycle to the next, the local average of v_{AG} may be forced to follow a required waveform and to synthesize a sinusoidal output waveform, the duty ratio should be varied according to eqn 11.11,

$$D = 0.5(1 + M \sin \omega t) \tag{11.11}$$

where ω is the angular frequency of the waveform to be synthesized, and the parameter M is known as the *depth of modulation* or the *modulation index*. With a modulation index of zero the duty ratio will be constant, whilst a modulation index of unity will result in the duty ratio varying over the full range of $0 \leq D \leq 1$. Substituting eqn 11.11 into eqn 11.10 gives the local average output voltage in eqn 11.12,

$$\bar{v}_{AG} = \frac{V_{in}}{2} M \sin \omega t \tag{11.12}$$

which has the required sinusoidal form, the amplitude being $(V_{in}/2)M$. The modulation index therefore controls the amplitude of the synthesized output.

Equation 11.12 shows that with a modulation index of unity the amplitude of the output waveform is limited to $V_{in}/2$. A modulation index of greater than unity, known as *over-modulation*, allows the amplitude of the fundamental component of the output to be increased, although eqn 11.12 breaks down in this region. However, over-modulation has the disadvantage of introducing low-order harmonics in the output spectrum. Alternatively the technique of third harmonic injection may be used to increase the amplitude of the fundamental component of the output by around 15 per cent without degrading the frequency spectrum. This is achieved by adding a small amount of third harmonic to the modulating waveform. The addition of the third harmonic does not affect the line-to-line output waveforms of a three-phase inverter since the third harmonic is a common mode component in three-phase systems.

A variety of methods have been developed for the practical implementation of sinusoidal PWM and the determination of the required duty ratio variation. Naturally sampled PWM is intuitively the most simple, and involves comparing the sinusoidal modulating waveform with a triangular carrier signal at the switching frequency. The analogue implementation of this technique is fraught with offset and drift problems, and, being analogue in nature, the method does not readily lend itself to digital implementations. Instead, methods such as regularly sampled PWM have been developed, which use sample and hold techniques to simplify the calculations within a processor. Most recently, the method of space vector PWM has become common for three-phase inverters. This uses a single rotating vector to represent the required three-phase output waveforms, and the transistor switching patterns for each of the three legs are calculated simultaneously, resulting in a very efficient implementation, furthermore the third harmonic injection to increase the maximum fundamental output is inherently provided.

11.5   High-frequency power supplies

High-frequency power supplies or *switched-mode power supplies* are widely used to derive low-voltage dc supplies from the ac utility, and operate by first rectifying the ac input to provide high-voltage dc, which is then converted into high-frequency ac, typically at around 100 kHz, and then passed through a physically small high-frequency transformer, before finally being rectified and smoothed to form the output. Power levels range from a few watts to a few kilowatts.

Traditionally, at these low power levels the input rectifier consists of a single-phase bridge and large smoothing capacitor, however in recent years this arrangement has become unacceptable for all but the very lowest power levels due to the high harmonic content of the currents drawn by such a rectifier circuit. The input current to the rectifier consists of narrow pulses at the peaks of the ac voltage waveform. Instead of adding large low-frequency filters at the rectifier input, active high-power factor rectifiers are increasingly being used to provide high-quality, near-sinusoidal input currents by using high-frequency converter techniques. A common example, based on the boost converter, is illustrated in Fig. 11.18.

A full-wave rectified voltage waveform is presented as the input to the boost converter, which, through the use of an input current feedback loop, is controlled to offer a constant input resistance, that is, to draw an input current that follows the full-wave rectified voltage. As a result the input power factor is almost unity. However, a

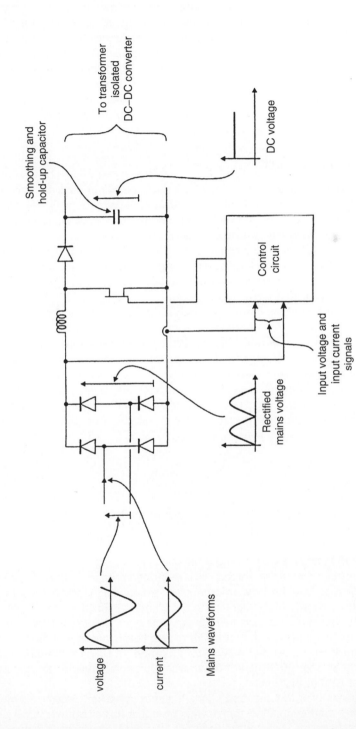

Fig. 11.18 Single-phase high-power factor rectifier

small high frequency filter may be required at the input to provide a path for the high frequency inductor ripple current. To ensure proper operation, the boost converter output voltage must be maintained above the peak of the ac input, which is achieved by sizing the output capacitor to accommodate the twice mains frequency pulsation in power flow that occurs through the boost converter, and also by modulating the amplitude of the boost converter input current to compensate for changes in the power drawn from the power supply output.

High-frequency coupled dc–dc converter circuits that are commonly used in switched mode power supplies are the *forward converter* and the *flyback converter*, and these are briefly reviewed in the following sections.

11.5.1   Forward converter

This is a transformer-isolated derivative of the buck converter, and the two circuits operate in a very similar manner. In the two-transistor version of the circuit, shown in Fig. 11.19, the devices operate in synchronism with a duty ratio D. The waveforms show a steady-state switching cycle, the transistors being turned on at the origin. The inductor current is assumed to be continuous with a small triangular ripple component. The waveforms show the primary voltage and current and the secondary current.

When the transistors turn on, the dc input voltage is applied to the primary winding. The voltage is reflected across to the secondary winding multiplied by the turns ratio and acts to forward bias the series diode and reverse bias the freewheel diode; the inductor voltage is therefore given by $NV_{in}-V_o$, causing a linear rise in the inductor current during the transistor on-time. The inductor current flows through the transformer secondary and is reflected through the turns ratio into the primary, transferring energy from the source. In addition the transformer magnetizing current rises linearly from zero.

When the transistors turn off, the inductor current diverts to the freewheel diode, the inductor voltage is equal to $-V_o$ and the inductor current falls linearly. By equating the positive and negative inductor volt-seconds, the expression for the voltage conversion ratio is given in eqn 11.13:

$$\frac{V_o}{V_{in}} = ND \tag{11.13}$$

The conversion ratio expression is similar to that of the buck converter, but is multiplied by the transformer turns ratio.

When the transistors turn off, the stored energy in the transformer core acts to maintain the flow of magnetizing current, the conduction path being through the two diodes in the primary circuit and the dc source. As a result, the primary voltage reverses, the magnetizing current falls linearly, returning the stored magnetic energy to the dc source. Since the reverse voltage applied to the primary is equal in magnitude to the forward voltage, the current decays to zero in a time equal to the transistors on time. Therefore, for the magnetizing current to fall to zero before the start of the next cycle, the transistor duty ratio must be limited to a maximum of 0.5. The limitation on the duty ratio results in poor utilization of the transformer, power flowing through it only for up to 50 per cent of the time.

For applications of a few hundred watts, a single transistor circuit is sometimes used with a third winding to allow transformer reset, but this circuit has the disadvantage that the transistor off-state voltage becomes $2V_{in}$. For higher powers, around 1 kW and above, a four-transistor full-bridge circuit (single-phase inverter) is used to drive

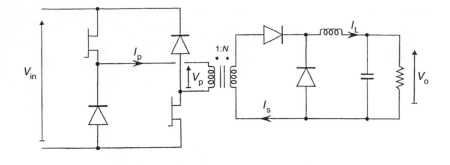

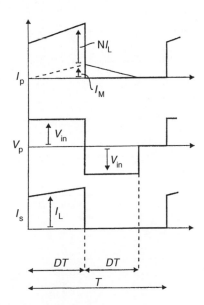

Fig. 11.19 Two-transistor forward converter

the transformer. This allows a true bidirectional voltage waveform to be impressed on the primary and enables power to flow through the transformer for almost 100 per cent of the cycle.

11.5.2 Flyback converter

This is often preferred at power levels below 100 W, where its simple circuit topology with only one magnetic component makes it the most economic solution. At higher power levels the circuit simplicity is out-weighed by the disadvantages of high ripple current in the output filter capacitor, high reverse voltages across the devices and poor utilization of the wound component.

The circuit and waveforms are shown in Fig. 11.20. Here, V_p is the primary winding voltage, I_p the primary winding current, I_s the secondary winding current, and I_c is the filter capacitor current. The currents in the magnetic component are assumed to be zero at the start of the cycle.

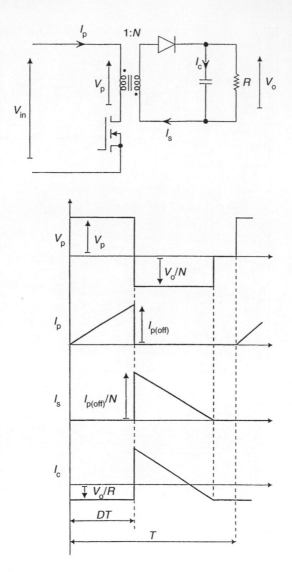

Fig. 11.20 Flyback converter

The turn-on instant is shown at the origin and when the transistor is conducting, the input voltage is impressed across the primary winding, causing a linear increase in current as energy is stored in the core and the flux increases. While the transistor is on, the secondary winding cannot conduct since the diode is reverse biased by a voltage $NV_{in} + V_o$.

The primary current is rapidly reduced to zero as the transistor turns off, and to maintain the core flux a current starts to flow in the secondary winding. The secondary current then falls linearly to zero as the stored magnetic energy is transferred to the load. It is important to note that the alternate conduction of the primary and secondary is quite different from the normal operation of a transformer, and results in the entire energy that passes through the circuit each cycle being stored momentarily in the core. The component is therefore known as a coupled inductor rather than a transformer, and due to the energy storage requirement, the coupled inductor tends to be significantly larger than a transformer designed for the same power and frequency.

The voltage conversion ratio for the converter may be obtained by equating the input and output energy per cycle as in eqn 11.14,

$$\text{Energy through put per cycle} = \frac{L_p I_{p(off)}^2}{2} = \frac{L_p}{2}\left[\frac{V_{in}DT}{L_p}\right]^2 = \frac{V_o^2 T}{R} \qquad (11.14)$$

where L_p is the self-inductance of the primary winding and D is the transistor duty ratio. Re-arranging gives eqn 11.15.

$$\frac{V_o}{V_{in}} = \frac{D}{\sqrt{2L_p/(RT)}} \qquad (11.15)$$

The conversion ratio is a linear function of D, but also depends upon load, operating frequency and the primary inductance. Surprisingly the turns ratio N does not appear in eqn 11.15; this is because the converter operates by transferring a fixed packet of energy to the output each cycle, which is independent of N.

Table 11.1 International and national standards relating to power electronics

IEC	EN/BS	Subject of standard	N. American
60065	EN 60065	Safety requirements for electronic apparatus for household use	
60335-2	EN 60335-2	Safety of household and similar apparatus	
60601-1	EN 60601-1	Medical electrical equipment – Pt 1: general requirements for safety	
60950-1	EN 60950-1	Safety of IT equipment	CSA 22.2 no 950-95 UL 1950
	EN 61000	General emission standard (section 14.3.1)	
	EN 61000-3-2	Emissions: mains harmonic current	
61010-1	EN 61010-1	Safety requirements for electrical equipment for measurement, control and laboratory use Pt 1: general requirements	
61204		LV power supply devices, dc output – performance and safety	
	EN 61558-2-7	Safety of power transformers and power supply units	
	EN 61800-3	EMC of adjustable speed drives	
		Industrial control equipment	UL 508
		Medical and dental equipment	UL 544
		Power units other than class 2	UL 1012
		LV video products without CRT displays	UL 1409
		Medical electrical equipment – Pt 1: general requirements for safety	UL 2601-1
		Electrical equipment for laboratory use – Pt 1: general requirements	UL 3101-1

11.6   Standards

There is a wide range of standards covering the performance of power electronic equipment, particularly with regard to electromagnetic compatibility and safety. Some standards are generic whilst others apply to specific classes of equipment. Examples of the most common standards are listed in Table 11.1.

References

11A.   Mohan, Undeland and Robbins: *Power Electronics: Converters, Applications and Design*, Third Edition, John Wiley, 2003.

Chapter 12

Batteries and fuel cells

R.I. Deakin
Professor J.T.S. Irvine
The University of St. Andrews

12.1   Introduction

The battery is at present the most practical and widely used means of storing electrical energy. The storage capacity of a battery is usually defined in ampere-hours (Ah); energy is strictly defined in kilowatt-hours (kWh) or joules, but since the voltage of a particular battery system is normally fixed and known, the Ah definition is more convenient. The terms *battery* and *cell* are often interchanged, although strictly a battery is a group of cells built together in a single unit.

Batteries can be classified into *primary* and *secondary* types.

A primary battery stores electrical energy in a chemical form which is introduced at the manufacturing stage. When it is discharged and this chemically stored energy is depleted, the battery is no longer serviceable. Applications for primary batteries are generally in the low-cost domestic environment, in portable equipment such as torches, calculators, radios and hearing aids.

A secondary or rechargeable battery absorbs electrical energy, stores this in a chemical form and then releases it when required. Once the battery has been discharged and the chemical energy depleted, it can be recharged with a further intake of electrical energy. Many cycles of charging and discharging can be repeated in a secondary battery. Applications cover a wide range. In the domestic environment secondary batteries are used in portable hand tools, laptop computers and portable telephones. Higher powered applications in industry include use in road and rail vehicles and in standby power applications. The capacity of secondary battery systems ranges from 100 mAh to 2000 Ah. Their useful life ranges from 2 to 20 years; this will depend, among other things, upon the number of charge–discharge cycles and the type and construction of battery used.

A *fuel cell* is an energy conversion device that is closely related to a battery. Both are electrochemical devices for the conversion of chemical to electrical energy. In a battery the chemical energy is stored internally, whereas in a fuel cell the chemical energy (fuel and oxidant) is supplied externally and can be continuously replenished. A *semi-fuel cell* is an intermediate between these, being a battery where the chemical energy can be replenished externally, but not continuously. The energy storage capacity of a fuel cell is a function of the fuel cell and its fuel (or possibly oxidant). A *reversible fuel cell* operates essentially as a secondary battery.

12.2   Primary cells

The majority of primary cells now in use fall into one of eight types:

- zinc carbon
- alkaline manganese
- mercury oxide
- silver oxide
- zinc air
- 3 V lithium primary
- lithium thionyl chloride
- 1.5 V lithium primary

Each of these is briefly reviewed in the following sections.

12.2.1   Zinc carbon

This battery is based upon Leclanché cells with zinc as negative and MnO_2 as the active positive electrode with carbon as current collector. In its latest form, this has been in use since 1950. The basic cell operates with a voltage range of 1.2 V to 1.6 V, and cells are connected in series to form batteries such as AAA, AA, C, D, PP3 and PP9. Typical construction of a PP9 battery is shown in Fig. 12.1.

Zinc carbon batteries are low in cost, they have an operating temperature range from −10°C to 50°C, and the shelf life at a temperature of 20°C is up to three years. They are ideally suited to applications with intermittent loads, such as radios, torches, toys, clocks and flashing warning lamps.

12.2.2   Alkaline manganese

The alkaline manganese battery differs from the zinc carbon in that the electrolyte is a potassium hydroxide solution as opposed to an ammonium chloride solution and that the configuration is inverted with zinc on the outside. A nickel-plated steel can is used as the positive current collector. The alkaline manganese battery became commercially available in its present form in the late 1950s. The basic cell operates with a voltage range of 1.35 V to 1.55 V, and cells are connected in series to form the standard battery sizes N, AAA, AA, C, D and PP3. Typical construction of a single-cell battery is shown in Fig. 12.2.

Cost is in the medium-low range. The operating temperature range is −15°C to 50°C, and the shelf life at 20°C is up to four years. Alkaline manganese batteries are suited to a current drain of 5 mA to 2 A; this makes them ideally suited to continuous duty applications which can include radios, torches, cd players, toys and cameras.

12.2.3   Mercury oxide

This type has been in commercial use since the 1930s. The single cell operates with a narrow voltage range of 1.2 V to 1.4 V; it is only available as a single button cell, the construction of which is illustrated in Fig. 12.3.

Cost is in the medium range. The operating temperature range is 0°C to 50°C, and the shelf life at 20°C is up to four years. Mercury oxide button cells are suitable for low current drains of 0.1 mA to 5 mA, with either continuous or intermittent loads.

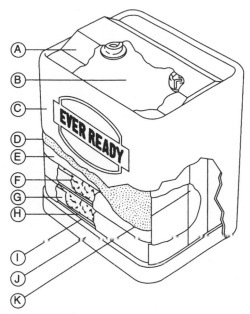

D   Wax coating. This seals any capillary passages between cells and the atmosphere, so preventing the loss of moisture.
E   Plastic cell container. This plastic band holds together all the components of a single cell.
F   Positive electrode. This is a flat cake containing a mixture of manganese dioxide and carbon black or graphite for conductivity. Ammonium chloride and zinc chloride are other necessary ingredients.
G   Paper tray. This acts as a separator between mix cake and the zinc electrode.
H   Carbon coated zinc electrode. Known as a duplex electrode, this is a zinc plate to which is adhered a thin layer of highly conductive carbon which is impervious to electrolyte.
I   Electrolyte impregnated paper. This contains electrolyte and is an additional separator between the mix cake and the zinc.
J   Bottom plate. This plastic plate closes the bottom of the battery.
K   Conducting strip. This makes contact with the negative zinc plate at the base of the stack and is connected to the negative socket at the other end.

A   Protector card. This protects the terminals and is torn away before use.
B   Top plate. This plastic plate carries the snap fastener connectors and closes the top of the battery.
C   Metal jacket. This is crimped on to the outside of the battery and carries the printed design. The jacket helps to resist bulging, breakage and leakage and holds all components firmly together.

Fig. 12.1 Zinc carbon PP9 battery construction (courtesy of Energiser)

Their small size and narrow voltage range are ideally suited for use in hearing aids, cameras and electric instruments.

12.2.4   Silver oxide

This has been in commercial use since the 1970s. The single cell operates with a very narrow range from 1.50 V to 1.55 V. Construction of a silver oxide button cell is shown in Fig. 12.4.

Cost is relatively high. They have an operating temperature range from 0°C to 50°C and a shelf life of up to three years. In button cell form, they are suitable for a current drain of 0.1 mA to 5 mA, either continuous or intermittent. Their small size and very narrow voltage range makes them ideal for use in watches, calculators and electrical instruments.

12.2.5   Zinc air

This battery has been in commercial use since the mid-1970s. A single cell operates with a voltage range of 1.2 V to 1.4 V. The construction of a zinc air button cell is shown in Fig. 12.5. Because one of the reactants is air, the battery has a sealing tab

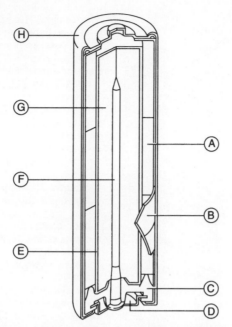

A Cathode pellet. This is of synthetic manganese dioxide, with graphite for conductivity, pressed into cylindrical pellets. The pellets make an interference fit with the can, ensuring good electrical contact. The graphite is the cathode current collector.
B Steel can. This acts as the cell container. The performed stud at the closed end of the can functions as the cell positive terminal.
C Closure. The hard plastic cell closure forms the seal at the open end of the can and carries the bottom cover/current collector assembly.
D Bottom cover. This metal plate acts as the negative terminal of the cell. It is welded to the current collector to ensure good electrical contact.
E Separator. This is of non-woven synthetic material.
F Anode current collector. This is a metal 'nail'.
G Zinc paste anode. The anode is of amalgamated high purity zinc powder made into a paste with the electrolyte.
H Shrink label. This covers the outside of the cell and carries printed design, and provides insulation of the side of cell.

Fig. 12.2 Construction of an alkaline manganese battery (courtesy of Energiser)

which must be removed before it is put into service. Following removal of the tab, the cell voltage will rise to 1.4 V.

Cost is in the medium range. Operating temperature is from –10°C to 50°C and the shelf life is almost unlimited while the seal is unbroken. They are produced as a button cell, and are suitable for current drain from 1 mA to 10 mA, with either intermittent or continuous loads. The service life is relatively short due to high self-discharge characteristics, but end of life can be accurately estimated because of the flat discharge current. The general performance and the life predictability make zinc air batteries ideal for use in hearing aids.

12.2.6   3 V lithium primary batteries

These typically have a lithium metal negative electrode with manganese dioxide or fluorinated graphite as positive electrode using a lithium salt dissolved in a non-aqueous

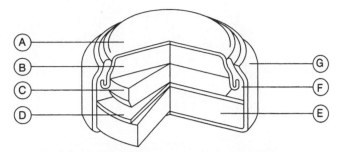

A   Cell cap. The plated steel cap functions as the negative terminal of the cell. The inside of the cap is laminated with copper.
B   Zinc anode. This is of high purity amalgamated zinc powder.
C   Absorbent pad. This is of a non-woven material and holds the alkaline electrolyte.
D   Separator. This is a synthetic ion-permeable material.
E   Mercuric oxide cathode. The cathode is a pellet of mercuric oxide plus graphite for conductivity and sometimes manganese dioxide. It is compressed into the can.
F   Sealing grommet. This plastic grommet both seals the cell and insulates the positive and negative terminals.
G   Cell can. The can is of nickel plated steel and functions as the positive terminal of the cell.

Fig. 12.3 Mercury oxide button cell (courtesy of Energiser)

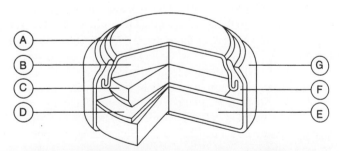

A   Cell cap. The plated steel cap functions as the negative terminal of the cell. The inside of the cap is laminated with copper.
B   Zinc anode. This is a high purity amalgamated zinc powder.
C   Absorbent pad. This pad is of a non-woven material and holds the alkaline electrolyte.
D   Separator. This is a synthetic ion-permeable membrane.
E   Cathode. This is a pellet of silver oxide plus graphite for conductivity. It is compressed into the can.
F   Sealing grommet. This plastic grommet both seals the cell and insulates the positive and negative terminals.
G   Cell can. The nickel plated steel can acts as a cell container and as the positive terminal of the cell.

Fig. 12.4 Silver oxide button cell (courtesy of Energiser)

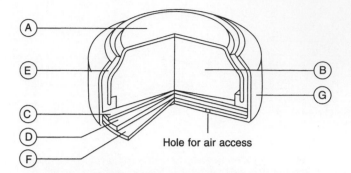

Hole for air access

A Cell cap. The plated steel cap functions as the negative terminal
   of the cell.
B Zinc anode. This is a high purity amalgamated zinc powder,
   which also retains the alkaline electrolyte.
C Separator. A synthetic ion-permeable membrane.
D Cathode. This is a carbon/catalyst mixture with a wetproofing
   agent on a mesh support, and with an outer layer of gas-
   permeable hydrophobic PTFE.
E Sealing grommet. This plastic grommet seals and insulates the
   positive and negative terminals, and seals the cathode to the base.
F Diffusion membrane. This permeable layer distributes air from
   the access holes uniformly across the cathode surface.
G Can. This plated steel can forms a support for the cathode, acts
   as a cell container and the positive terminal of the cell. The
   holes in the can permit air access to the catalyst cathode.

Fig. 12.5 Zinc air cell (courtesy of Energiser)

polar solvent such as ethylene carbonate. This type has been available since 1975, but it has only become commercially available in the past few years. The single cell operates with a relatively high voltage, in the range 2.8 V to 3.0 V. A typical construction is shown in Fig. 12.6.

Cost is relatively high. The operating temperature covers a wide range, from −30°C to 50°C, and the shelf life is up to ten years. The batteries are suitable for continuous or intermittent loads; in the form of a button cell they are suitable for 0.1 mA to 10 mA current drain, or as a cylindrical cell they can deliver from 0.1 mA to 1 A. They have a high discharge efficiency down to −20°C, and are suitable for applications in instruments, watches and cameras.

12.2.7 Lithium thionyl chloride

This type utilizes a liquid positive electrode. The single cell has a high voltage in the range 3.2 V to 3.5 V, with an open-circuit voltage of 3.67 V. A cross section of an example is shown in Fig. 12.7.

Cost is relatively high. The operating temperature range is one of the highest available, from −55°C to 85°C, and the shelf life is up to 10 years. These batteries are suitable for intermittent or continuous duties, with a current drain from 15 mA up to 1.5 A. Their high energy density (>700 Whdm^{-3}), wide temperature range and high discharge efficiency make these batteries particularly well suited to memory back-up applications in programmable logic controllers, personal computers and alarm systems.

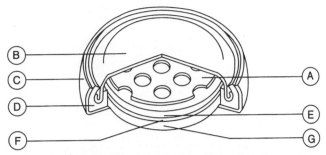

A Current collector. This is a sheet of perforated stainless steel.
B Stainless steel top cap. This functions as the negative
   terminal of the cell.
C Stainless steel cell can. This functions as the positive
   terminal of the cell.
D Polypropylene closure. This material is highly impermeable to
   water vapour and prevents moisture entering the cell after it
   has been sealed.
E Lithium negative electrode. This is punched from sheet
   lithium.
F Separator containing electrolyte. The separator is of non-
   woven polypropylene cloth and contains electrolyte, a
   solution of lithium perchlorate in a mixture of propylene
   carbonate and dimethoxyethane.
G Manganese dioxide positive electrode. The cathode is made
   from a highly active electrolytic oxide.

Fig. 12.6 Lithium manganese dioxide cell (courtesy of Energiser)

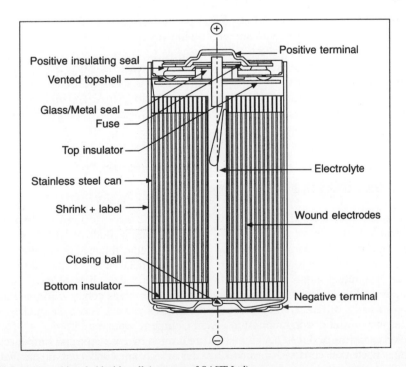

Fig. 12.7 Lithium thionyl chloride cell (courtesy of SAFT Ltd)

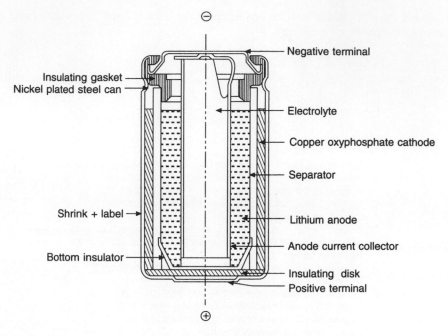

Fig. 12.8 Lithium copper oxyphosphate cell (courtesy of SAFT Ltd)

12.2.8   1.5 V lithium primary batteries

These utilize iron sulphide or copper oxide positive electrodes. The single cell has a voltage range 2.1 V to 2.5 V, with an open-circuit voltage of 2.7 V. A sectional view of a typical unit is shown in Fig. 12.8.

The batteries are expensive because of their high energy density. They operate over a temperature range −40°C to 60°C and they have a shelf life of up to 10 years. They are suitable for continuous loads from 37 µA up to 25 mA. The high energy density and wide temperature range make these batteries specially suitable for applications in computers.

12.3   Secondary cells based upon aqueous electrolytes

Until the last few years of the twentieth century, the main forms of secondary or rechargeable battery were lead acid and nickel cadmium, both of which have been in use for about 100 years. Whilst lead acid has largely retained its market, nickel cadmium has been largely superseded by nickel metal hydride and lithium ion batteries. The latter, along with lithium batteries dominates a host of new consumer applications. Lithium cells will be treated separately as their construction and design are significantly different from the aqueous-based systems; however, nickel metal hydrides are similar in construction to nickel cadmium batteries.

Nickel cadmium and lead acid batteries are formed by connecting a number of cells in series. Each cell consists of vertical plates which are connected in parallel and are divided into a positive group and a negative group. Adjacent positive and negative plates are insulated from each other by separators or rod insulators which have to provide the following functions:

- obstruct the transfer of ions between the plates as little as possible
- keep in place the active material
- enable the escape of charging gases from the electrolyte

All the plates and separators together form a *plategroup* or *element*.

The insulator in nickel cadmium batteries is formed from vertical plastic rods, which may be separate or as grids which are inserted between the plates. Corrugated or perforated PVC is also used as a separator. The active material is kept in place by steel pockets.

In lead acid batteries, the separators are normally microporous sheets of plastic, which are usually combined with corrugated and perforated spacers; porous rubber separators are also used. In some designs, the separator completely surrounds the positive or negative plate. The selection of separator is very important, and it depends upon the plate design and the use for which the battery is intended. In batteries intended for short discharges, such as the starting of the diesel engine, thin separators and spacers are used, whereas batteries designed for long discharges, as in standby power or emergency lighting, have thick separators and spacers. Batteries that require extra strength to withstand mechanical stresses often have a glass fibre separator against the positive plates in order to minimize the shedding of active material from these plates.

The outermost plates are of the same polarity on both sides. In lead acid batteries the outside plates are normally negative; in nickel cadmium batteries they are normally positive.

The plategroups are mounted in containers which are filled with electrolyte. The electrolyte is dilute sulphuric acid in lead acid batteries, and potassium hydroxide in nickel cadmium batteries. In a fully charged lead acid battery the electrolyte density is between 1.20 and 1.29. In nickel cadmium batteries it is between 1.18 and 1.30 irrespective of the state of charge.

The containers may house a single cell or they may be of a monoblock type, housing several cells. The top (or lid) may be glued or welded to the container, or it may be formed integrally with the container. In the lid is a hole for the vent, which is normally flame arresting, and the cells are topped up with electrolyte through the vent hole.

Below the plate there is an empty space which is known as the *sludge space*. This is provided so that shredded active material and corrosion product from the supporting structure can be deposited without creating any short circuits between the plates. The sludge space in nickel cadmium batteries is smaller than that in lead acid units because there is no corrosion of the supporting steel structure and no shredding of active material; it is usually created by hanging the plategroup from the lid. In lead acid batteries the sludge space may be created by hanging the plates from the lid or container walls (this is common in stationary batteries), by standing the plates on supports from the bottom (this is common for locomotive starting batteries) or by hanging the positive plates with the negative plates standing on supports (this is also used for stationary batteries).

From the plates, the current is carried through plate lugs, a connection strap or bridge and a polebolt. The bolts pass through the lid and are sealed against the lid with gaskets. Above the lid, the cells are connected to form the battery. Monoblocks may have connectors arranged directly through the side of the partition cell wall, in which case the polebolts do not pass through the lid, or the connectors may be located between the cell lid and an outer lid. The monoblocks made in this way are completely insulated and easy to keep clean.

The main forms of aqueous electrolyte secondary cell now commercially available are:

- nickel cadmium – sealed
- nickel cadmium – vented
- nickel metal hydride
- lead acid – pasted plate
- lead acid – tubular
- lead acid – Planté
- lead acid – valve-regulated sealed (VRSLA)

Each of these is described separately in sections 12.3.1 to 12.3.7, and a summary of the main features of the four lead acid types and nickel cadmium is given in Table 12.1.

12.3.1   Sealed nickel cadmium

This type utilizes a cadmium positive and a nickel oxyhydroxide negative with alkaline electrolyte and has been commercially available since the early 1950s. The single cell operates across a voltage range 1.0 V to 1.25 V and is ideally suited to a heavy continuous current drain of up to eight times the nominal ampere-hour capacity of the battery. The low internal resistance of the cell makes it ideal for heavy current discharge applications in motor-driven appliances such as portable drills, vacuum cleaners, toys and emergency systems. Batteries of this type are capable of being recharged hundreds of times using a simple constant current charging method; however the electrochemistry has a memory effect and capacity is progressively lost if the batteries are completely discharged and then fully recharged.

Table 12.1 Comparison of features and performance of the main types of secondary battery

Battery type	Lead acid					Nickel cadmium
	Pasted plate		Tubular	Planté	VRSLA	
	Lead antimony	Lead calcium				
Cycle duty	Good	Poor	Very good	Suited to shallow discharge duty	Suited to shallow discharge duty	Suited to shallow discharge duty
Maximum temperature	45°C	35°C	45°C	45°C	40°C	45°C
Gas generation and maintenance	Low	Very low	Low	Very low	Negligible No topping up needed	Low to moderate
Volume/Ah indicator	50	50	45	100	35	55
Relative cost	70	70	90	100	80	200–300

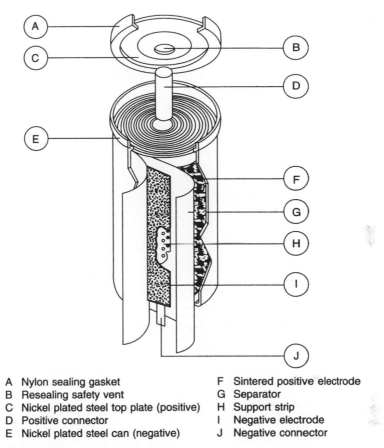

A Nylon sealing gasket
B Resealing safety vent
C Nickel plated steel top plate (positive)
D Positive connector
E Nickel plated steel can (negative)

F Sintered positive electrode
G Separator
H Support strip
I Negative electrode
J Negative connector

Fig. 12.9 Sealed nickel cadmium cell (courtesy of Energiser)

The cells tolerate a wide temperature range from −30°C to 50°C, and they have a shelf life of up to eight years at 20°C. Typical construction of a sealed nickel cadmium cell is shown in Fig. 12.9. The cell is available commercially in a variety of standard sizes including AAA, AA, C, D, PP3 and PP9.

12.3.2 Vented nickel cadmium

Using the same electrochemistry as the sealed variant, vented cells use 'pocket' plates, which are made from finely perforated nickel-plated steel strip filled with active material. The pocket plates are crimped together to produce a homogenous plate. Translucent plastic or stainless steel containers are used. An example is shown in Fig. 12.10.

The batteries are robust and they have the benefits of all-round reliability; they are resistant to shock and extremes of temperature and electrical loading. They can be left discharged without damage and require very little maintenance. Their cycling ability is excellent and can offer a service life of up to 25 years.

Vented nickel cadmium batteries are used typically in long-life applications where reliability through temperature extremes is required and where physical or electrical abuse is likely. These applications include railway rolling stock, off-shore use, power

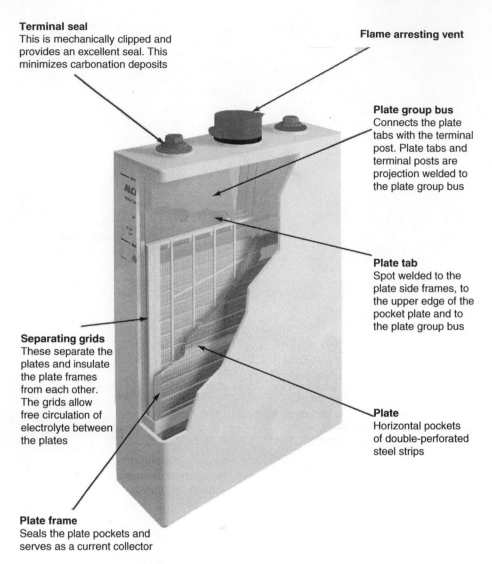

Terminal seal
This is mechanically clipped and
provides an excellent seal. This
minimizes carbonation deposits

Flame arresting vent

Plate group bus
Connects the plate
tabs with the terminal
post. Plate tabs and
terminal posts are
projection welded to
the plate group bus

Plate tab
Spot welded to the
plate side frames, to
the upper edge of the
pocket plate and to
the plate group bus

Separating grids
These separate the
plates and insulate
the plate frames
from each other.
The grids allow
free circulation of
electrolyte between
the plates

Plate
Horizontal pockets
of double-perforated
steel strips

Plate frame
Seals the plate pockets and
serves as a current collector

Fig. 12.10 Vented nickel cadmium battery (courtesy of ALCAD)

system switch tripping and closing, telecommunications, uninterruptible power
supplies, security and emergency systems and engine starting.

12.3.3   Nickel metal hydride

The sealed nickel metal hydride battery has become commercially available in the past
few years and market share has rapidly risen to make this one of the major technologies. A single cell operates with a voltage range of 1.1 V to 1.45 V, and a nominal
voltage of 1.2 V. The construction of a metal hydride cell is illustrated in Fig. 12.11.

These batteries are relatively expensive, but they have no toxicity problems on
recycling. Their energy density is in the region of 60–70 Wh/kg, which is higher than
the sealed nickel cadmium battery. The operating temperature range is –20°C to 60°C

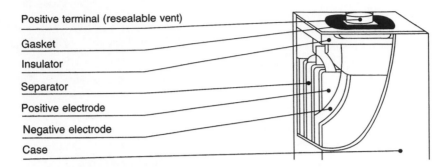

Positive terminal (resealable vent)

Gasket

Insulator

Separator

Positive electrode

Negative electrode

Case

Fig. 12.11 Nickel metal hydride battery (courtesy of SAFT Ltd)

and the shelf life is up to five years. Nickel metal hydride batteries are suitable for applications with milliampere current drains from 0.2 to 3 times the nominal ampere-hour capacity of the battery. The high energy density and specific power (perhaps 250 W kg $^{-1}$) makes this type of battery ideally suited to applications for memory back-up and portable communications. Nickel metal hydride batteries will endure recharging up to 500 times, although the method of charging is slightly more complicated than for sealed nickel cadmium systems.

12.3.4 Lead acid – pasted plate

The positive plates of pasted plate cells are made by impressing an oxide paste into a current-collecting lead alloy grid, in which the paste then forms the active material of the plate. The paste is held in position by a long interlocking grid section. In addition to the long-life microporous plastic separator, a glass-fibre mat is used. This becomes embedded in the face of the positive plate holding in place the active material and so prolongs the life of the battery.

Cells can be manufactured with lead–antimony–selenium alloy plates. These require very little maintenance, offer good cycling performance and are tolerant to elevated temperatures. Alternatively, lead–calcium–tin alloy plates may be used; these offer even lower maintenance levels.

Pasted-plate lead acid batteries are the lowest in purchase costs; their service life is in the range 2 to 20 years and the maintenance requirements are low, with extended intervals between watering. They are compact and provide high power density with excellent power output for short durations. Typical applications are in medium-to-long duration uses where initial capital cost is a primary consideration. These include telecommunications, UPS, power generation transmission, switch tripping and closing, emergency lighting and engine starting.

12.3.5 Lead acid – tubular

In a tubular cell, an example of which is shown in Fig. 12.12, the positive plate is constructed from a series of vertical lead alloy spines or fingers which resemble a comb. The active material is lead oxide; this is packed around each spine and is retained by tubes of woven glass fibre which are protected by an outer sleeve of woven polyester or perforated PVC. This plate design enables the cell to withstand the frequent charge–discharge cycles which cause rapid deterioration in other types of lead acid cell.

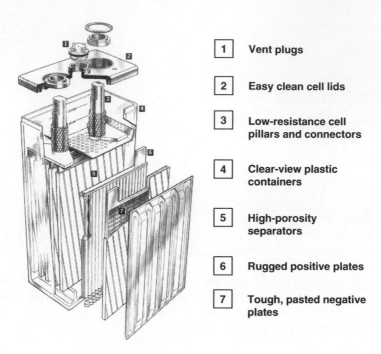

1	Vent plugs
2	Easy clean cell lids
3	Low-resistance cell pillars and connectors
4	Clear-view plastic containers
5	High-porosity separators
6	Rugged positive plates
7	Tough, pasted negative plates

Fig. 12.12 Tubular cell lead acid battery (courtesy of Invensys)

The advantages of the tubular cell are excellent deep cycling characteristics, service life of up to 15 years and a compact layout which gives a high power per unit volume. The low antimony types also offer extended watering intervals.

Typical applications are where the power supply is unreliable, and where discharges are likely to be both frequent and deep. These include telecommunications, UPS, emergency lighting and solar energy.

12.3.6 Lead acid – Planté

The distinguishing feature of the Planté cell is the single pure lead casting of the Planté positive plate. The active material is formed from the plate surface, eliminating the need for mechanical bonding of a separately applied active material to a current collector. This makes the Planté cell the most reliable of all lead acid types. An example of the construction of Planté batteries is shown in Fig. 12.13.

Ultra High Performance Planté cells have a thinner plate. This gives the highest standards of reliability for UPS and other high-rate applications. The advantages of the Planté cell are extremely long life (which can be in excess of 20 years for High Performance cells and 15 years for Ultra High Performance cells) and the highest levels of reliability and integrity. Constant capacity is available throughout the service life, an assessment of condition and residual life can readily be made by visual inspection. Usage at high temperatures can be tolerated without significantly compromising the life expectancy, and maintenance requirements are very low.

Typical applications are long-life float duties in which the ultimate in reliability is required. These include critical power station systems, telecommunications, UPS, switch tripping and closing, emergency lighting and engine starting.

1. Vent plugs
2. Cell lids
3. Cell pillars and connectors
4. Bar guard
5. Negative plates
6. Separators
7. Planté positive plates
8. Plastic containers

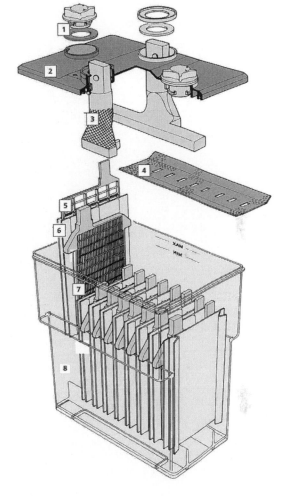

Fig. 12.13 Lead acid Planté cells (courtesy of Invensys)

12.3.7   Lead acid – valve regulated sealed (VRSLA)

In sealed designs, lead calcium or pure lead is used in the grids. The separator is a vital part of the design because of its influence on gas recombination, and the amount of electrolyte that it retains; it often consists of a highly porous sheet of microfibre. Examples of sealed cells are shown in Fig. 12.14.

Besides these sealed units, there is a large variety of so-called maintenance-free batteries which rely more upon carefully controlled charging than upon the gas recombination mechanism within the cells. Non-antimonial grids and pasted plates are used, and in most units the electrolyte is immobilized or gelled. The batteries are provided with vents, which open at a relatively low overpressure.

The benefits are a long life, no topping up, no acid fumes and no requirement for forced ventilation. There is no need for a separate battery room and the units can be located within the enclosure of an electronic system. In addition, these batteries are lighter and smaller than the traditional vented cells.

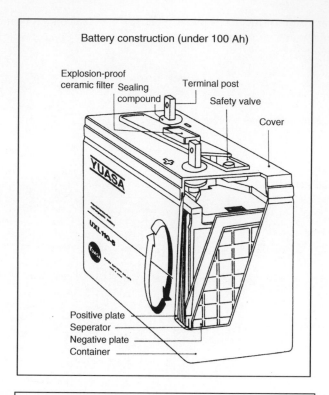

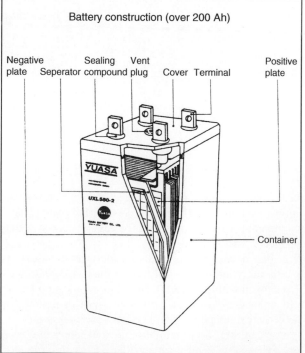

Fig. 12.14 Examples of sealed lead acid batteries (courtesy of Yuasa)

Typical applications include main exchanges for telecommunications, PABX systems, cellular radio, microwave links, UPS, switch tripping and closing, emergency lighting and engine starting.

12.4   Secondary lithium ion cells

Rechargeable lithium batteries proved much more difficult to develop than their primary analogues; however, the resultant technology is certainly the most important advance in energy storage in the last fifty years. Lithium, which is one of the lightest elements is also highly electronegative, offering both high potential and capacity. This high electronegativity also necessitates the use of non-aqueous solvents in the electrolyte. To date, lithium metal has not proved suitable for use in rechargeable batteries due to dendrite formation, although work on polymer electrolytes seeks to resolve this problem. Thus, currently available rechargeable lithium batteries tend to use lithium carbon as the negative and metal oxides as the positive electrode. Rechargeable lithium batteries are available in a wide variety of sizes as both lithium ion and lithium polymer (gel type) with most sizes up to about 100Ah currently being available.

12.4.1   Lithium ion batteries

These batteries consist of three active elements, a carbon negative electrode, into which lithium inserts reversibly, a non-aqueous electrolyte immobilized in a porous separator and a positive metal oxide electrode, into also which lithium inserts reversibly, (see Fig. 12.15). The term *rocking chair battery* has often been used to emphasize the interchange of inserted lithium between the electrodes on cycling. Three forms of

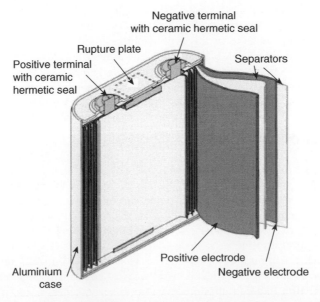

Fig. 12.15 Structure of a lithium-ion battery used for satellite and aerospace applications – the LSE and LVE product series (courtesy GS Yuasa Ltd.)

metal oxide dominate commercial batteries, these being layered lithium cobalt oxides, lithium nickel oxides and spinel lithium manganese oxide. The cobalt oxide was the first to achieve commercial success and is still the most widely used. Both nickel and cobalt oxides offer 3.6 V operation in combination with carbon. Although nickel is cheaper than cobalt, its more complex electrochemistry has led to the majority of manufactures preferring cobalt, but lithium manganese oxide is now challenging cobalt oxide for the dominant market position. Although the operating cell voltage is lower at 3.0 V, the considerable advantages in terms of availability, cost and low toxicity more than compensate.

Lithium ion batteries offer high energy density (125 Whkg^{-1}, 300 Whdm^{-3}), high operating voltage (3.6 V), high cyclability with up to 1000 cycles being possible and a rapid recharge (e.g. 2 hours) is possible. There is no memory effect on charge/discharge as found in nickel cadmium and the batteries are much safer than lithium primary batteries. Self-discharge on standing is less than 10 per cent per month, which is acceptable for a rechargeable battery. Although special high-voltage stable non-aqueous electrolytes have been developed, there are still some issues with stability if careful control of charging is not maintained, and the top-of-charge voltage is exceeded. Charging is carried out at constant current until top-of-charge is reached at about 80 per cent of capacity, then voltage is held constant as the current decays to a limiting value. Normally cells are only sold as part of an integrated pack that contains control electronics and the protection circuit.

12.4.2   Lithium polymer batteries

Considerable efforts have been made to commercialize lithium polymer batteries utilizing a polymer such as polyethylene oxide instead of an organic solvent to dissolve the lithium salt in the electrolyte. Such systems would be able to safely utilize lithium metal as an electrode, which would considerably increase capacity. Unfortunately the electrical resistance of the polymer is still slightly too high for wide-scale commercialization, but some important advances are being made. A slightly different compound based on a gel electrolyte has been more successful and most commercial polymer lithium batteries are of this type. Here, a liquid non-aqueous electrolyte is encapsulated in a polymer gel, typically polyvinylidine fluoride or PVDF. Apart from the immobilized electrolyte, such gel-based polymer batteries are very similar to more conventional lithium ion batteries. The polymer construction facilitates a range of innovative concepts based upon winding and folding to the various elements.

12.5   Fuel cells

Like a battery, a fuel cell converts the energy of a chemical reaction directly into electrical energy. The process here involves the oxidation of an external fuel, which is normally a hydrogen-rich gas, and the reduction of an oxidant, which is usually atmospheric oxygen. Electrons are passed from the fuel electrode to the oxidant electrode through the externally connected load, and the electrical circuit is completed by ions that cross an electrolyte to produce water.

The fuel cell has a number of advantages over the conventional heat engine and shaft-driven generator for the production of electrical power. The generation efficiency is much higher in a fuel cell at scales even up to several megawatts, it has a higher power density, lower vibration characteristics, and reduced emission of pollutants. An individual fuel cell operates at a dc voltage of about 1 V. Cells must therefore be

Table 12.2 Comparison of different fuel cell types

	Electrolyte	Operating temperature (°C)	Development status	Applications	Suitable fuel
SOFC	Solid oxide	750–1000	100 kW (tubular)	CHP	Partially reformed natural gas, hydrogen, carbon monoxide, etc
			5–50 kW (planar)	Power generation APU	
MCFC	Molten carbonate	630–650	2 MW	Power generation CHP	Carbon monoxide, hydrogen
PAFC	Phosphoric acid	190–210	11 MW	CHP	Hydrogen with traces of carbon monoxide
				Power generation	
PEMFC	Solid polymer	70–90	250 kW	Transport CHP	Ultrapure hydrogen
AFC	Alkaline	50–200	Developed	Space	Ultrapure hydrogen, no CO_2 tolerance
				Transport	

connected to stacks of series and parallel connection in order to deliver the voltage and power required for many applications. Another complication is that many fuel cells cannot operate effectively using a raw hydrocarbon fuel; most types require a reformer, which converts the hydrocarbon into a hydrogen-rich gas suitable for passing directly into the fuel cell. The main types of fuel cell are best classified according to temperature of operation and fuel requirement, as in Table 12.2.

The two main types of fuel cells being developed at present are the *Solid Polymer Electrolyte Fuel Cell (PEFC)* and the *Solid Oxide Fuel Cell (SOFC)*. These two main classes are distinguished by the type of electrolyte they use. This in turn determines their operating temperature.

12.5.1   The Polymer Electrolyte Fuel Cell (PEFC)

An example of a PEFC is shown in Fig. 12.16. The solid polymer fuel cell is based upon a proton conducting polyfluorosulphonic acid membrane with finely dispersed platinum electrodes upon a porous carbon matrix. Machined carbon or steel plates are utilized as bipolar plates to form series stacks. The PEFC operates at temperatures below 100°C with humidified gases. It is a potentially clean method of generating electricity which is silent, robust and efficient. Applications for the PEFC cover the domestic, commercial and industrial range, but it will probably be best suited to small-scale Combined Heat and Power (CHP) applications, and for transport duties. The PEFC is often described with reference to power density, which ranges from 0.25–1.0 kW/l. Some PEFCs are commercially available for demonstration or early application at 1–5 kW for CHP and up to 25 kW for transport. Expected lifetime is still only a few thousand hours for the stack, but system lifetime is expected eventually to be in the range of 8–20 years.

Fig. 12.16 The solid polymer fuel cell (courtesy of Loughborough University)

12.5.2 The Solid Oxide Fuel Cell (SOFC)

The SOFC is illustrated in Fig. 12.17. It typically comprises a nickel zirconia cermet anode, yttria stabilized zirconia electrolyte, lanthanum strontium manganite cathode and a chrome-based alloy interconnect. The SOFC operates at much higher temperatures than the PEFC, in the region of 850°C to 1000°C. It offers the possibility of high electrical efficiency together with high-grade exhaust heat; important applications for the SOFC are therefore in the combined heat and power field and in other generation applications where a significant heat load is present.

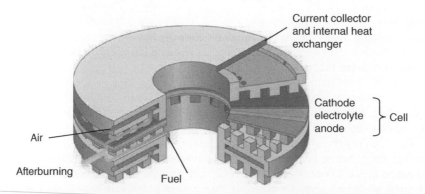

Fig. 12.17 Combined heat and power solid oxide fuel cell (courtesy Sulzer Hexis)

Advantages of the SOFC are very high electrical efficiency, the ability to utilize conventional fuels such as natural gas with limited processing and the facility to produce steam or hot water from the high temperatures that are available. Typical power ratings are in the range 150–250 kW and life expectancy is 5–20 years, with adequate maintenance. The first demonstration systems in the range 1–2 kW (electrical output) have recently become commercially available, normally in combined heat and power with about 10 kW (thermal output).

12.6 Battery charging

12.6.1 Small commercial batteries (up to 10 Ah)

The rechargeable batteries used in portable equipment are mainly lithium ion, nickel cadmium or nickel metal hydride. Recharging of these batteries can be carried out by the following methods:

- transformer-rectifier
- switched-mode power supply unit
- capacity charger

The transformer-rectifier circuit is used to reduce ac mains voltage to a lower dc voltage. The charge delivered to the battery can be regulated using a constant current, a constant voltage, constant temperature or a combination of these. The transformer in this system provides the inherent advantage of isolation between the mains and low-voltage circuits.

In the switched-mode power supply unit, the mains voltage is rectified, switched at a high frequency (between 20 kHz and 1 mHz), converted to low voltage through a high-frequency transformer, rectified back to dc, which is then regulated to charge the battery in a controlled manner. Although more complex than the transformer-rectifier circuit, this technique results in a smaller and more efficient charging unit.

In a capacity charger the mains is rectified and then series coupled with a mains-voltage capacitor with current regulation directly to the battery packs. This makes for a small and cheap charging unit, but it has to be treated with care, since the capacitor is not isolated from the mains.

In addition to these three methods, several manufacturers have responded to the need for shorter recharge periods and better battery condition monitoring by designing specific integrated circuits that provide both the charging control and the monitoring.

12.6.2 Automotive batteries

The main types for recharging automotive batteries are:

- single battery, out of the vehicle
- multiple batteries, out of the vehicle
- starter charging, battery in vehicle

Transformer/reactance is predominantly used in each of these applications, with either a ballast in the form of resistance, or resistance to control the charging current. More sophisticated chargers have constant voltage and current-limiting facilities to suit the charging of 'maintenance free' batteries. A typical circuit is shown in Fig. 12.18.

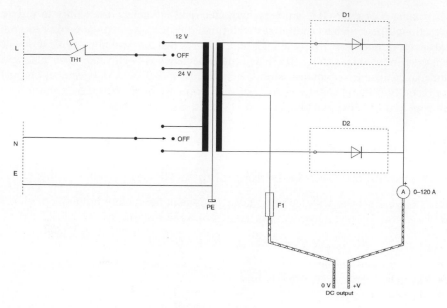

Fig. 12.18 Typical circuit for automotive battery charging

Vented automotive batteries are usually delivered to an agent in the dry condition. They have to be filled and charged by the agent. A multiple set of batteries connected in series is usually charged for a preset time in this case, and this requires the use of a bench-type charger. Bench chargers are rated from 2 V to 72 V, with a current capability ranging from 10–20 A dc (mean). Examples are illustrated in Fig. 12.19. The bench-type charger may alternatively be left on charge indefinitely using a constant-voltage, current-limited charger.

Starter charging is used to start a vehicle which has a discharged battery. A large current is delivered for a short period, and starter charges have usually a short-term rating. They can be rated at 6 V, 12 V or 24 V, delivering a short-circuit rated starting current from 150 A to 500 A dc (rms) and a steady-state output of 10 A to 100 A dc.

Output voltage is from 1.8 V to 3.0 V per cell. Voltage ripple may be typically up to 47 per cent with a simple single-phase transformer-rectifier, but the ripple will depend upon output voltage, battery capacity and the state of discharge. Voltage regulation is important in starter charging applications since high-voltage excursions can damage sensitive electrical equipment in the vehicle.

Since these chargers are usually short-time rated, some derating may be required if they have to operate at a high ambient temperature.

Safety features are built into automotive battery chargers in order to avoid the risk of damage to the battery, to vehicle wiring or to the operator. The normal safety features include:

- reverse polarity protection
- no battery – short-circuit protection
- thermal trip – abuse protection

Fig. 12.19 Examples of bench-type battery chargers (courtesy of Deakin Davenset Rectifiers)

12.6.3 Motive power

Motive power or traction battery chargers are used in applications where the batteries provide the main propulsion for the vehicle. These applications include fork lift trucks, milk floats, electrical guided vehicles, wheelchairs and golf trolleys. The requirement is to recharge batteries which have been discharged to varying degrees, within a short period (7–14 hours). Both the battery and the charger may be subject to wide temperature variations. A well-designed charger will be simple to operate, will automatically compensate for fluctuations in main voltage and for differences between batteries arising from such factors as manufacture age and temperature, and will even tolerate connection to abused batteries which may have some cells short-circuited.

The most common type of charger is the *modified constant potential* or *taper charger* shown in Fig. 12.20. In all but the smallest chargers, the ballast resistor shown in this circuit is replaced by a reactance; this reactance may be in the form of a choke connected in series with the primary or secondary windings of the transformer, or more usually it is built into the transformer as leakage reactance.

While the battery is on charge, the voltage rises steadily from 2.1 V per cell to 2.35 V per cell, at which point the battery is approximately 80 per cent charged, and gassing begins. Gassing is the result of breakdown and dissipation of the water content in the electrolyte. Charging beyond this point is accompanied by a sharp rise in voltage, and when the battery is fully charged the voltage settles to a constant voltage, the value of which depends upon a number of factors, including battery construction, age and temperature.

During the gassing phase (above 2.35 V per cell) the charging current must be limited in order to prevent excessive over heating and loss of electrolyte. The purpose of

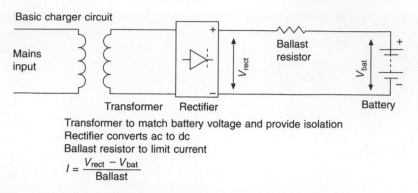

Basic charger circuit

Mains input

Transformer   Rectifier

V_{rect}

Ballast resistor

V_{bat}

Battery

Transformer to match battery voltage and provide isolation
Rectifier converts ac to dc
Ballast resistor to limit current

$$I = \frac{V_{rect} - V_{bat}}{Ballast}$$

Fig. 12.20 Modified constant potential, or taper charger

the ballast resistor in the charger circuit is to reduce the charging current as the battery voltage rises, hence the name taper charger. By convention, the current output from the charger is quoted at a voltage of 2.0 V per cell, and the proportion of this current which is delivered at 2.6 V per cell is defined as the *taper*. A typical current limit recommended by battery manufacturers is one-twelfth of the battery capacity (defined as the 5-hour rate in ampere-hours) at the mean gassing voltage of 2.5 V per cell. A disadvantage of the high-reactance taper charging system is that the output may be very sensitive to changes in the input voltage; the charge termination method and the required recharge time must therefore be taken into account while sizing the charger.

Because of the inefficiencies of energy conversion, particularly due to the heating and electrolysis during the gassing phase, the energy delivered by the motive power charger during recharging is 12–15 per cent higher than the energy delivered by the battery during discharge.

To recharge a battery fully in less than 14 hours a high rate of charge is necessary and the termination of charging when the battery is fully charged must be controlled. The two types of device for termination of charge are voltage–time termination and rate-of-charge termination.

A voltage–time controller detects the point at which the battery voltage reaches 2.35 V per cell, and then allows a fixed 'gassing' time for further charging, which is usually 4 hours. This method is not suitable for simple taper chargers with nominal recharge times of less than 10 hours because of the variation in charge returned during the time period as a result of mains supply fluctuations. If a short recharge time is required with a voltage–time controller then a *two-step taper charger* is used, in which a higher charging current is used for the first part of the recharge cycle. When the voltage reaches 2.35 V per cell, the timer is started as discussed here, and the current during the further charging period is reduced by introducing more ballast in the circuit.

The *rate-of-charge* method of termination has predominated in large chargers during the past decade offering benefits to both the manufacturer and the user. In the rate-of-charge system the battery voltage is continuously monitored by an electronic circuit. When the battery voltage exceeds 2.35 V per cell, the rate of rise of the battery voltage is calculated and charging is terminated when this rate of rise is zero, that is when the battery voltage is constant. This method can be used with single-rate taper chargers with recharge periods as short as 7 hours because of the higher precision of termination. In order for a rate-of-charge termination system to operate satisfactorily, there must be compensation for the effects of fluctuations in mains supply. A change in mains voltage results in a proportional change in the secondary output voltage from the charger transformer; if uncompensated this will cause a change in charging current and therefore in battery voltage. For a 6 per cent change in mains voltage, the charging current may change a much as 20 per cent and the battery voltage may change by 3 per cent.

Many batteries have the facility for *freshening* or *equalizing*.

Freshening charge is supplied to the battery after the termination of normal charge in order to compensate for the normal tendency of a battery to discharge itself. A freshening charge may be continuous low current, or trickle charging, or it may be a burst of higher current applied at regular intervals.

Equalizing charge is supplied to the battery in addition to the normal charge to ensure that those cells which have been more deeply discharged than others (due, for instance, to tapping off a low-voltage supply) are restored to a fully charged state.

A *controlled charger* is a programmable power supply based on either thyristor phase angle control or high-frequency switch mode techniques. The main part of the recharge cycle is usually at constant current and the power taken by the charger is therefore constant until the battery voltage reaches 2.35 V per cell. Many options are available for the current–voltage profile during the gassing part of the recharging cycle, but all of these profiles deliver a current which is lower than the first part of the cycle.

Voltage drop in the cable between the charger and the battery is important because the charge control and termination circuitry relies upon an accurate measurement of the battery voltage. It is not normally practical to measure the voltage at the battery terminals because the measuring leads would be either too costly or too susceptible to damage, and it is common practice to sense the voltage at the output of the charger and to make an allowance for the voltage drop in the cables. Alteration to the length or cross section of these cables will therefore cause errors, especially with low-voltage batteries.

Motive power chargers are typically available from 6 V (three cells) to 160 V (80 cells) with mean dc output currents from 10 A to 200 A. A typical circuit is shown in Fig. 12.21.

The output current is rated at 2.0 V per cell; the taper characteristics set by the transformer reactance then results in 25 per cent output current at typically 2.65 V per cell. Rating is not continuous, and derating to 80 per cent is typical to take into account the taper characteristic. Consideration must be given to this if a multiple shift working pattern is to be adopted.

Voltage ripple is typically 15–25 per cent for single-phase chargers and 5–15 per cent for three-phase chargers. The precise level of ripple will vary with time and it will depend upon mains voltage, battery capacity and the depth of the battery.

12.6.4   Standby power applications

Typical standby power applications include emergency lighting, switch tripping, switch closing and telecommunications. The main functional requirements for the battery charger in these cases are:

- to ensure that the state of charge of battery is maintained at an adequate level, without reducing battery life or necessitating undue maintenance
- to ensure that the output voltage and current of the complete system are compatible with the connected electrical load

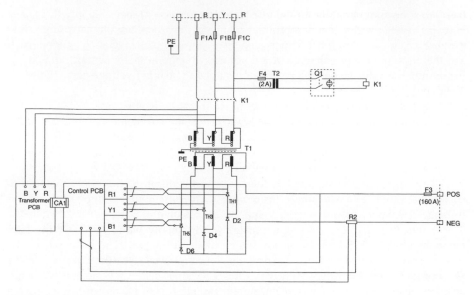

Fig. 12.21 Typical circuit diagram for a motive power charger

- to ensure after a discharge that the battery is sufficiently recharged within a specified time to perform the required discharge duty
- to provide adequate condition monitoring, to the appropriate standards

Assuming initially that the battery is fully charged, the simple option is to do nothing. A charged battery will discharge if left disconnected from a load and from charging equipment, but if the battery is kept clean and dry this discharge will be quite slow. For some applications, open-circuit storage is therefore acceptable.

For most applications, however, there is a need for battery charge to be maintained. The current–voltage characteristic is not linear, and a small increase in charging voltage will result in a large-scale increase in current. Nevertheless, it is always possible to define a voltage which, when applied to a standby power battery, will maintain charge without excessive current, and the charging current flowing into the battery has only to replace the open-circuit losses in the battery, which are usually small.

Once these open-circuit losses have been made up, any additional current flowing is unnecessary for charging purposes and is normally undesirable. In vented cells it causes overheating and gassing and eventually, if not checked, damage and loss of capacity of the cell. In sealed cells there can be overheating, in extreme cases expulsion of gases through the pressure vent and ultimately, a loss of capacity.

On the other hand, if the battery voltage is allowed to fall too much, the open-circuit losses will not be replaced and the battery will slowly discharge.

The charging voltage has therefore to be controlled carefully for best battery maintenance. The usual limits are within ±1 per cent of the ideal voltage, which is normally termed the *float voltage*. The float voltage has a negative temperature coefficient, which must be accounted for when batteries are to operate in exceptionally hot or cold environments. Float voltages for the major types of standby power cell are shown in Table 12.3.

Table 12.3 Typical cell voltages for systems with limited load voltage excursions

Cell type	Cell voltage (V)			
	End of discharge	Float	Refresh	Boost
Vented nickel cadmium	1.1	1.45	1.55	1.7
Vented lead acid	1.8	2.25	2.45	2.7
Sealed lead acid	1.8	2.27	N/A	N/A

For vented cells there are, however, circumstances under which the float voltage should be exceeded. Batteries that are new or have suffered abuse will benefit from a vigorous gassing up to the *boost voltage* shown in Table 12.3. Batteries which have stood on float charge with no discharge–charge cycle for many months will benefit from a *refresh charge* with gassing, at the refresh voltages which are also shown in Table 12.3.

Few dc standby power systems can be designed without taking account of *limits on the load voltage*. For good battery operation it is necessary to charge at the float voltage (or sometimes a higher level), but it is also necessary to discharge the battery to a sufficiently low voltage if the full capacity is to be released from the cells. These considerations impose fundamental limits that define the maximum excursion of the system output voltage. Table 12.4 shows the minimum and maximum voltages which are reasonable in a 50 V standby power system for each for the three major battery types. Systems with other voltages will require excursions which are in direct proportion.

It can be seen from Table 12.4 that, allowing for the discrete steps in voltage when changing the number of cells, it is not possible to achieve a voltage excursion of less than about ±12 per cent under conditions of float alone; the excursion limits are larger in refresh operation, and still larger with boost.

Table 12.4 shows that the sealed lead acid cell offers minimum overall voltage variation because of the absence of the larger excursions due to refresh and boost charges. If it is not possible to use a sealed lead acid system, another alternative is to disconnect the load for the full boosting operation; this should normally be necessary only at the time of system installation in any case.

A voltage regulator should be included in the system if closer limits of voltage variation are required. Diode regulators are now reliable and widely used; they operate by

Table 12.4 Load voltage ranges in a 50 V standby power system

Function	Minimum volt/cell (V)	Maximum volt/cell (V)	50 V system		
			No. of cells	Minimum voltage (V)	Maximum voltage (V)	±Voltage excursion (%)
Vented lead acid cells						
Float	1.80	2.25	25	45.5	56.3	11.3
Refresh	1.80	2.45	24	43.2	58.8	15.6
Boost	1.80	2.70	22	39.6	59.4	19.8
Vented nickel cadmium cells						
Float	1.10	1.45	39	42.9	56.6	13.7
Refresh	1.10	1.55	38	41.8	58.9	17.1
Sealed lead acid						
Float	1.80	2.27	25	45.0	56.8	11.8

switching banks of series-connected diodes in and out as the battery voltage slowly varies. It is important to ensure that switching in the regulator occurs only as a result of changes in battery voltage, and not as a result of load changes; if the regulator responds to changes which occur as a result of load change, excursions outside the specific voltage limits may occur because of the delays in the process of switching the diode bank. Ensuring that this distinction is made normally requires a computer simulation for all but the simplest regulators. Control of the switching of the diode bank is best achieved by a programmable controller, especially if additional complicated relay-type logic is required in the system. Diode regulators are large and their heat dissipation is substantial. Higher efficiency regulators using actively switched devices are becoming available, but they are at present limited to relatively low power applications.

Condition monitoring is now included in most dc systems in order to warn of excessive battery voltage excursions. The applications are diverse, but the main features are:

- *high voltage detection:* this is necessary in order to prevent a fault on the supply system from damaging the battery or load circuit
- *low voltage detection:* this warns of load failure due to insufficient voltage, and it is also needed to trigger the disconnection of sealed batteries which may be damaged by excessive discharging
- *charge failure:* this is needed in order to stimulate action to restore the ac supply or to prepare for disconnection of the load
- *earth leakage:* this is needed where an unearthed load system is used, for safety and for avoidance of double faults

Communication of a fault is through volt-free contact on a relay, which usually signals to the monitoring centre using a 110 V or other voltage supply.

It has been seen in Table 12.4 that there are restrictions on the choice of charging voltage. The preference for many applications is to limit the voltage to the float voltage, and while it may be necessary to increase the charging voltage to speed up the recharging, the voltage should be returned to the float voltage as soon as possible. At this float voltage level, all the cell types discussed will be recharged to about 80 per cent of their nominal capacity. Assuming the charger current is at the adequate level shown in Table 12.5, recharge to 80 per cent capacity will be achieved in the times indicated in the table.

To recover the remaining 20 per cent of the charge is more difficult, and different techniques are necessary for the three cell types. A vented lead acid battery will be fully recharged at float voltage in about 72 hours, but if the charging voltage is boosted to 2.7 V per cell, a full recharge will take about 14 hours. For a vented nickel cadmium cell full charge will never be achieved without increasing the voltage above float level. Exact times may vary between cell types, but typically a refresh charge at 1.55 V per

Table 12.5 Charging time and current necessary to recharge to 80 per cent of capacity using float voltage

Cell type	Charging current	Charging time
Vented lead acid	7% of capacity	14 hours
Vented nickel cadmium	20% of capacity	6–8 hours (depending on type)
Sealed lead acid	10% of capacity	9 hours

cell will give full charge in about 200 hours on the highest performance cells, and boosting to 1.7 V per cell will reduce this time to 9–10 hours. Sealed lead acid cells will reach full charge after about 72 hours at float voltage; an increase of charge voltage to 2.4 V per cell will reduce charging time to about 48 hours, but there is no way in which this can be significantly reduced further.

If a fast recharge is essential, an alternative is to oversize the battery; for instance if 100 Ah capacity is required with a full recharge within 8 hours, then a battery with 125 Ah capacity could be installed.

Most cells will protect themselves from excess charging current, provided that the voltage is limited to the float voltage, but above this level excessive charging current can damage the cell. For vented lead acid cells the 7 per cent of capacity shown in Table 12.5 is recommended as an upper limit. For sealed lead acid cells, the recommended upper limit is 50 per cent of capacity. Nickel cadmium cells normally require a minimum charge current of 20 per cent of capacity, as indicated in Table 12.5, but a lower limit of 10 per cent of capacity is recommended when boosting in order to avoid excessive gassing and electrolyte spray.

Standby power chargers are available in a wide range of capacities to suit many applications. Typical dc outputs are 6, 12, 24, 30, 48, 60, 110, 220 and 240 V, with dc mean output current ranging from 1 A to 1000 A. DC output current is rated at 100 per cent of the output current at the full specified voltage. Regulation of the output is generally within ±1 per cent for an input voltage change of ±10 per cent and a load current change of 0–100 per cent. A typical circuit is shown in Fig. 12.22. Although standby power systems are continuously rated, some derating may be necessary for operation in tropical climates if this was not originally specified.

Differing levels of output smoothing can be incorporated into the charging system, depending upon the application. General applications require a maximum of 5 per cent ripple, but for telecommunications supplies, specifications are based on CCITT telecommunications smoothing, which requires 2 mV phosphometrically weighted at 800 Hz. The key components of an installation are shown schematically in Fig 12.23.

12.7   Battery monitoring

The monitoring of battery condition is becoming more important as remote operation and reduced maintenance requirements are increasingly specified in standby power systems. The options which are available are summarized in the following sections.

12.7.1   Load-discharge testing

The basis of this test is to discharge the battery into a selected load for a preset time, after which the battery voltage is checked. In the majority of cases this will require that the battery is taken off line. The load-discharge test is very reliable, but it requires a special resistance load bank, and it is labour-intensive and time-consuming. Figure 12.24 shows an example of the type of equipment that is required.

12.7.2   dV/dt-load testing

The technique here is to connect a fixed known load to the battery for a short period and record the battery voltage over this period. By comparing the recorded load voltage with data from the battery manufacturer, estimations can be made regarding the condition of the battery and its ability to perform.

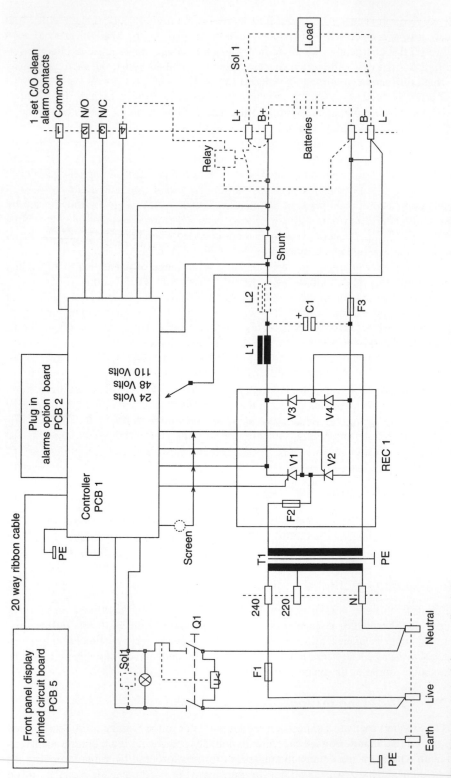

Fig. 12.22 Typical circuit for a standby charger

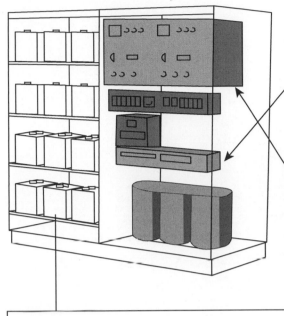

The rectifier/battery charger

The rectifier/charger converts incoming ac mains into dc to provide a stable, constant voltage output with automatic current limiting. The rectifier–charger automatically floats and boost charges the battery simultaneously providing the required dc output to the load.

Control and distribution

Distribution panels typically provide several fused outlets, monitoring of busbar voltage and current, earth fault status together with busbar and interbusbar controls. Depending upon application, the control and distribution equipment may be contained within the charger cubicle (as shown) or may be housed in a separate matching enclosure.

The battery

The battery is a key element in a secure power supply and must be selected with care. The type of battery selected will depend on many factors including: reliability, operating temperature, cost, life, standby time required, maintenance parameters, ventilation and available space. Depending on the type of battery, it can be supplied on a stand or in a ventilated cabinet. The battery should always be located as close as possible to the dc power supply to minimize line voltage drop.

The main types of battery used are:

- valve regulated lead acid; minimal gassing so can be used in electronics enclosure, maintenance free, relative installed cost medium, typical life 10 yrs.
- flat plate lead acid: gas, with adequate ventilation and periodic maintenance required. Relative installed cost low, typical life 10–15 yrs.
- Planté lead acid: gas, with adequate ventilation and periodic maintenance required. Highest levels of reliability. Relative installed cost high, typical life 20–25 yrs.
- nickel cadmium: gas, with adequate ventilation and periodic maintenance required. Ideal or arduous conditions. Relative installed cost very high, typical life 20–25 yrs.

Fig. 12.23 Key elements of a standby power battery unit

12.7.3 Computer monitoring of cell voltages

In this case, measurement leads are taken from each cell to a central monitoring computer. The voltage of each cell is recorded through charge and discharge of the battery. Although cell voltage is not a direct indication of residual capacity, trends can be observed and weak cells can be identified.

12.7.4 Conductance monitoring

This method has increased in popularity in recent years and several manufacturers now offer a standard monitoring product based upon the technique. Figure 12.25 shows typical equipment. Monitoring the conductance of individual cells over their working life can give an indication of impending failure, allowing preventive maintenance to be

Fig. 12.24 Equipment for load-discharge testing (courtesy of Deakin Davenset Rectifiers)

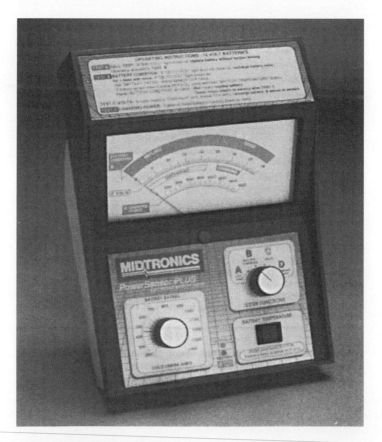

Fig. 12.25 Conductance monitoring equipment

carried out. A particular advantage is that the readings can be reliably used whatever be the state of the charge of the battery.

12.8   Installation, testing and commissioning

Larger lead acid or nickel cadmium installations should be sited in a correctly constructed room, vehicle, cubicle or rack which allows ready access for maintenance and ensure adequate ventilation of batteries and battery chargers. The battery should not be located close to a source of heat such as a transformer or heater. Battery racks should adhere to recommendations regarding spacing between cells, and drip trays and insulators should be fitted where applicable. All cell connections should be cleaned, coated with no-oxide grease and tightened to the suppliers' specification. A typical installation is shown in Fig. 12.26.

The rating of the ac mains input cable to the battery charger should be based on the peak current that will be drawn by the charger at the lowest mains voltage. The rating of the dc cables from the battery charger and in the battery should be no lower than the maximum protection fuse rating, and attention should be paid to the overall voltage drop in the system, especially on lower voltage dc systems. The cables should be adequately protected, and segregation may be necessary on systems with large dc current.

It may be necessary to charge batteries for several days before their full capacity is reached, and in some cases a special initial commissioning charge level may be required.

Fig. 12.26 Typical battery installation (courtesy of Invensys)

Tests should be set up so that the results enable the operation of the system to be compared with the original specification. For larger systems, the manufacturer should propose a set of witness tests to be performed at the works, followed by final tests to prove capacity and duty on site. A typical test schedule should be designed to establish:

- the ability of the battery charger to support the standing load, where applicable
- the ability of the battery charger to recharge a fully discharged battery in the required time, whilst simultaneously supporting the standing load, again where applicable
- the ability of the battery charger to maintain regulation within specified limits throughout the required range of load and input conditions
- correct operation of the monitoring features and instrumentation
- the ability of the battery and charger system to supply the load for the required period during a power failure

12.9   Operation and maintenance

12.9.1   Primary cells

Primary cells cannot be recharged and once their energy is depleted they should be disposed off in the manner recommended by the manufacturer. The cells should be examined occasionally during their life for signs of leakage or physical deformation. Chemical hazards can arise if batteries are misused or abused and in extreme cases, if there is a risk of fire or explosion.

12.9.2   Secondary cells

Small secondary cells should be examined periodically under both charging and discharging conditions. As the battery ages its ability to accept recharge and hold its capacity will decrease. This ageing is usually shown by the dissipation of heat during charging as a result of the inability to accept charge, whilst on discharge a reduction of capacity will be observed. Excessive overcharging will cause leakage of electrolyte or deformation of the cell.

Cells should be handled carefully, with the operator wearing the correct protective clothing. When handling vented cells, supplies of saline solution (for eye washing) and clean water should be readily available. If a cell is to be filled, only purified water should be used when mixing the electrolyte.

The maximum storage time for cells which are filled and nominally charged is 6 months for nickel cadmium, 12 months for valve-regulated lead acid batteries (at 20°C) and 8 weeks for vented lead acid batteries. Cells which are stored for longer than this should be periodically charged and the electrolyte level should be checked, where applicable.

Whether filled or empty, the battery may contain explosive gases, and no smoking, sparks or flames should be permitted in the vicinity of the installation.

Although the voltage at any point in a battery system can be reduced by the removal of inter-cell connectors, the cells are electrically live at all times and cannot be de-energized or isolated in the conventional sense. When connecting cells together, insulated tools should be used wherever possible to avoid accidental short circuits and sparks. Tools should be cleaned before use if they are to be used on both lead acid and nickel cadmium batteries, since acid will destroy a nickel cadmium battery.

On large installations it is good practice to operate a system of *cell log sheets*. Such sheets or books are normally supplied by the cell manufacturer. Completion of these sheets will require the measurement of voltage, ambient temperature and, if applicable, the specific gravity of each cell at regular intervals. Cell electrolyte level (where applicable) and the tightness of cell connections should also be checked at these intervals.

Faults to be watched out for are:

- loose connections in the cells or in the charger system
- low or high electrolyte level (where applicable)
- debris or electrolyte spillage on the top cells which may lead to short circuits
- excessive loss of electrolyte (where applicable) due to overcharging or battery ageing
- cells overheating because of their inability to accept charge
- loss of capacity due to undercharging (specific gravity should be checked where applicable)
- loss of charger regulation due to a control circuit fault
- ac or dc fuses operated
- excessive loading on the battery or on the charger

12.10 Standards

There are many standards covering various types of battery and battery charging systems. The key IEC recommendations together with equivalent BS and EN standards and related North American standards are summarized in Table 12.6.

Table 12.6 Comparison of international, regional and national standards for batteries and battery charges

IEC	EN	BS	Subject	N. American
428		5142	Specification for standard cell	
896	60896-1	6290-1	General specification for lead acid cells	
896	60896-2	6290-2	Planté cells	IEEE 450/484
896	60896-3	6290-3	Pasted cells	IEEE 450/484
896	60896-4	6290-4	Value-regulated sealed cells	IEEE 1188
1056	61056-1	6745	Portable value regulated lead acid cells	
		3031	Sulphuric acid for use in lead acid batteries	
		4974	Specification of water for lead acid batteries	
		7483	Specification for lead acid batteries in light vehicles	
254	60254	2550	Specification for lead acid traction batteries	
		6287	Code of practice for safe operation of traction cells	
95	60095	3911	Lead acid starter batteries	
		6133	Code of practice for safe operation of lead acid stationary batteries	

(contd)

Table 12.6 (*contd*)

IEC	EN	BS	Subject	N. American
623	60623	6260	Nickel cadmium single cells	IEEE 1106
285	60285	5932	Nickel cadmium cylindrical cells	
622	60622	6115	Nickel cadmium single cells (prismatic)	ANSI C 18.2 M
		6132	Code of practice for safe operation of alkaline secondary cells and batteries	
1044	61044	EN 61044	Operation charging of lead acid traction batteries	
	DIN 41774		Traction battery chargers – taper characteristics	
	DIN 41773		Traction battery chargers – characteristics	
952	60952	EN 60952	Aircraft batteries	
335-2-29	60335-2-29	3456-2-29	Domestic battery chargers	UL 1564 ANSI 1564
146	60146	4417	Specification for converters	UL 458 UL 1012 UL 1236 UL 1310
950	60950	7002	Information technology equipment	UL 1950

References

12A.   May, G.J., *Journal of Power Sources*, **42**, 1993, pp. 147–153.
12B.   May, G.J., *Journal of Power Sources*, **53**, 1995, pp. 111–117.
12C.   *Rechargeable Batteries Handbook*, Butterworth-Heinemann, Oxford, UK, 1992.
12D.   Dell, R.M. and Rand, D.A.J., *Understanding Batteries,* RSC, Cambridge, 2001.

Chapter 13

The power system

Dr B.J. Cory
Professor M.R. Irving
Brunel University

13.1   Introduction

All countries now have a power system which transports electrical energy from generators to consumers. In some countries several separate systems may exist, but it is preferable to interconnect small systems and to operate the combination as one, so that economy of operation and security of supply to consumers is maximized. This integrated system (often known as the 'grid') has become dominant in most areas and it is usually considered as a major factor in the well-being and level of economic activity in a country.

All systems are based on alternating current, usually at a frequency of either 50 Hz or 60 Hz. The 50 Hz is used in Europe, India, Africa and Australia, and 60 Hz is used in North and South America and parts of Japan.

Systems are traditionally designed and operated in the following three groupings:

- the source of energy – *generation*
- bulk transfer – *transmission*
- supply to individual customers – *distribution*

13.2   Generation

Generators are required to convert fuels (such as coal, gas, oil and nuclear) and other energy sources (such as water, wind and solar radiation) into electrical power. Nearly all generators are rotating machines, which are controlled to provide a steady output at a given voltage. The main types of generator and the means of control are described in Chapter 5.

The total power output of all operating generators connected to the same integrated system must at every instant be equal to the sum of the consumer demand and the losses in the system. This implies careful and co-ordinated control such that the system frequency is maintained, because the majority of generators in an ac power system are synchronous machines and their rotors, which produce a magnetic field, must lock into the rotating magnetic field produced by alternating currents in the stator winding. Any excess of generated power over the absorbed power causes the frequency to rise, and a deficit causes the frequency to fall. As the demand of domestic, commercial and industrial consumers varies, so the generated power must also vary, and this is normally managed by transmission system control which instructs some generators to maintain a steady output and others (particularly hydro and gas turbine plant) to 'follow' the load; load 'following' is usually achieved by sensitive control of the input, dependent upon frequency. It is desirable to run the generating plant such that the overall cost of supplying the consumer at all times is a minimum, subject to the various constraints which are imposed by individual generator characteristics.

In de-regulated, or privatized power systems this is achieved by competition among generators combined with additional regulated markets for ancillary services and use of the transmission system.

13.2.1   Distributed and renewable generation

The worldwide imperative to reduce greenhouse gases, particularly CO_2, and to secure energy supplies for the long-term future has prioritized the development of electricity generation from renewable energy sources. Renewable generators and other high-efficiency schemes, such as *Combined Heat and Power (CHP)* are relatively small in capacity compared with large thermal power stations. Consequently, these generators are often embedded or distributed in the network at voltages, such as 11 kV or 33 kV.

As the penetration of renewables into the system increases there are major issues for the planning and operation of the power system. Some renewables, such as biomass can be regarded as providing firm capacity, or others, such as tidal power may be predictable but periodic, but most, including wind, wave and photovoltaic, have to be regarded as intermittent. Hydroelectric power is also a renewable and apart from the run-of-river plant, it offers a valuable energy storage capability. The various plant characteristics can have a significant impact for the system operator, especially in determining the required level of spinning reserve, and/or demand management, required within the system to cover the increased intermittence of supply. At present, the higher capital cost of renewable generation needs to be compensated through government subsidy to seek the total levels of renewable energy desired.

Distributed Generation (DG) also poses serious technical issues for the distribution network. These include power quality problems, such as harmonic current injection, and the inability of many DGs to 'ride-through' voltage dips (thereby exacerbating the problem). The DGs also affect fault levels in networks, either by contributing excess fault current in the case of directly connected induction generators or by not contributing sufficient fault current, where generators are connected through power electronic converters. Further issues include the possibility of bi-directional power flow in low-voltage networks and whether 'islanded' sections of the network could become a safe and acceptable operational option in the future. These technical issues are certainly solvable using present day technology, but they pose an interesting and important challenge for power system engineers.

13.3   Transmission

Many large generators require easy access to their fuel supply and cooling water, so they cannot necessarily be sited close to areas of major consumption. Environmental constraints may also preclude siting close to areas of consumption. A bulk power transmission system is therefore needed between the generators and the consumers.

Large generating plant produces output ranging from 100 MW to 2000 MW and for economic reasons this normally operates with phase-to-phase voltages in the range

10 kV to 26 kV. In order to reduce transmission losses so that transmission circuits are economic and environmentally acceptable, a higher voltage is necessary. Phase-to-phase transmission voltages of up to 765 kV are used in sparsely populated large countries, such as Brazil, USA and Canada, but 380–400 kV is more prevalent in Europe. The standard voltages recommended by IEC are 765 kV, 500 kV, 380–400 kV, 345 kV, 275 kV, 220–230 kV, 135–138 kV and 66–69 kV.

Most transmission circuits are carried overhead on steel pylons. An example is shown in Fig. 13.1. They are suspended from insulators which provide sufficient insulation and air clearance to earth to prevent flashovers and danger to the public. A typical suspension-type insulator is shown in Fig. 13.2. Each country has tended to have its own acceptable tower and conductor design. At higher voltages, Aluminium Conductor Steel Reinforced (ACSR) conductor is used, a core of steel strands providing the required strength. A typical cross section for an ACSR conductor is shown in Fig. 13.3. For voltages over 200 kV two or more conductors per phase are used. This results in lower losses because of the large conductor cross section and lower radio interference and corona because of the lower voltage stress at the conductor surface.

In cases where an overhead line route is impossible because of congestion in an urban area or for environmental amenity reasons, buried cables may be employed, but the cost is 15–20 times higher than that of an equivalent overhead line. On sea

Fig. 13.1 Transmission line tower – a 400 kV double-circuit line

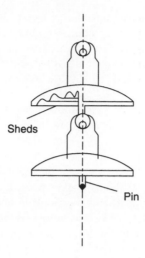

Suspension-type insulator

Fig. 13.2 Suspension-type insulator

crossings an underwater cable is the only solution, but these are often dc, for reasons explained in section 13.3.1.1.

A high-voltage transmission system interconnects many large generators with areas of high electricity demand; its reliability is paramount, since a failure could result in loss of supply to many people and to vital industry and services. The system is therefore arranged as a network so that the loss of one circuit can be tolerated. This is shown in Fig. 13.4. In many countries, three-phase lines are duplicated on one tower, in which case a tower failure might still result in a partial blackout. Mixed-voltage systems are often carried on a single tower, but this is not the practice in the UK.

In order to achieve flexibility of operation, circuits are marshalled at substations. The substations may include transformers to convert from one voltage level to another,

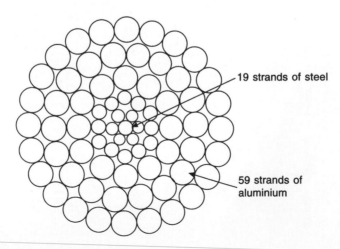

Fig. 13.3 Typical cross section of an ACSR overhead line conductor

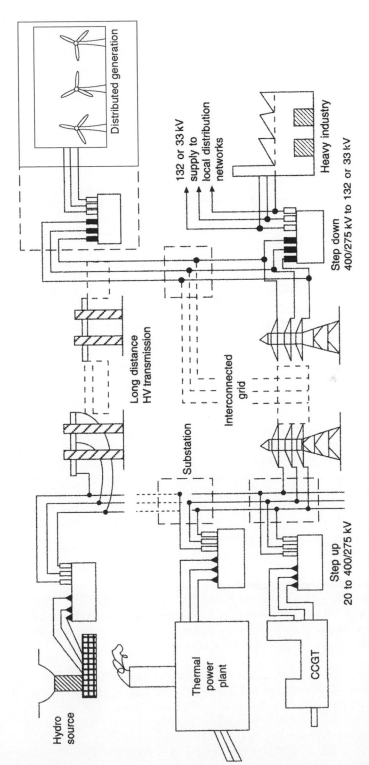

Fig. 13.4 Generation and transmission network (UK voltages and practice)

and switchgear to switch circuits and interrupt faults. Substations are normally outdoors and they occupy an extensive secure area, although compact indoor substations using SF_6 (as described in section 7.5.3.2) have become more prevalent recently because of their improved reliability in adverse weather and a more compact layout.

An interconnected transmission network can comprise many substations which are all remotely controlled and monitored to ensure rapid reconnection after a disturbance or to enable maintenance.

13.3.1   Principles of design

The two main requirements of transmission systems are:

- the interconnection of plant and neighbouring systems to provide security, economic operation and the exchange of energy on a buy–sell basis
- the transport of electrical energy from remote generation to load centres

These objectives are met by selection of the most economical overhead line design, commensurate with the various constraints imposed by environmental and national considerations. The design and approval process for new lines can take many years; following public enquiries and judicial proceedings before planning, permission is granted and line construction begins. Typical objections to new overhead lines, particularly in industrialized countries are:

- the visual deterioration of open country areas
- the possibility that electromagnetic field propagation may cause interference with television, radio and telecommunications, with an increasing awareness of public health issues
- emission of noise by corona discharges, particularly at deteriorating conductor surfaces, joints and insulation surfaces
- the danger to low-flying aircraft
- a preference for alternative energy supply, such as gas and local environment-friendly generation including solar cells and wind power, or measures for reducing electricity demand, such as better thermal insulation in houses and commercial premises, lower energy lighting, natural ventilation in place of air conditioning, and even changes of lifestyle

Power system planners are required to show that extensive studies using a range of scenarios and sensitivities have been carried out, and that the most economical and least environmentally damaging design has been chosen. Impact statements are also required in many countries to address the concerns regarding the many issues raised by local groups, planning authorities and others.

Technically, the key issues which have to be decided are:

- whether to use an overhead line or an underground cable
- the siting of substations and the size of substation required to contain the necessary equipment for control of voltage and power flow
- provision for future expansion, for an increase in demand, and in particular for the likelihood of tee-off connections to new load centres
- the availability of services and access to substation sites, including secure communications for control and monitoring

13.3.1.1   HVDC transmission

Direct current is being used increasingly for the high-voltage transport of electrical energy. The main reasons for using dc in preference to ac are:

- dc provides an asynchronous connection between two ac systems which operate at different frequencies, or which are not in phase with each other. It allows real power to be dispatched economically, independently of differences in voltage or phase angle between the two ends of the link
- in the case of underground cables or undersea crossings, the charging current for ac cables would exceed the thermal capacity of the cable when the length is over about 50 km, leaving no capacity for the transfer of real power. A dc link overcomes this difficulty and a cable with a lower cross section can be used for a given power transfer

Where a line is several hundred kilometres long, savings on cost and improvements in appearance can be gained in the dc case by using just two conductors (positive and negative) instead of the three conductors needed in an ac system. Some security is provided should one dc conductor fail, since an emergency earth return can provide half power. The insulation required in a dc line is equivalent only to that required for the peak voltage in an ac system, and lower towers can therefore be used with considerable cost savings.

Against these reductions in cost with a dc line must be set the extra cost of solid state conversion equipment at the interfaces between the ac and dc systems, and the corresponding harmonic correction and reactive compensation equipment which is required in the substations. It is normally accepted that the break-even distance is 50 km for cable routes and 300 km for overhead lines; above these distances dc is more economical than ac.

As many ac systems already exist, and because the trading of energy across national and international boundaries is becoming more prevalent, dc transmission is being increasingly chosen as the appropriate link. An added advantage is that a dc infeed to a system allows fast control of transients and rapid balancing of power in the case of loss of a generator or other supply, and it does not contribute to the fault level of the receiving system. It is important here that when a short circuit occurs on the ac system side, the current that flows can be safely interrupted by the ac-side circuit breaker.

The accepted disadvantages of a dc infeed, apart from the already-mentioned extra cost of conversion equipment, are the lack of an acceptable circuit breaker for flexible circuit operation, and the slightly higher power losses in the conversion equipment compared with an equivalent ac infeed.

With careful design of the transmission and conversion components, the reliability of a dc and ac systems is comparable.

A typical HVDC scheme providing a two-way power flow is shown schematically in Fig. 13.5. Each converter comprises rectifying components in the three-phase bridge connection, and each of the rectifying components consists of a number of thyristors connected in series and parallel. Increasingly, Gate Turn Off (GTO) thyristors are being used because of the greater control they allow. The current rating of each bridge component can be up to 200 A at 200 kV and bridges may be connected in series for higher voltages up to 400 kV or even 600 kV± to earth, each bridge being supplied from a three-phase converter transformer. By triggering the thyristors, the current flow through the system can be controlled every few electrical degrees, hence rapid isolation

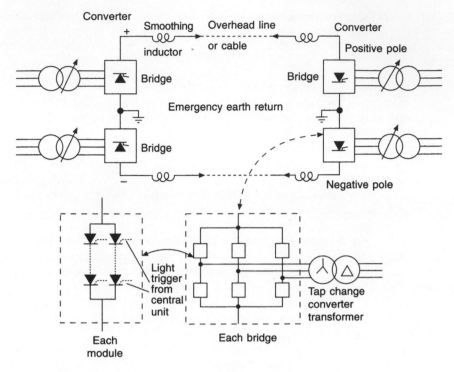

Fig. 13.5 Complete HVDC scheme, showing converters and dc link

can be achieved in the event of a fault on the system. Similarly, by delay triggering, the current can be easily controlled, the direct voltage being best set by tapchange on the converter supply transformers. The triggering of the thyristors may be by a light pulse which provides voltage isolation. *Inversion*, which enables power flow from the dc system to the ac system, depends on the ac back-emf being available with a minimum fault level in the receiving systems, so inversion into an isolated system is not possible unless devices with turn-off capability are available.

A further feature of dc converter substations is the need for ac transmission filters to produce an acceptable ac sinusoidal waveform following the infeed or outfeed of almost square-wave blocks of current. Such filters for $6n \pm 1$ harmonics (where n is the number of bridges and substations) can add 25 per cent to the cost of a substation, although they can also be used to provide some of the VAr generation which is necessary to control the power factor of the inverter.

13.3.1.2 AC system compensation

As ac power systems become more extensive at transmission voltages, it is desirable to make provision for flexible operation with compensation equipment. Such equipment not only enables larger power flows to be accommodated in a given rating of ac circuit, but also provides a means of routing flows over the interconnected system for economic or trading purposes.

Compensation equipment consisting of fixed or variable inductance and capacitance can be connected in series with the circuit, in which case it must be rated to carry the circuit current, or it may be connected in shunt, and used to inject or absorb *reactive power* (VArs) depending upon requirements. In the same way, that real power injected into the system must always just balance the load on the system and the system losses at that instant, so too must the reactive power achieve a balance over regions of the system.

Transmission circuits absorb VArs because their conductors are inductive, but they also generate VArs because they have a stray capacitance between phases and between phase and earth. The latter can be particularly important with high-voltage cables. The absorption is proportional to I^2X, where I is the current and X is the reactance of the circuit, and the generation is proportional to V^2B, where V is the voltage and B is the susceptance of the circuit. When I^2X is equal to V^2B at all parts of the circuit, it is found that the system voltage is close to the rated value. If V^2B is greater than I^2X, then the system voltage will be higher than the rated value, and vice versa. Designers therefore need to maintain a balance over the foreseeable range of current as loads vary from minimum to maximum during the day, and over the season and the year.

Compensation may be provided by the following three main methods:

- *series capacitors connected in each phase* to cancel the series inductance of the circuit. Up to 70 per cent compensation is possible in this way.
- *shunt inductance* to absorb excessive VArs generated by the circuit stray capacitance or (exceptionally) to compensate for the leading power factor of a load.
- *shunt capacitance* to generate VArs for the compensation of load power factor or excessive VArs absorbed under heavy current flow conditions on short overhead line circuits.

A combination of these arrangements are possible, especially in transmission substations where no generators are connected; generators are able to generate or absorb VArs through excitation control (see section 5.3.2).

13.3.1.3   FACTS devices

Recently, *Flexible AC Transmission Systems (FACTS)* devices have become available. In these, the amount of VAr absorption by inductors is varied through the control of thyristors connected in series with the inductor limbs. A typical shunt controllable unit is shown in Fig. 13.6. This is known as a *Shunt Variable Compensator (SVC)*. Other FACTS devices are variable series capacitors, variable phase-shifters and *Universal Power Controllers (UPC),* in which energy is drawn from the system in shunt and injected back into the system in series at a controlled phase angle by means of GTO thyristors.

The FACTS devices offer instantaneous control over voltage, current and impedance. Reactive power can be generated or absorbed and the flow of real power can be varied between alternative paths. The FACTS devices can therefore be used for voltage control, power flow control between parallel paths (particularly following a circuit outage), and by fully utilizing the instantaneous capability they can stabilize intermachine or interarea power oscillations. The disadvantages of FACTS are the additional cost, including filters to reduce the associated injection of undesirable harmonic currents into the grid.

Although a wide variety of FACTS devices have been proposed and investigated, at present the SVC is the main device which has achieved widespread adoption. The SVC offers a modern, cost-effective alternative to the rotating synchronous condenser

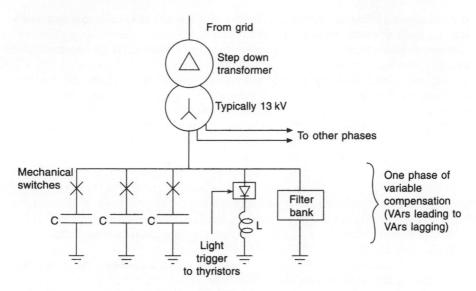

Fig. 13.6 Schematic of a Shunt Variable Compensator (SVC)

(an over-excited synchronous generator), providing a dynamic reactive power source for voltage control in parts of a grid remote from synchronous generators.

13.3.2   System operation

A transmission system may be *vertically integrated*, in which case the generating plant belongs to the same utility, or more commonly it may be *unbundled*, in which case it has only transmission capacity, with no generation plant. In either case, the main tasks for the transmission operator are to maintain a constant frequency and voltage for all consumers, and to operate the system economically and securely. Security in this context means maintaining voltage within limits, staying within a prescribed stability margin and operating all circuits within their thermal rating. This requires adequate monitoring of all the transmission components, with sufficient communication and control facilities to achieve these desired goals. Most transmission systems will, therefore, have a co-ordinating room and possibly a number of manned outstations for local or regional devolvement of responsibility.

For frequency control, some of the synchronized generators are equipped with sensitive governors which use a frequency signal rather than a speed signal. The output of these generators is dependent upon the balancing power required to achieve a steady frequency over the whole system. The transmission system operator, backed up by computer forecasts of load variations and knowledge of the available plant and their offer prices, may have the authority to instruct generators to start up or to shut down (*unit commitment*) and to set their output (*loading* or *dispatching*) so that over a prescribed hourly, daily or weekly period they generate energy to meet the consumer demand at the minimum overall cost. In the UK however, the system operator only has the authority to select offers and bids from generators and energy purchasers to effect a *balancing market*, whereas the bulk unit commitment and dispatch of generation and demand is accomplished by a *bilateral market*. In a bilateral market, generators contract directly with energy purchasers and are responsible for their own output scheduling. The balancing market operates over a short time period (one hour in the UK) imposing any adjustments necessary to obtain balanced supply and demand and technical satisfaction of any transmission system constraints.

There is a considerable scope for minimization of the losses in an interconnected system through the control of the compensation devices described in section 13.3.1.2. This control is guided by the use of optimal load flow programs, security assessments and calculations of transient stability margin. One of the main concerns is to arrange patterns of generation, including some plant which may otherwise be uneconomic, to maintain voltage despite outages of circuits and other components for maintenance, extension and repair. Safety of utility personnel and the operation of the system to avoid risk to the public is at all times paramount.

13.4   Distribution

An example of a three-phase distribution system is illustrated in Fig. 13.7. In the UK, voltages of 132 kV, 110 kV, 66 kV, 33 kV and 11 kV are typically used to provide primary distribution, with a 380–415 V three-phase and neutral low-voltage supply to smaller consumers, such as residential or smaller commercial premises, where 220–240 V single-phase to neutral is taken off the three-phase supply. Distribution voltages in continental Europe are typically 110 kV, 69 kV and 20 kV, but practice varies from country to country. In the USA, voltages of 138 kV, 115 kV, 69 kV, 34.5 kV, 13.2 kV and 4.16 kV are employed.

The transformer stepping down from the primary distribution to the low-voltage supply may be pole-mounted or in a substation, and it is close to the consumers in order to limit the length of the low-voltage connection and the power losses in the low-voltage circuit. In a national power system, many thousands of transformers and their associated circuit breakers or fuses/protective devices are required for distribution to low-voltage circuits, in contrast to high-voltage transmission and primary distribution systems, where the number of substations is in the hundreds.

The progressive introduction of small-scale distributed generation (DG) is now a major issue for distribution networks. These networks, which have been designed for one-way-traffic of energy from transmission levels down to consumers, may have to be modified to accept the possibility of reversed flow caused by the DG exceeding local demand. This has a significant impact on the protection systems required.

It will be noted from Fig. 13.7 that the primary and low-voltage distribution systems are connected in a radial configuration. Circuit loops between adjacent substations are avoided because these can lead to circulating currents, which may increase the power losses and create difficulty in protection schemes. However, tie circuits between adjacent lines and cables are available to reconfigure the network when a portion of the low-voltage circuit is out of service for maintenance or because of failure. These tie circuits are controlled by a normally open switch which can be closed manually within a few minutes, although an increasing trend is for automation of this operation by *Supervisory Control and Data Acquisition (SCADA)* systems.

In urban and suburban areas, much of the primary and low-voltage distribution system is underground, with readily accessible substations sited in cellars or on small secure plots. Industrial sites may also have a number of substations incorporated into buildings or secure areas; these may be controlled by the works engineer or operated and maintained by an electricity distribution company.

In rural areas and in more dispersed suburban areas, many three-phase overhead lines operating at 10–15 kV or 27–33 kV are supported for many miles on poles which

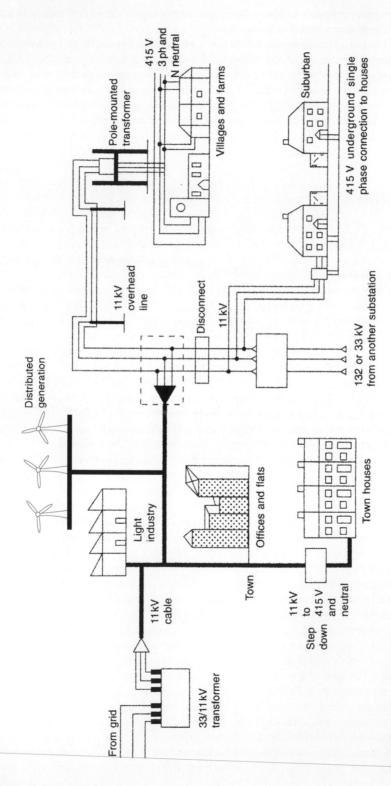

Fig. 13.7 Distribution network

may be of wood, concrete or steel lattice. The 380–415 V three-phase supply is taken from these lines through a small pole-mounted fused input/output transformer. If the maximum load to be taken is below about 50 kW, the supplies for homes or farmsteads may be derived from a single-phase 10–15 kV supply. Typically, a rural primary distribution system supplies up to 50 step-down transformers spread over a wide region. The lines in such a system are vulnerable to damage by tree branches, snow and ice accumulation and lightning strikes and it therefore has lower reliability than underground systems in urban areas. Considerable ingenuity has been applied to protection of this type of system with the use of auto-reclosing supply circuit breakers and automatic reconnection switches, which are described in section 7.4.2.2.

It is now a common practice in developed countries to monitor the primary distribution system down to 10–15 kV and to display alarm, voltage and power-flow conditions in a control room; and in the event of an incident, repair crews are despatched quickly. Repairs to the low-voltage system are still dependent, however, on consumers notifying a loss of supply.

The proper earthing of distribution systems is of prime importance in order that excessive voltages do not appear on connections to individual consumers. It is the practice in UK and some other countries to connect to earth the neutral conductor of the four-wire system *and* the star point of the low-voltage winding on the step-down transformer, not only at the transformer secondary output but also at every load point with a local meter and protective fuse. This is known as the *Protective Multiple Earth (PME)* system, which is described in relation to cable technology in section 9.3.1.3. It is designed to ensure that all metallic covers and equipment fed from the supply are bonded so that dangerously high voltages do not hazard lives.

13.4.1   System design

It is essential that a distribution system is economical in operation and easy to repair, and its design should enable reconnection of a consumer through adjacent feeders to supply substations in the event of failure or outage of part of the system.

Copper or aluminium conductors with a cross section of 150 mm² to 250 mm² are typically used at the lowest voltages, and these are arranged so that the maximum voltage drop under the heaviest load conditions is no more than 6 per cent; alternatively the voltage at the connection to every consumer must not rise more than 6 per cent under light load conditions. Local adjustment is achieved by off-load tapchanging (usually ± 2.5 per cent or ± 5 per cent), and voltage in the primary circuits (usually 11 kV and 33 kV in the UK) is controlled by on-load tapchangers under Automatic Voltage Control. The construction and operation of tapchangers is described in more detail in section 6.2.5. Reinforcement or extension on the LV network is usually arranged through the installation of a new primary feed point with a transformer, rather than by upgrading the LV network.

Primary underground networks now employ cross-linked polyethylene (XLPE) three-phase cables, as described in sections 9.2.4 and 9.3.1.3. The latest designs incorporate fibre-optic strands for communication purposes. Cables require careful routing and physical protection to minimize the risk of inadvertent damage from road and building works in the vicinity.

Overhead lines are usually of a simple flat three-phase configuration which avoids conductor clashing in high winds and ice precipitation. An example is shown in Fig. 7.20. Impregnated wood poles are normally used. These provide some degree of

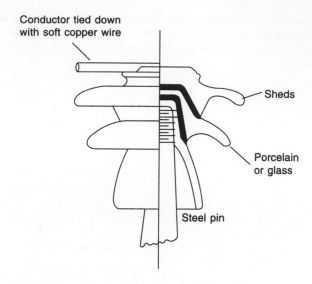

Fig. 13.8 Pin-type insulator

insulation to ground, and they can be quickly replaced in the event of collapse or decay. Insulators are usually of the cap-and-pin type, an example of which is illustrated in Fig. 13.8. Surge diverters and arcing horns provide a considerable measure of over-voltage protection, particularly where many kilometres of exposed line are fed from a primary substation. Arcing horns consist of a carefully positioned air gap between each conductor and earth. They are designed to flash over at a particular voltage when a potentially dangerous surge occurs on the system. Since there is no provision for extinguishing the resulting arc in this event, it is preferable, although more expensive, to provide a surge diverter at strategic positions in the overhead system. A surge diverter consists of a zinc oxide resistance between the live conductor and earth; this presents a very high resistance at normal voltages, but a low value at overvoltages, thereby conducting surge current safely to earth.

Primary substations consist usually of two or three step-down tapchanging trans-formers, with common busbars which may be duplicated for reliability and circuit separation. Remotely controlled interconnecting circuit breakers are installed to provide operating flexibility. Such substations are continuously monitored and metered, the data being brought back to a distribution control centre for display and archiving. A modern tendency is to reduce the number of control centres and to increase the geographical area covered by each through the use of computers and clever alarm handling and by analysis with powerful software which may in the future incorporate artificial intelligence.

13.4.2 System operation

The most important requirement in distribution control is good communication with district personnel and maintenance crews in order to ensure that maintenance and repair is efficiently, safely and swiftly carried out. Ready telephone access by the public and other consumers is also necessary so that dangerous situations or supply failures can be easily reported to engineers and technicians in the field.

The main task of the distribution controller is, therefore, to monitor equipment alarms and to ensure that rapid and effective action is taken. Schedules of equipment outages and maintenance have to be planned and effected with the minimum of disruption to consumers.

13.5   De-regulation and privatization

The electrical energy supply sector has undergone dramatic changes since the beginning of the 1990s. *De-regulation*, or *privatization*, has been adopted throughout the world as a new market-oriented approach. Before privatization, power systems were generally owned by national governments and operated as vertically integrated organizations including generation, transmission and distribution. Privatized structures vary considerably, with the power system being split geographically and/or vertically into separate companies. Competition between generators is always possible, and energy supply contracts to end-users can be open to competition, but the transmission and distribution networks are an inherent monopoly with access and charging for their use overseen by a government regulator. The private companies have access to capital at prevailing interest rates, avoiding government spending restriction (or encouragement) that is a feature of government-owned utilities. It is generally believed that competition can drive energy efficiency up and electricity prices down. However, there is still some debate about how effectively market forces alone can stimulate the correct level of planning and construction of new generation and other plant at a level which ensures satisfactory security of supply for the longer term. This difficulty is exacerbated by the technical inability to store significant quantities of electrical energy. An area of doubt is therefore whether short-term electricity price fluctuations provide a sufficient economic signal to drive the long-term decision of a private company to build new plant, based on their analysis of risk and reward. If a de-regulated system does fail to provide the level of energy supply security expected by modern communities it might not always be clear where the 'obligation to supply' rests.

13.6   Future trends

Power systems are continually evolving and with the increasing capability of computers and software, systems are becoming more precisely controlled for economy of operation and security.

The future is certain to include a greater proportion of primary energy from renewable sources. Generators are likely to be distributed throughout the power system and there may be a very large number of mass-produced very small generators. The mode of operation of the power system will need to become more flexible and is likely to be based on local intelligent control. Operation of the distribution network as an active system can be anticipated, with routine flow reversal and continued operation under islanded conditions becoming feasible. It is expected that generation and load controllers will need to respond to local network conditions, for example reducing the output of relevant generators to cover the period of a circuit outage. This flexible response must be autonomous, to allow unmanned operation, and it would need to be controlled in a local decentralized manner.

We seem to be at the beginning of a new era of major change in power systems plant, technology and management. Great challenges and opportunities for electrical engineers lie ahead.

References

13A.   Weedy, B.M. and Cory, B.J. *Electric Power Systems* (4th edn), Wiley, 1987.
13B.   *EHV Transmission Line Reference Book*, Edison Electric Institute, USA, 1968.
13C.   *Modern Power Station Practice*, Volume L, (3rd edn), Pergamon, 1991.
13D.   Jenkins, N. et al. *Embedded Generation*, Institution of Electrical Engineers, 2000.

Chapter 14

Power quality and electromagnetic compatibility

G.W.A. McDowell and A.J. Maddocks
ERA Technology Ltd

14.1 Power quality

14.1.1 Introduction

The term *power quality* seeks to quantify the condition of the electrical supply. It not only relates largely to voltage but also deals with current and it is largely the corrupting effect of current disturbances upon voltage. Power quality can be quantified by a very broad range of parameters, some of which have been recognized and studied for as long as electrical power has been utilized. However, the advent of the term itself is more modern and it has created a useful vehicle for discussing and quantifying all factors that can describe supply quality.

Power quality is yet another means of analysing and expressing electromagnetic compatibility (EMC), but in terms of the frequency spectrum, power quality characterizes mainly low-frequency phenomena. Perhaps because of this and because of the manner in which it affects electrical equipment, power quality has largely been dealt with by engineers with electrical power experience rather than those with an EMC expertise. In reality, resolving power problems can benefit from all available expertise, particularly since power quality disorders and higher frequency emissions can produce similar effects.

In 1989, the European Community defined the supply of electricity as a product, and it is therefore closely related to the provisions and protection of the EMC Directive (89/336/EEC), but in drawing a comparison between electricity and other manufactured product it is essential to recall a significant difference. Electricity is probably unique in being a product which is manufactured, delivered and used at the same time. An electricity manufacturer cannot institute a batch testing process for example and pull substandard products out of the supply chain. By the time electricity is tested it will have been delivered and used by the customer whether it was of good quality or not.

14.1.2 What determines power quality

Arguably, power quality is not an absolute quantity. Many users of electrical power would consider their electrical supply to be of adequate quality, provided the output of the process it is used to supply is of a high standard. For example, if the desktop computer functions without interruption, temporary resets or voltage-related component failures, then the quality of its electrical supply would be considered perfectly acceptable. Industry would not consider a supply voltage with repeated temporary 100 ms dips to be of poor quality if it did not have an affect on the manufacturing processes.

So power quality is a perception driven by the tolerance (*susceptibility* being the more usual EMC parlance) of the electrical devices that depend upon it. Therefore, the power quality parameters that have evolved to describe acceptable limits are largely an expression of equipment tolerance.

Importantly, although utilities may declare the power quality limits, their attempts to meet this may not be achieved at all times and there is a finite probability that power quality will sometimes fall outside these limits. If the number of times that a supply falls outside the declared power quality limits is more than can be economically tolerated due to the adverse effects upon plant and equipment, then mitigating measures should be taken.

14.1.3 Supply characteristics

The complexity introduced by the length of supply distribution circuits, environmental factors, loading characteristics and the natural degradation of the distribution equipment means that power quality can differ significantly with location.

In general, power quality tends to degrade progressively, the greater the distance is from the source of generation. This is because the supply impedance is lowest near the terminals of the generating systems or where interconnections exist between several sources, and from such points the supply impedance increases along lines and cables, and through transformers. In addition, the predominantly radial nature of power distribution means that load current fluctuations tend to have a more pronounced effect closer to where the load is concentrated.

The power quality at the *Point of Common Coupling (PCC)* with the public electricity supply network defines the quality of the electricity supply to the consumer. For many years the electricity supply industry in the UK has applied planning limits and set standards through engineering recommendations, such as G5/4 (references 14A and 14B) for harmonics. In recent years CENELEC has created EN 50160 (reference 14C) which defines the standards of power quality that is common in Europe for medium-voltage (less than 35 kV) and low-voltage (less than 1 kV) systems. EN 50160 effectively defines the electrical supply environment that European consumers can expect to experience rather than setting the standards for supply. It has usefully added to the awareness of power quality by encompassing within one document the broad spectrum of power quality parameters defined in Table 14.1. It has not set tighter standards, but merely characterized those which already exist.

In order to set immunity levels for equipment, it is first necessary to define the electrical environment. This is defined by IEC 61000-2-2 (reference 14D) which sets the compatibility levels for low-frequency conducted disturbances on low-voltage networks. These levels are less stringent than EN 50160. Therefore, not surprisingly, the margin between voltage characteristic levels (as defined by EN 50160) with the end-user equipment immunity (as defined by IEC 61000-6-1 and IEC 61000-6-2) shows that the level of immunity for most power quality phenomena would be insufficient to protect end-user equipment adequately. In practice, the actual level of disturbance is likely to be equal or less than the immunity levels defined by generic standards, and the likelihood of a disturbance affecting the compliant end-user equipment is greatly reduced.

14.1.4 Key parameters

The parameters that are commonly used to characterize supplies are listed in Table 14.1 together with the typical tolerance limits which define acceptable norms. Within Europe these power quality limits are defined by the EN 61000 series of

Table 14.1 Summary of power quality levels defined by EN 50160

Phenomena	EN 50160 (LV)
Power frequency (50 Hz)	*Interconnected systems* ±1% (95% of week) +4% (absolute level) −6% (absolute level)
Supply voltage variations on 230 V nominal	±10% (95% of week based on 10 min samples, rms)
Rapid voltage changes	±5% Frequent ±10% Infrequent
Flicker	P_k= 1.0 (95% of week)
Supply voltage dips	*Majority* Few 10 s Duration <1 s Depth <60% *Some locations* Few 1000 per year of < 15% depth
Short interruptions	20–500 per year Duration 1 s of 100% depth
Long interruptions	10–50 per year Duration >180 s of 100% depth
Temporary power frequency overvoltage	<1.5 kV
Transient overvoltages	*Majority* <6 kV *Exceptionally* >6 kV
Supply voltage unbalance	*Majority* <2% (95% of the week) *Exceptionally* >2%, <3% (95% of the week)
Harmonic voltage distortion	THD <8% (95% of the week)
Interharmonic voltage distortion	Under consideration
Mains signalling	95 to 148.5 kHz at up to 1.4 Vrms (not in MV)

standards in order to be compatible with the susceptibility limits set for equipment. The more a supply deviates from these limits, the more likely it is that malfunction could be experienced in terminating equipment. However, individual items of equipment will have particular sensitivity to certain power quality parameters while having a wider tolerance to others. Table 14.2 provides examples of equipment and the power quality parameters to which they are particularly sensitive.

Table 14.2 shows a preponderance of examples with a vulnerability to voltage dips. Of all the power quality parameters, this is probably the most troublesome to the manufacturing industry; and in the early 1970s, as the industry moved towards a reliance on electronic rather than electromagnetic controls, it was commonly observed how much more vulnerable the industrial processes were to supply disturbances.

Supply distortion (characterized by harmonics) is another power quality parameter that has received enormous attention, with many articles, textbooks and papers written on the subject. However, the modern practices that will be discussed later have reduced the degree to which this currently presents a problem.

Other parameters tend to be much less problematic in reality, although that is not to say that perceptions sometimes suggest otherwise. Voltage surge and transient overvoltage in particular are often blamed for a wide range of problems.

Table 14.2 Examples of sensitivity to particular power quality parameters

Equipment type	Vulnerable power quality parameter	Effect if exceeded	Range
Induction motor	Voltage unbalance	Excessive rotor heating	<3%
Power factor correction capacitors	Spectral frequency content. This is usually defined in terms of harmonic distortion	Capacitor failure due to excessive current flow or voltage. Most sensitive if resonance occurs	<10% voltage THD in non-resonant conditions. In resonant conditions, even an individual component of 1% could cause failure
PLCs	Voltage dips	Disruption to the programmed functionality	Varies, but <60% volts for 30 ms could trigger failure
Computing systems	Voltage dips	Disruption to the programmed functionality	Varies, but <60% volts for 30 ms could trigger failure
Variable speed drives, motor starters and attracted armature control relays	Voltage dips	Disruption to the control system causing shutdown	Varies, but <60% for 30 ms could trigger failure
Power transformers	Spectral frequency content of load current. This is usually defined in terms of harmonic distortion	Increased losses leading to excessive temperature rise	At full load, the acceptable additional losses will be minimal
Devices employing phase control, such as light dimmers and generator automatic voltage regulator (AVRs)	Alteration in waveform zero crossing due to waveform distortion, causing multiple crossing or phase asymmetry	Instability	Will depend upon the rate at which multiple transitions occur
Motor driven speed-sensitive plant	Induction and synchronous motor shaft speed are proportional to supply frequency. Some driven loads are sensitive to even small speed variations	The motors themselves are tolerant of small speed variations. At high supply frequencies (>10%) shaft stresses may be excessive due to high running speeds	Limits depend on the sensitivity of the driven load

However, very often this is a scapegoat when the actual cause cannot be identified. Even when correlation with switching voltage transients is correctly observed, the coupling introduced by poor wiring installations or bad earth bonding practices can be the real problem. Unlike the other power quality parameters, voltage transients have a high frequency content and will couple readily through stray capacitance and mutual inductance into neighbouring circuits. Coupling into closed conductor loops that interface with sensitive circuits such as screens and drain wires can easily lead to spurious events.

14.1.5   Common problems

14.1.5.1   Harmonic voltage and current distortion

Equipment which uses power electronic switching techniques to control intake power is a common source of current distortion. Examples include variable speed drives, uninterruptible power supplies and switch mode power supplies. The dominant frequency components tend to be of low orders with 3rd, 5th ,7th, 11th and 13th being common. The presence of these components in current and voltage waveforms introduces additional losses in the conductors and magnetic circuits of cables, transformers, motors and generators. The frequency components can also excite resonance in *Power Factor Correction (PFC)* capacitors.

The following mitigating measures can be applied:

- specification of frequency content in current loading and applied voltage on purchase specifications
- use of K-factor rated transformers
- use of detuned PFC banks
- use of double-rated neutrals (or higher) in systems with high 3rd harmonic loading
- use of passive or active filters to reduce harmonic content

Voltage and current waveform distortion is also a consequence of distorting loads. Mitigating measures for waveform distortion include:

- specification of waveform distortion tolerance requirements in purchase specifications
- use of equipment which does not rely on phase measurement by point-on-wave such as zero-crossing detection
- use of rms rather than peak sensing
- use of 3-phase rather than 1-phase sensing for devices such as AVRs
- use of permanent magnet generators with generator AVRs
- use of generators with 2/3 pitch windings to help minimize harmonic impedance and thus minimize voltage distortion

Low frequency harmonic distortion is also often accompanied by high frequency non-harmonic components. These are produced by rapid switching and are a feature of virtually all power electronic systems. For this reason, it is important that earthing and bonding be applied in order to minimize impedance. Stranded conductors of large cross-sectional area installed with parallel paths in a mesh arrangement are examples of good practice.

14.1.5.2   Voltage dips

Voltage dip immunity in end-user equipment continues to be a major quality issue
in Europe and is only gradually improving with the implementation of IEC and
CENELEC standards. As a result, equipment malfunction due to transient voltage dips
continues to lead to complaints about adverse power quality. In some regions of
Europe it may even be perceived that power quality is reducing due to lack of voltage
dip immunity alone. The reality is more likely to be that such perception is influenced
by an increase in the numbers of susceptible equipment, or changes in the application
of equipment, despite the fact that dip immunity complies with present day EMC
standards. However, the IEC and CENELEC standards offer a framework for specification and testing that could eventually benefit both the electricity supply industry and
end users.

The common cause of voltage dips is system faults within public electricity distribution networks. Severe voltage dips will also occur during fault clearance on end-user
distribution networks. Thus, although it is common to use circuit protection to isolate
on-site faults, the widespread disturbing effect of the accompanying voltage dip is
often overlooked. References 14E and 14F discuss voltage mitigation measures some
of which are briefly referred.

Mitigating measures to be applied are:

*   specification of dip-tolerant equipment at the time of purchase
*   use of mitigation measures based on FACTS techniques, such as the *Dynamic
    Voltage Restorer (DVR)*. This can be applied at the PCC to support the
    supply voltage during network faults. It is expensive but can offer an effective
    solution.
*   use of constant-voltage transformers to improve voltage regulation without the
    need for energy storage
*   application of battery, diesel rotary or flywheel storage to provide energy
    ride-through for critical plant
*   use of mechanically latched rather than electrically held relays and contactors
    on critical plant

14.1.6   Monitoring

Monitoring is essential when tackling equipment malfunction which is thought to
be power quality related. It is also a good policy to benchmark the quality of any new
supply arrangements that support important business processes so that the relevance
of power quality conditions, or changing conditions, can be assessed at a later time.
There are a large number of power quality monitors available (reference 14G) and their
selection should target the particular power quality parameters of importance. It must
be decided whether data logging is required, or a more simple detection of power quality parameters that exceed predefined thresholds. Some monitors offer the opportunity
to achieve both functions simultaneously and then to examine retrospectively with
regard to thresholds or standards. Large data storage capability using hard disk drives
or static memory has made the monitoring activity much easier to apply but can lead
to an enormous amount of data to inspect.

Power quality monitors are packaged in portable and fixed patterns. Fixed pattern
versions are becoming more common and offer the opportunity to collect data and
observe trends throughout the life of the installation. Permanently installed monitors

can also be internet connected to allow the data to be downloaded and to allow expert opinion to be called upon for analysis.

Any power quality survey should be carefully planned with clear objectives. The monitor and its sensors (such as current transformers) must be selected to provide the appropriate operating range and response bandwidth. Also the physical constraints of conductor diameter and distances, together with safe working, must be considered when planning how the power quality monitor is to be applied.

Time correlation is a powerful tool in relating power quality disturbances to process problems. Therefore, it is necessary to establish a time log, in which the operators can enter the time and details of events. It is not unusual to discover power quality characteristics that were unknown but bear no relationship to the problem being investigated.

14.2   Electromagnetic compatibility

14.2.1   Introduction

Electromagnetic Compatibility (EMC) is achieved when co-located equipment and systems operate satisfactorily, without malfunction, in the presence of electromagnetic disturbances. For example, the electrical noise generated by motor-driven household appliances, if not properly controlled, is capable of causing interference to domestic radio and TV broadcast reception. Equally, microprocessor-based electronic control systems need to be designed to be immune to the electromagnetic fields from hand-held radio communication transmitters, if the system is to be reliable in service. The issues covered by EMC are quality of life, spectrum utilization, and operational reliability, through to safety of life, where safety-related systems are involved.

The electromagnetic environment in which a system is intended to operate may comprise a large number of different disturbance types, emanating from a wide range of sources including:

- mains transients due to switching
- radio frequency fields due to fixed, portable and mobile radio transmitters
- electrostatic discharges from human body charging
- powerline surges, dips and interruption
- power frequency magnetic fields from power lines and transformers

In addition to having adequate immunity to all these disturbances, equipment and systems should not adversely add electromagnetic energy to the environment above the level that would permit interference-free radio communication and reception.

14.2.1.1   Sources

The essence of all EMC situations is contained in the simple source, path, receptor model shown in Fig. 14.1.

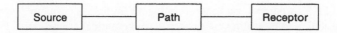

Fig. 14.1 Source, path, receptor model

Sources comprise electromechanical switches, commutator motors, power semi-conductor devices, digital logic circuits and intentional radio frequency generators. The electromagnetic disturbances they create can be propagated via the *path* to the *receptor*, such as a radio receiver, which contains a semiconductor device capable of responding to the disturbance, and causing an unwanted response, i.e. interference.

For many equipments and systems, EMC requirements now form part of the overall technical performance specification. The EU's EMC Directive, 89/336/EEC, was published in 1992 and came into full implementation on 1 January 1996. All apparatus placed on the market or taken into service must, by law, comply with the Directive's essential requirements, that is it must be immune to electromagnetic disturbance representative of the intended environment and must generate its own disturbance at no greater than a set level that will permit interference-free radio communication. The Directive refers to relevant standards which themselves define the appropriate immunity levels and emission limits. More information on the EMC Directive and its ramifications is available in reference 14H.

14.2.1.2   Coupling mechanisms

The path by which electromagnetic disturbance propagates from source to receptor comprises one or more of the following:

- conduction
- capacitive or inductive coupling
- radiation

These paths are outlined in Fig. 14.2.

Coupling by conduction can occur where there is a galvanic link between the two circuits, and dominates at low frequencies where the conductor impedances are low. Capacitance and inductive coupling takes place usually between reasonably long co-located parallel cable runs. Radiation dominates where conductor dimensions are comparable with a wavelength at the frequency of interest, and efficient radiation occurs.

For example, with a personal computer, the radiation path is more important for both emission and immunity at frequencies above 30 MHz, where total cable lengths are of the order of several metres. Designers and installers of electrical and electronic equipment need to be aware that all three coupling methods exist so that the equipment can be properly configured for compatibility.

14.2.1.3   Equipment sensitivity

Analogue circuits may respond adversely to unwanted signals in the order of millivolts. Digital circuits may require only a few 100 mV disturbance to change state. Given the high levels of transient disturbance present in the environment, which may be in the order of several kilovolts, good design is essential for compatibility to occur.

14.2.2   Simple source models

For many EMC situations, such as coupling by radiation, effective prediction and analysis are achieved by reference to simple mathematical expressions. For nearly all products experiencing EMC problems, the equipment will work perfectly in

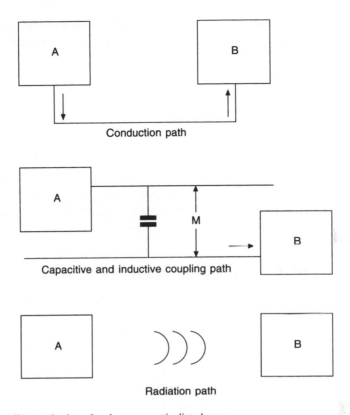

Fig. 14.2 Coupling mechanisms for electromagnetic disturbance

the development laboratory and only when it is subjected to external electromagnetic disturbance do other facets of its characteristics emerge. Under these circumstances, circuit conductors are considered as antennas capable of both transmitting and receiving radio energy. The circuit can usually be assessed as either a short monopole antenna, for instance where one end of the conductor is terminated in a high impedance, or as a loop where both ends are terminated in a low impedance.

For the monopole equivalent at low frequencies and at distances greater than a wavelength, the field strength E, at distance d (in metres) is given by eqn 14.1:

$$E = \frac{60\pi \times I \times h}{\lambda \times d} \text{ (V/m)} \tag{14.1}$$

where I is the current in amperes, h is the length of the conductor and λ is the wavelength ($= 300 \div$ frequency in megahertz), both in metres. It can be seen that the field strength is greater for shorter distances and higher frequencies (shorter wavelengths).

For loop radiators at low frequencies, the field is given by eqn 14.2:

$$E = \frac{120\pi^2 \times I \times A \times n}{d \times \lambda^2} \text{ (V/m)} \tag{14.2}$$

where n is the number of turns and A is the area of the loop. The field strength is greater at shorter distances and higher frequencies.

At high frequencies where the conductor lengths are comparable with a wavelength, a good approximation of the field at a distance can be taken if the source is considered as a half-wave dipole. The field is then given by eqn 14.3:

$$E = \frac{7\sqrt{P}}{d} \text{(V/m)} \tag{14.3}$$

where P is the power in watts available in the circuit.

14.2.2.1   Receptor efficiency

To estimate the degree of coupling in the radiated path, empirical data give values of induced currents of about 3 mA for an incident electromagnetic wave of 1 V/m. This relationship can be used to good effect in converting the immunity test levels in the standards into an engineering specification for induced currents impressed at an input port due to coupling via an attached cable.

14.2.3   Signal waveforms and spectra

For many digital electronic systems, the main concerns in emission control at low frequencies are associated with power line disturbance generated by switch mode power supplies. These devices switch at a relatively high rate, in the order of 30–100 kHz, and produce a line spectrum of harmonics spreading over a wide frequency band as shown in Fig. 14.3.

Without mains filtering, the emission levels from the individual harmonics are considerably in excess of the common emission limits. Care is needed in the sourcing of these subassemblies to ensure that they are compliant with the relevant standards.

At higher frequencies, noise from digital circuits switching at very high rates (clock frequencies in excess of 100 MHz are not uncommon) couples via external cables and

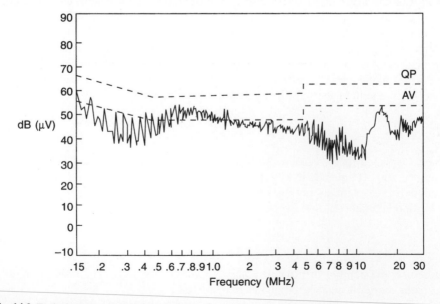

Fig. 14.3 Typical conducted emission spectrum of a switch mode power supply

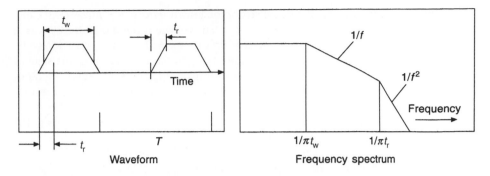

Fig. 14.4 Waveform and frequency spectrum of a digital signal

radiates in the VHF band (30–300 MHz) or it may radiate directly from circuit boards in the UHF band (300–1000 MHz). A clock oscillator has a waveform and a spectrum of the type shown in Fig. 14.4.

It can be seen that the turning points in the spectrum are $(1/\pi) \times$ the pulse width t_w, above which frequency the spectrum decreases is inversely proportional to frequency, and $(1/\pi) \times$ the rise time t_r. At frequencies greater than $1/\pi t_r$ the spectrum decays rapidly in inverse proportion to the square of the frequency. Thus for longer pulse durations, the high-frequency content is reduced; if the *rise time* is slow then the content is further reduced. For good emission control, slower clock speeds and slower edges are better for EMC. This is contrary to the current trends where there are strong performance demands for faster edges and higher frequency clocks.

Many equipment malfunction problems in the field are caused either by transient disturbances, usually coupled onto an interface cable, or by radar transmitters if close to an airfield. The disturbance generated by a pair of relay contacts opening comprises a series of short duration impulses at high repetition rate, as illustrated in Fig. 14.5.

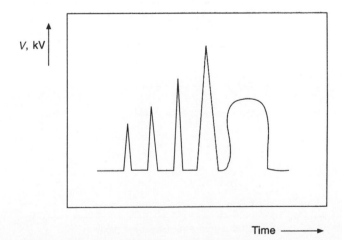

Fig. 14.5 Voltage across opening relay contacts

As the contacts separate the energy stored in the circuit inductance is released, causing a high voltage to occur across the contacts. The voltage is often sufficient to cause
a discharge and a spark jumps across the gap. This is repeated until the gap is too wide.
When coupled into a digital electronic circuit, the disturbance can change the state of
a device and interference in the form of an unwanted operation or circuit 'lock-up'
occurs. Control is exercised by ensuring that the receptor circuit has adequate immunity to this type of disturbance.

The radar transmission is one example of a modulated radio frequency signal, and
this can often cause interference even in low-frequency electronic circuits. The radio
frequency energy is rectified at the first semiconductor junction encountered in its
propagation path through the equipment, effectively acting as a diode rectifier or
demodulator. The rectified signal, an impulse wave in the case of a radar, is thus
processed by the circuit electronics and interference may result if the induced signal is
of sufficient amplitude, see Fig. 14.6.

Most equipment is designed and constructed to be immune to radio frequency fields
of 3 V/m (or 10 V/m for industrial environments) at frequencies up to 1 GHz. Many
radars operate at frequencies above 1 GHz (for example 1.2 GHz, 3 GHz and 9.5 GHz)
and equipment close to an airfield may suffer interference because it is not designed

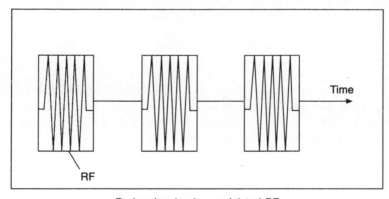

Radar signal pulse modulated RF

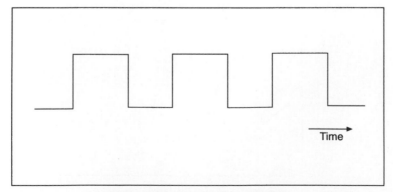

Radar signal after circuit rectification

Fig. 14.6 Radar signal

to be immune. Under these circumstances, architectural shielding is required, often comprising the use of glass with a thin metallic coating which provides adequate light transmittance and, more importantly, effective shielding at these high radio frequencies.

14.2.4   EMC limits and test levels

14.2.4.1   Emissions

The radio frequency emission limits quoted in EMC standards are usually based on the level of disturbance that can be generated by an apparatus or system such that radio or TV reception in a co-located receiver is interference-free.

Conducted emission limits are set to control the disturbance voltage that can be impressed on the voltage supply shared by the source and receptor where coupling by conduction occurs. Limits currently applied in European emission standards are shown in Fig. 14.7. The Class A limits are appropriate for a commercial environment where coupling between source and receptor is weaker than in the residential environment, where the Class B limits apply. The difference in the limits reflects the difference in attenuation in the respective propagation paths.

The limits shown apply when using the 'average' detector of the measuring instrument, and are appropriate for measuring discrete frequency harmonic spectral line emissions. A 'quasi-peak' detector is employed to measure impulsive noise for which the limits are up to about 10 dB higher, i.e. more relaxed. The standards often require that the 'average' limit is met when using the 'average' detector, and the 'quasi-peak' limit is met using this 'quasi-peak' detector. Nearly all modern EMC measuring instrumentation contains provision for measuring using both detectors to the required degree of accuracy.

Typical limit levels are in the order of 1–2 mV for Class A and 0.2–0.6 mV for Class B. These are quite onerous requirements given that some devices, such as triacs for motor speed control may be switching a few hundred volts peak.

Radiated emission limits are derived from a knowledge of the field strength at the fringe of the service area, from typical signal to noise ratios for acceptable reception, and also from applying probability factors where appropriate. Typical limits are shown in Fig. 14.8.

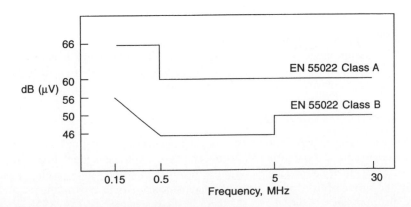

Fig. 14.7 Conducted emission limits (average detector)

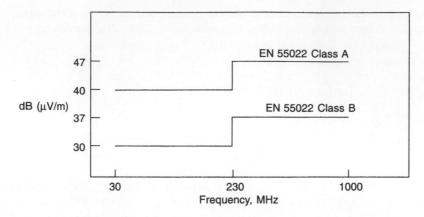

Fig. 14.8 Radiated emission limits

Below 230 MHz the Class B limits are equivalent to a field strength of 30 μV/m at a distance of 10 m. Many items of information technology equipment have clock frequencies in the range 10–30 MHz, for which the harmonics in the ranges up to a hundred megahertz or more are effectively wanted signals, required to preserve the sharp edges of the digital waveform. This harmonic energy has to be contained within the apparatus by careful PCB design and layout and/or shielding and filtering.

Low-frequency limits for harmonic content are derived from the levels of disturbance that the supply networks can tolerate, and are expressed either as a percentage of the fundamental voltage or as a maximum current. Even harmonics are more strongly controlled than odd harmonics, since the even harmonics indicate the presence of a dc component. Similarly, flicker limits are set by assessing the effects on lighting of switching on and off heavy power loads, such as shower heaters.

14.2.4.2   Immunity

Immunity test levels are set to be representative of the electromagnetic environment in which the equipment is intended to operate. In European standards, two environments are considered, the residential, commercial and light industrial environment and the heavy industrial environment. The distinction between the two is not always clear but most equipment suppliers and manufacturers are knowledgeable on the range of disturbances that must be considered for their product to operate reliably in the field and can make the appropriate choice without difficulty. The key factor is whether heavy current switching and/or high-power radio frequency sources are present. If they are, the more severe 'industry' levels should be selected.

(a) Electrostatic discharge (ESD)

Electrostatic charge is built up on a person walking across a carpet or by other actions where electric charge separation can occur. The charge voltage is much higher for synthetic materials in dry atmospheres with low relative humidity. Although charge potentials in the order of 10–15 kV may be encountered in some environments such as hotels, the standards bodies have selected an air discharge level of 8 kV as being representative of a broad range of circumstances. The ESD event is very fast with a sharp edge having a rise time of about 1 ns and a duration of about 60 ns. This generates a spectrum which extends into the UHF bands and therefore presents a formidable test for many types of equipment.

(b) *Electrical fast transients*

This is the disturbance type adopted by the standards bodies to be representative of the showering arc discharge encountered across opening relay contacts. Figure 14.9 shows the general waveform of the disturbance applied in the harmonized European standards. These transients are applied directly to the mains power conductor, or to interface and input/output cables via a capacitive coupling method to simulate the effects of co-located noisy power conductors. The voltage levels applied on the mains supply are 1 kV for the residential environment and 2 kV for the industrial environment.

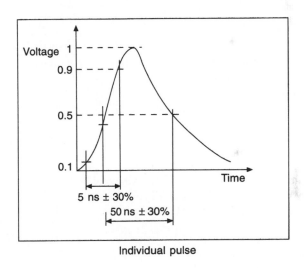

Individual pulse

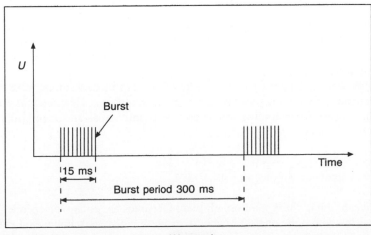

Wavetrain

Fig. 14.9 Waveform of fast transient

(c) *Radio Frequency (RF) fields*

At low frequencies (below 80 MHz) the interaction of incident electromagnetic waves with receptor systems can be simulated effectively by a simple induced voltage impressed either with respect to ground via a network known as a CDN, or longitudinally via a transformer (bulk current injection). At higher frequencies (80–1000 MHz) the field is applied directly to the equipment under test, usually by setting up a calibrated transmitting antenna situated at a separation distance of about 3 m. The tests are performed in shielded enclosures in order that the RF energy can be controlled and no external interference occurs.

In the more recent standards, the walls of the screened rooms are lined with absorbing material to provide a reasonably uniform field within the chamber. The applied waveform is modulated with either a 1 kHz tone to simulate speech or a pulse train to represent digital cellular radio transmission.

(d) *Dips, surges and voltage interruptions*

Many other types of disturbances are present in the environment. These are the slow-speed types which are usually associated with power switching, lightning pulses and power failure. They are usually simulated by specialist disturbance generators. In many cases, immunity to these disturbances is achieved by good design of the mains power supply in the equipment under test. This is discussed fully in section 14.1.5.2.

(e) *Magnetic fields*

Equipment containing devices sensitive to magnetic fields should be subjected to power frequency (50 Hz) magnetic fields. Typical levels are 3 A/m for the residential, commercial and light industrial environment and 30 A/m in the more severe industrial environment.

14.2.5   Design for EMC

14.2.5.1   *Basic concepts*

The preferred and most cost-effective approach to the achievement of electromagnetic compatibility is to incorporate the control measures into the design. At the design inception stage some thought should be given to the basic principles of the EMC control philosophy to be applied in the design and construction of the product. The overall EMC design parameters can be derived directly or determined from the EMC standards to be applied, either as part of the procurement specification or as part of the legal requirements for market entry. There are usually two fundamental options for product EMC control:

- shielding and filtering (see Fig. 14.10(a))
- board level control (see Fig. 14.10(b))

For the *shielding and filter* solution, all external cables are either screened leads, with the screens bonded to the enclosure shield, or unscreened leads connected via a filter. The basic principle is to provide a well-defined barrier between the inner surface of the shield facing the emissions from the PCBs and the outer surface of the shield

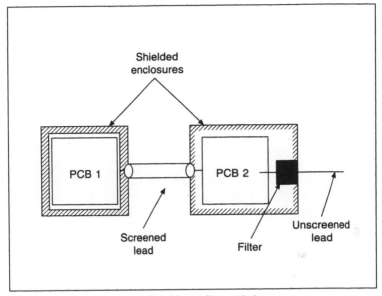

(a) Shield and filter solution

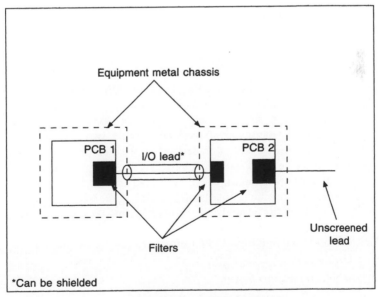

(b) Board level control solution

Fig. 14.10 Options for product EMC control

interacting with the external environment. This solution requires the use of a metal or metal-coated enclosure. Where this is not possible, or not preferred, then the *board level control* option is appropriate, where the PCB design and layout provides inherent barriers to the transfer of electromagnetic energy. Filters are required at all the cable interfaces, except where an effective screened interface can be utilized.

14.2.5.2 Shielding

For design purposes, effective shielding of about 30–40dB can be provided at high frequencies by relatively thin metal sheets or metal coatings on plastic. The maximum shielding achievable is associated with the apertures, slots and discontinuities in the surface of the shield enclosure. These may be excited by electromagnetic energy and can resonate where their physical length is comparable with a wavelength, significantly degrading the performance of the shield. The following basic rules apply:

- the maximum length of a slot should be no greater than 1/40 of the wavelength at the highest frequency of concern
- a large number of small holes in the shield gives better performance than a small number of large holes
- the number of points of contact between two mating halves of an enclosure should be maximized
- mains or signal line filters should be bonded to the enclosure at the point of entry of the cable

14.2.5.3 Cable screens termination

For maximum performance, cable shields should be terminated at both ends with a 360° peripheral (i.e. glanded) bond. This is not always possible and in some cases it is undesirable because of the associated ground loop problem (Fig. 14.11). Noise currents, I, in the ground generate a voltage V between the two circuits A and B which can drive a high current on the outer surface of the interconnecting screen, thus permitting energy to be coupled into the internal system conductors.

In many applications involving the use of long conductors in noisy environments, the simplest solution is to break the ground loop by bonding the screen at one end only, as shown in Fig. 14.12. Here the ground noise voltage is eliminated but the shield protects only against electric fields and capacitive coupling. Additional or alternative measures, such as opto-isolation are required if intense magnetic fields are present.

14.2.5.4 PCB design and layout

Control at board level can be achieved by a careful design of the board involving device selection and track layout. As discussed in section 14.2.2, emissions from the PCB tracks may be reduced at high frequencies if the devices are chosen to have slow switching rates and slow transition (i.e. long rise and fall) times.

Device selection can also improve immunity by bandwidth control. The smaller the bandwidth, the less likely it is that high-frequency disturbances will be encountered within the pass-band of the circuit. Although rectification of the disturbance may occur in the out-of-band region, the conversion process is more inefficient and higher immunity usually results.

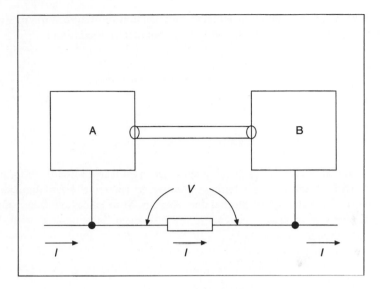

Fig. 14.11 The ground loop problem for screen terminated at both ends

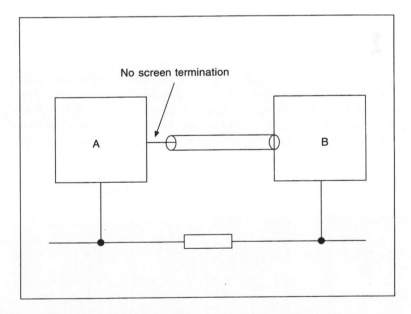

Fig. 14.12 Ground loop broken by termination of the screen at one end only

The tracks on PCBs can act as antennas and the control methods involve reducing their efficiency. The following methods can be applied to good effect:

- reduce the area of all track loops
- minimize the length of all high-frequency signal paths
- terminate lines in resistors equal to the characteristic impedance
- ensure that the signal return track is adjacent to the signal track
- remove the minimum amount of copper on the board, i.e. maximize the surface area of the $0\,V$ (zero-volt ground) and VCC (power) planes

The latter two points are generally achieved where a multilayer board configuration is employed. These measures are highly effective at reducing board emissions and improving circuit immunity to external disturbances. Multilayer boards are sometimes considered relatively expensive but the extra cost must be compared with the total costs of other measures that may be required with single or double-sided boards, such as shielding, filtering and additional development and production costs.

14.2.5.5 Grounding

Grounding is the method whereby signal returns are managed, and it should not be confused with earthing which deals with protection from electrical hazards. Grounding is important at both the PCB level and circuit interconnection level. The three main schemes are shown in Fig. 14.13.

In the series ground scheme, noise from circuit A can couple into circuit C by the common impedance Z. This problem is overcome in the single-point ground, but the scheme is wasteful of conductors and not particularly effective at high frequencies where the impedance of the grounding conductors may vary, and potential differences can be set up.

The ideal scheme is the multipoint ground. Generally, single-point grounding is used to separate digital, analogue and power circuits. Multipoint grounding is then used whenever possible within each category of analogue or digital. Some series ground techniques may be employed where the coupled noise levels can be tolerated. Usually the overall optimum solution is derived from good basic design and successive experimentation.

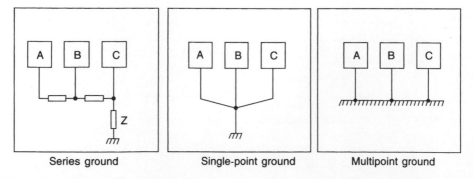

  Series ground   Single-point ground   Multipoint ground

Fig. 14.13 Three main grounding schemes

14.2.5.6 Systems and installations

The general principles discussed above may be applied at a system or installation level. Guidance on this topic is available in references 15 J and 15 K.

14.2.6 Measurements

14.2.6.1 Emissions

Conducted emission tests comprise measurements of voltage across a defined network which simulates the RF impedance of a typical mains supply. These *Line Impedance Stabilizing Networks (LISNs)* also provide filtering of the supply to the *Equipment Under Test (EUT)* and are also known as artificial mains networks or isolating networks. The EUT is connected to the LISN in a manner which is representative of its installation and use in its intended environment. Figure 14.14 shows the general arrangement.

The EUT is configured in a typical manner with peripherals and inputs/outputs attached, and operated in a representative way which maximizes emissions.

Radiated emissions are made by measuring the field strength produced by the EUT at a defined distance, usually 3 m or 10 m. The measurements are made on an open area test site which comprises a metallic ground plane, over a flat surface with no reflecting objects and within a defined ellipse.

The ground plane should cover a larger area than the test range, for example a 6×9 m area would be ideal for a 3 m range, and a 10×20 m area for a 10 m range. The EUT is situated 1 m above the ground plane on an insulating support (unless it is floor-standing equipment) and a calibrated antenna is placed at the required test distance from the EUT. At any emission frequency, such as the harmonic of the clock oscillator in a PC, the receiver is tuned to the frequency and the antenna height is raised between 1 m and 4 m in order to observe the maximum field strength radiated by the product. (The net field strength is the sum of the direct and ground-reflected waves and it varies with height.) The EUT is also rotated about a vertical axis in order to measure the maximum radiation in the horizontal plane.

Measuring instruments for both conducted and radiated emission measurements comprise spectrum analysers or dedicated measuring receivers. The spectrum analyser

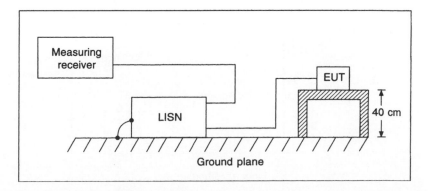

Fig. 14.14 Measurement of conducted emissions

usually has to be modified to have a stage for preselection which prevents overload and damage in the presence of impulsive noise, and it may require additional external pulse-limiting protection when performing conducted emissions measurements with a LISN. Both instruments usually have facilities of computer control by the IEEE bus, avoiding the necessity for manual operation. When using spectrum analysers, it is important to check for overload or spurious emissions by ensuring that the observed indication on the display reduces by 10 dB when an additional 10 dB RF attenuation is introduced at the front end of the analyser.

The EMC measuring receivers are designed to meet the stringent requirements of Publication 16 of CISPR (Committee International Special Perturbations Radioélectrique), a subdivision of the IEC. This sets out specifications for input impedance, sensitivity, bandwidth, detector function and meter response, such that the reproducibility of the tests can be guaranteed.

14.2.6.2 Immunity

Electrostatic discharge tests are made with an ESD 'gun', set to the desired voltage which is equivalent to the human charge potential, and having well-defined charge and discharge characteristics. The ESD is applied to all user-accessible parts of the EUT. The operation of the equipment is thus observed for any malfunction. Immunity to the ESD event is improved by minimizing the ESD energy that can enter the enclosure containing the electronics. The ideal solution is either a good shielded enclosure with small apertures and good bonding between sections, or a totally non-conducting surface. Generally it is difficult to design a product which completely satisfies either solution, but designers should attempt to steer towards one or the other. Measurements of immunity to RF fields are made in a shielded enclosure, the modern types being lined with absorbing materials, such as ferrite tiles on at least five of the six inner surfaces. The EUT is subjected to radiation from an antenna situated in the near vicinity as shown in Fig. 14.15. The field is pre-calibrated to the required level of field strength specified in the appropriate standard, prior to the introduction of the EUT into the chamber. The RF is swept slowly from 80–1000 MHz and any equipment misoperations are noted, the performance level of the EUT having been defined prior to the start of the test.

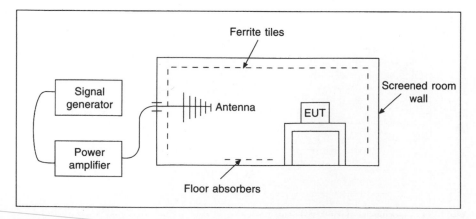

Fig. 14.15 RF field immunity-test arrangement

Transients, surges, dips and interruptions tests are performed with dedicated test instrumentation which fully satisfies the requirements of the relevant standards. Generally the tests are much quicker to perform than the RF field test, and information on the EMC performance of the EUT can be gathered rapidly.

For ESD tests and fast transient tests, the equipment should carry on working after the application of the disturbance without any loss of data. For the RF field test, there should be no loss of performance outside that specified by the manufacturer at any time during the test. For dips and surges etc. provided the equipment works satisfactorily, after a manual reset, it will be deemed to have passed the test.

14.3   Standards

The three types of EMC standards in current use in Europe are *basic* standards, *generic* standards and *product-specific* standards. Basic standards contain the test methods and test levels at limits but do not specify a product type. Product standards contain comprehensive details on how the product should be configured and operated during the test and what parameters should be observed. Generic standards apply in the absence of a product standard and are relevant to all products which may be operated within a defined environment. Both product standards and generic standards may refer to basic standards for their test methods. The important standards in current use in Europe are listed. Those listed in section 14.3.1 and 14.3.2 which are applicable to a particular product or system may be applied by the manufacturer to demonstrate compliance with the EMC Directive.

14.3.1   Generic standards

EN 61000-6-3: Emissions: residential, commercial and light industrial environment
EN 61000-6-4: Emissions: industry environment
EN 61000-6-1: Immunity: residential, commercial and light industrial environment
EN 61000-6-2: Immunity: industrial environment

14.3.2   Important product standards

EN 50011: Emissions: industrial, scientific and medical equipment
EN 50013: Emissions: radio and TV equipment
EN 55020: Immunity: radio and TV equipment
EN 50014-1: Emissions: household appliances and portable tools
EN 55014-2: Immunity: household appliances
EN 50015: Emissions: lighting equipment
EN 61547: Immunity: lighting equipment
EN 50022: Emissions: information technology equipment
EN 55024: Immunity: information technology equipment
EN 50091-2: Uninterruptable Power Supplies (UPS)
EN 55103: Professional audio and video equipment
EN 60947: Low-voltage switchgear and controlgear
EN 61131-2: Programmable controllers
EN 61326: Measurement and control instruments
EN 61800-3: Electrical power drives
Other product standards are emerging at a high rate which will mean that reliance on generic standards will diminish significantly in the future.

14.3.3   Basic standards

EN 61000-3-2: Emissions: mains harmonic current
EN 61000-3-3: Emissions: voltage fluctuation and flicker
EN 61000-4-2: Immunity: ESD
EN 61000-4-3: Immunity: RF fields
EN 61000-4-4: Immunity: fast transients
EN 61000-4-5: Immunity: surges
EN 61000-4-6: Immunity: conducted RF voltages
EN 61000-4-8: Immunity: magnetic fields
EN 61000-4-11: Immunity: dips and interruptions

References

14A.   Engineering recommendation G5/4. *Limits for Harmonics in the United Kingdom Electricity Supply System*: Electricity Association.

14B.   ETR 122 *Guide to the application of engineering recommendation G5/44* in the assessment of harmonic voltage distortion and connection of non-linear equipment to the electricity supply system in the UK; Electricity Association; 2002.

14C.   EN50160. *Voltage characteristics of electricity supplied by public distribution systems*, CENELEC, 2000.

14D.   IEC 61000-2-2, *Compatibility levels for low-frequency conducted disturbances and signalling in public low-voltage power supply systems*; 2002.

14E.   McDowell, G.W.A. *Protective measures against transient voltage depressions*. Protecting Electrical Networks and Quality of Supply in a De-Regulated Industry Leatherhead, UK: ERA Technology, 1995. p. 8 (Conference: London, UK, 15–16 February 1995).

14F.   Greig, E. *How to improve voltage dip immunity in industrial and commercial power distribution systems*. ERA Report 99-0632R, ERA Technology Ltd, 1999.

14G.   Lam, D. *A review of power quality monitors*. ERA Report 2002-0213, ERA Technology Ltd, 2002.

14H.   Marshman, C. *The Guide to the EMC Directive 89/336/EEC*, EPA Press, 1992, ISBN 095 173623X, 308 pp.

14I.   Maddocks, A.J., Duerr, J.H. and Hicks, G.P. *Designing for Electromagnetic Compatibility: A Practical Guide*, ERA Report 95-0030, ERA Technology Ltd, 1995, ISBN 0700805842, 147 pp.

14J.   Goedbloed, J. *Electromagnetic Compatibility,* Prentice Hall, 1992. ISBN 0132492938, 400 pp.

14K.   Ott, H.W. *Noise Reduction Techniques in Electronic Systems* (2nd edn). John Wiley, USA, 1988, ISBN 0471850683, 426 pp.

14L.   Paul, C. *Electromagnetic Compatibility*. John Wiley, USA, 1988, ISBN 0471850683, 426 pp.

Chapter 15

Electricity and potentially explosive atmospheres

G. Tortoishell
Associate Consultant
Sira Safety Compliance

15.1 Introduction

UK legislation has contained requirements for electrical installations in adverse environments in the Electricity Regulations of the Factories Act since 1908. Over the years these have been given more detail as legislation has been updated. Potentially explosive atmospheres, otherwise known as *hazardous areas* or *flammable atmospheres*, have been given particular emphasis because of the possible major consequences of inadvertent ignition. The disasters at Flixborough and Abbeystead are examples of the many incidents which have occurred, and continue to occur, worldwide. There have been significant changes to the UK legislation in recent years, largely as a result of EU directives dealing with the safety and health of workers. This chapter concentrates on the specific problem of potentially explosive atmospheres, the legal requirements and the practices for meeting them.

15.2 EU directives

The two EU directives dealing specifically with potentially explosive atmospheres are 1999/92/EC (reference 15A), which deals with the safety and health of workers exposed to potentially explosive atmospheres, and 94/9/EC (reference 15B), which covers the equipment for use in such atmospheres. The Chemical Agents 98/24/EC Directive (reference 15C) is also relevant as it contains requirements for flammable materials.

Directive 1999/92/EC covers the assessment of explosion risks, technical measures for explosion protection, organizational measures for explosion protection, coordination duties and explosion protection documentation.

Directive 94/9/EC covers the certification process for hazardous area equipment and the role of Notified Bodies. It contains *Essential Health and Safety Requirements (EHSRs)* for the equipment and is intended to allow equipment to cross EU boundaries without discrimination on the grounds of safety. It requires CE marking of equipment, unlike earlier directives for hazardous area equipment, which signifies that the equipment complies with all relevant directives and it covers all equipment intended for use in hazardous areas. Directive 94/9/EC is a 'New Style' directive which, with the introduction of the EHSRs, does not require replacement as equipment standards change.

These directives are usually referred to as the *ATmosphere EXplosible (ATEX)* 137 and 100a Directives respectively. Equipment Categories were introduced in the ATEX

100a directive and the corresponding zones of use (see section 15.4) are defined in the ATEX 137 directive.

15.3   UK legislation

The ATEX 137 Directive, along with the parts of the Chemical Agents Directive dealing with fire and explosion, were enacted into UK law as the *Dangerous Substances and Explosive Atmosphere Regulations 2002 (DSEAR)* (reference 15D), and both the ATEX 137 Directive and DSEAR came into force on 1 July 2003. Workplaces which were already in use on this date have until 1 July 2006 to comply. A key feature of DSEAR is area classification, which zones workplaces according to the probability of a potentially explosive atmosphere being present. The Health and Safety Executive have published Codes of Practice (references 15E to 15J) giving guidance on compliance with DSEAR.

The ATEX 100a Directive was enacted into UK law as the *Equipment and Protective Systems Intended for Use in Potentially Explosive Atmospheres Regulations 1996 (EPS)* (reference 15K). They came into force on 1 July 2003 and compliance was optional during the transitional phase between 1996 and 2003. Although DSEAR specifies the use of equipment complying with the EPS regulations, there is no requirement to change equipment in workplaces which were already in use on 1 July 2003.

DSEAR is directed at employers and the EPS regulations are directed at equipment manufacturers. The link between them is in DSEAR, where the categories of equipment defined in the EPS regulations are allocated to the various zones defined in DSEAR.

15.4   Area classification

Area classification defines hazardous areas on the basis of the probability of a flammable atmosphere being present. It is a legal requirement under DSEAR for any employer whose business involves flammable materials in any form, and it results in the division of the workplace into one or more zones. The levels of probability used in area classification are high, medium and low. Historically, dust and gas or vapour were dealt with separately which is reflected in different nomenclature for their zoning. Zones 0, 1 and 2 are for gas, vapour and mist; zones 20, 21 and 22 are for dust. The definitions for gas or vapour zones and dust zones are very similar and are paraphrased here for simplicity:

- Zone 0, Zone 20 – High probability (A flammable atmosphere is present continuously, for long periods or frequently)
- Zone 1, Zone 21 – Medium probability (A flammable atmosphere is likely in normal operation occasionally)
- Zone 2, Zone 22 – Low probability (A flammable atmosphere is unlikely in normal operation and, if it occurs, will exist only for a short time)

The zones do not reflect the ignition properties of the flammable materials; this is dealt with separately by their ignition characteristics. It should also be noted that the zones describe 'atmospheres', which are mixtures of flammable material with air under atmospheric conditions. Any other situations are not part of area classification but require consideration as part of the overall safety case for the workplace.

'Normal operation' is considered to be the actual, day-to-day operation of the workplace. It is not the idealized operation of the workplace as it was designed, and the area classification has to be reviewed regularly to accommodate any changes to the workplace and the work therein. 'Not likely in normal operation' raises the question of how unlikely an event does area classification consider. This is best illustrated by considering a pipeline with flanged joints carrying a flammable gas. For a properly designed, constructed and maintained pipeline, fracture of the pipeline would not be considered; this is considered as a catastrophic failure and dealt with by the overall risk assessment. Leakage at a flanged piping joint would, however, be considered.

There are circumstances where area classification is not appropriate and these will be revealed by the overall risk assessment. For example, small quantities of flammable materials in the workplace may represent such a small risk to workers that other control measures may suffice.

15.4.1   Codes of practice for area classification

Codes of practice for area classification usually recommend that area classification is carried out by a team which includes the electrical/control function. Electrical engineers working in industries within the scope of DSEAR should, therefore, have some familiarity with the process of area classification and its meaning. The codes of practice fall into three broad categories; those which deal with principles, those dealing with industry sectors and those covering specific companies, installations or flammable materials. There are too many codes of practice to include in an exhaustive list but the more common ones are mentioned here. Dealing with area classification principles are BS EN 60079-10:2003 (reference 15L) for gases and vapours, and BS EN 50281-3:2002 (reference 15M) for combustible dust. One of the most common industry codes of practice is the Institute of Petroleum Code (reference 15N) which covers gases, vapours and mists. Specific codes of practice include IGE/SR/25 (reference 15O) for natural gas installations and various codes of practice and guidance from the British Compressed Gases Association covering the storage, handling and use of compressed gases in transportable cylinders.

The process of area classification includes the following:

A. Identify all materials which are capable of forming a potentially explosive atmosphere in the conditions of use, including both normal and abnormal process conditions
B. Identify the actual and potential sources of release of those materials
C. Grade the sources of release according to the probability of the release forming a potentially explosive atmosphere (see Table 15.1)
D. Consider the effect of ventilation on the releases to determine the resulting zones and their extents (see Table 15.2)
E. Document the results in a report and produce area classification drawings (see Fig. 15.1)

Table 15.1 Grading sources of release

Continuous grade release	A release which is continuous or is expected to occur frequently or for long periods
Primary grade release	A release which can be expected to occur periodically or occasionally during normal operation
Secondary grade release	A release which is not expected to occur in normal operation and, if it does occur, is likely to do so only infrequently and for short periods

Table 15.2 The relationship between the grade of release, the ventilation and the resultant zone (applicable to gases and vapours only)

Grade of release	Ventilation						
	Degree[f]						
	High			Medium[d]			Low
	Availability[f]						
	Good	Fair	Poor	Good[d]	Fair	Poor	Good, Fair and Poor
Continuous	Zone 0 NE[a] Non-hazardous	Zone 0 NE[a] Zone 2	Zone 0 NE[a] Zone 1	Zone 0	Zone 0[e] + Zone 2	Zone 0[e] + Zone 1	Zone 0
Primary	Zone 1 NE[a] Non-hazardous	Zone 1 NE[a] Zone 2	Zone 1 NE[a] Zone 1	Zone 1	Zone 1[e] + Zone 2	Zone 1[e] + Zone 1	Zone 1 or Zone 0[c]
Secondary	Zone 2 NE[a] Non-hazardous	Zone 2 NE[a] Non-hazardous	Zone 2	Zone 2[b]	Zone 2[b]	Zone 2[b]	Zone 1 and even Zone 0[c]

Notes: (a) Zone 0 NE, 1 NE or 2 NE indicates a theoretical zone which would be of negligible extent under normal conditions.
(b) The zone 2 area created by a secondary grade of release may exceed that attributable to a primary or continuous grade of release; in this case, the greater distance should be taken.
(c) Will be zone 0 if the ventilation is so weak and the release is such that in practice an explosive gas atmosphere exists virtually continuously (i.e. approaching a 'no ventilation' condition).
(d) A 'Medium' degree of ventilation with 'Good' availability is equivalent to an outdoor location.
(e) '+' denotes 'surrounded by'.
(f) The definitions of 'degree' and 'availability' of ventilation are given in BS EN 60079-10:2002. Table 15.2 applies to gases and vapours only.

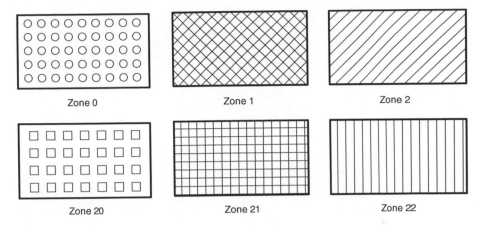

Fig. 15.1 Area classification drawing symbols

15.5   Hazardous area equipment

15.5.1   Methods of protection

Electrical equipment has been used in potentially explosive atmospheres for over one hundred years and several methods of protection have been developed during that time. The probability of the equipment becoming an effective source of ignition determines the zone of use. Equipment with a higher probability of being an effective source of ignition can be used only in the zones of lower hazard. The 94/9/EC Directive and the EPS regulations recognize this in specifying the corresponding category for the equipment. All the methods of protection use one or more of only four basic techniques which are summarized in Table 15.3.

Type of Protection n is another method of protection for electrical equipment which uses simplified versions of the methods above and is category 3 only (see Table 15.4). All the above methods of protection were originally developed for use with flammable gases and vapours. There is a further method of protection specifically for combustible dust and several of the methods of protection for gases and vapours have been adapted for use with combustible dust (see Table 15.5).

All the above methods of protection have standards for their design and construction which address most of the EHSRs of the ATEX directive, and hence the EPS regulations; the remaining EHSRs are concerned with instructions for use and other essential information for the end-user.

It is also possible to design and construct equipment which complies with the EHSRs with no reference, or only partial reference, to supporting standards. This equipment will carry only the marking specified in the directive without the identifying letters above.

15.5.2   Temperature classes and apparatus groups

Equipment for use in potentially explosive atmospheres must also be suitable for the ignition characteristics of the flammable material. IEC 79-20:1996 (reference 15P)

Table 15.3 Methods of Protection for electrical equipment

Basic technique	Method of protection	Identifying letters	ATEX category
The flammable atmosphere is excluded from contact with ignition sources	Pressurization: flammable atmosphere excluded from the interior of the equipment by internal air or inert gas pressure	p	2 3 (depending on action taken on loss of pressurization)
	Oil immersion: flammable atmosphere excluded from the interior of the equipment by filling with oil	o	2
	Powder filled: flammable atmosphere excluded from the interior of the equipment by filling with quartz granules	q	2
	Encapsulation: flammable atmosphere excluded from the interior of the equipment by encapsulation in solid material	m ma (extra requirements for the encapsulation)	2 1
Equipment has no ignition sources	Increased safety: the equipment has no ignition sources and a fault causing an ignition source is unlikely	e	2
Ignition sources are not ignition-capable	Intrinsic safety: the energy and power available to ignition sources are limited, in normal operation and under fault conditions, to values which are not capable of igniting a flammable atmosphere	ia (two independent faults considered) ib (one fault considered)	1 2
An ignition of flammable atmosphere within the equipment is contained	Flameproof enclosure: the enclosure withstands the pressure of an internal explosion without damage and prevents ignition of the surrounding flammable atmosphere	d	2

Table 15.4 Type of protection n

Technique	Similar method of protection	Identifying letters
No sources of ignition (non-sparking)	Increased Safety e	nA
Flammable atmosphere excluded from sources of ignition by complete sealing	Encapsulation m	nC
Power and energy limited at sources of ignition	Intrinsic Safety i	nL
Flammable atmosphere has restricted access to sources of ignition by sealing (restricted breathing)	None	nR
Flammable atmosphere excluded from sources of ignition by internal air or inert gas pressure	Pressurization p	nP

Table 15.5 Methods of protection for combustible dust

Technique	Identifying letters
Protection by enclosure	tD
Intrinsic safety	iD
Pressurization	pD
Encapsulation	mD

lists the properties of approximately 400 flammable gases and vapours. The relevant properties are:

- *Auto-ignition temperature (AIT)* – the temperature at which the most easily ignited mixture with air will ignite
- *Minimum ignition energy/minimum ignition current (MIE/MIC)* – the smallest amount of energy or current in a spark which will ignite the most easily ignited mixture with air
- *Maximum experimental safe gap (MESG)* – the largest gap between parts of a flameproof enclosure through which the internal explosion will not propagate to ignite a surrounding flammable atmosphere

The AIT is used to select equipment on the basis of the temperature of the equipment. The temperature of the equipment must not be greater than the AIT of the surrounding atmosphere. The equipment for use in a potentially explosive atmosphere will be marked with a temperature class (T class), a maximum temperature or both.

In Table 15.6, the maximum temperature of equipment is the temperature of the hottest part of the equipment to which the flammable atmosphere has unprotected access, for instance the external temperature of flameproof equipment but the internal temperature of increased safety equipment.

The ambient temperature in which the equipment is used affects the temperature of the equipment, so a maximum ambient temperature is also specified. The normal maximum ambient temperature is 40°C unless an alternative is marked on the equipment. Minimum ambient temperature is also specified; this is –20°C unless marked otherwise.

The MIE and MESG of flammable gases and vapours have been measured and fall into three distinct bands. This has been used to produce three sets of design criteria in the standards for non-mining equipment and a separate set for mining equipment.

Table 15.6 Ignition temperature

Temperature class	Maximum temperature of equipment or Minimum AIT of flammable material °C
T1	450
T2	300
T3	200
T4	135
T5	100
T6	85

These allow equipment to be designated as Group I for underground mining applications and Group II elsewhere. Within Group II, apparatus is designated as IIA, IIB or IIC according to the design criteria used, the IIC design criteria being the most stringent and that of IIA being the least stringent (see Table 15.8). The subgroups A, B and C are important only if the method of protection relies on limitation of energy or current, for instance Intrinsic Safety or Type of Protection nL, or on containment of an explosion such as a flameproof enclosure. Otherwise the equipment is allocated to the whole of Group II.

For convenience IEC 79-20 lists the Group and Temperature Class of equipment suitable for each flammable material.

15.5.3 Equipment marking

Prior to the 100a ATEX Directive, hazardous area equipment was marked with a 'Certification Coding', a certificate number and other information. A typical equipment label would contain:

- the manufacturer's name
- the name or type of the equipment
- the 'Distinctive Community Mark' ⟨Ex⟩
- a serial number
- the certification code
- the certificate number
- any 'Special Conditions' for safe use
- electrical ratings
- any other information required by the standards for the equipment

In a typical certification code, say EEx de IIB T3 (Tamb = 50°C), E means certified to European standards, Ex means equipment for a potentially explosive atmosphere, de means methods of protection, IIB means group and subgroup, T3 means temperature class and Tamb = 50°C means maximum ambient temperature. If the equipment is marked with a temperature in addition to, or in place of a Temperature Class, then this is used to denote the minimum ignition temperature of the gas or vapour.

In a typical certificate number such as SIRA01E1987X, SIRA is the name or logo of the certifying body, 01 is the year of certification, E is the revision status of the standards to which certified, 1 means the principal method of protection, 987 is the serial number of the certificate and X means that 'Special Conditions' for safe use exist.

The 'Distinctive Community Mark' ⟨Ex⟩ equipment complies with a hazardous area equipment directive. The 100a ATEX Directive introduces additional marking:

- the address of the manufacturer
- the year of manufacture
- the CE marking
- the number of the notified body responsible for quality assurance
- the ATEX Category

In a typical additional ATEX marking: CE   0518   ⟨Ex⟩   II 2 G

the meaning is: $C \epsilon$ Complies with all relevant Directives
0518 Notified body number
⟨Ex⟩ 'Distinctive Community Mark'
II Apparatus Group
2 ATEX category
G Suitable for gases and vapours (D denotes combustible dust)

Equipment certified to the EHSRs, without supporting standards, will have only the ATEX marking.

15.6 Equipment selection

Although further selection criteria may be relevant, such as the possibility of chemical attack, vibration, and high risk of mechanical damage, the four essential criteria for equipment selection are:

- the protection concepts and ATEX category are suitable for the intended zone of use (Table 15.7)
- the Apparatus Grouping is suitable for the intended flammable materials (Table 15.8, which applies only to gases and vapours)
- the Temperature Class or surface temperature is suitable for the ignition temperature of the flammable material (Table 15.6)
- the Ambient Temperature limits for the equipment are suitable for the intended location

A combustible dust has two ignition temperatures, one for the dust in the form of a cloud (T_{cloud}) and the other for the dust as a layer. These temperatures depend on the physical properties of the dust, such as particle size, moisture content, and the layer ignition temperature increases as the layer thickness increases. Consequently, large safety margins between the equipment temperature (T_{max}) and the dust ignition temperatures are used (Table 15.9). In Europe, dust layer ignition temperatures are usually measured with 5 mm thick layer ($T_{5\,\text{mm}}$) but in USA, a 12.5 mm layer is used.

If dust layers are expected to exceed 5 mm the derating graph given in BS EN 50281-1-2:1999 (reference 15Q) may be used. If dust layers are expected to exceed 5 mm and $T_{5\,\text{mm}}$ is also 250°C or less, special precautions are given.

Equipment for combustible dust applications has additional special ingress protection requirements, given in BS EN 50281-1-2:1999, and summarized in Table 15.10. The general criteria for ingress protection are covered in Chapter 10 and in section 16.2.2.1.

Table 15.7 Zones of use

Intended zone of use	ATEX category
0	1G
20	1D
1	1G or 2G
21	1D or 2D
2	1G, 2G or 3G
22	1D, 2D or 3D

Table 15.8 Apparatus grouping

Apparatus group	Gas/vapour group
II	All Group II
IIA	Group IIA only
IIB	Group IIA or IIB
IIC	Group IIA, IIB or IIC

Table 15.9 Dust ignition temperature

Dust cloud	Dust layer
$T_{max} = (2/3) \times T_{cloud}$	$T_{max} = T_{5\,mm} - 75\,K$

15.7 Equipment installation

The code of practice for the installation of hazardous area electrical equipment is BS EN 60079-14 (reference 15R). The guidance in this code of practice is additional to the requirements for non-hazardous area installations. Electrical equipment should, preferably, be located outside a hazardous area or in the lowest risk area available. It should be installed in accordance with its documentation, including any special requirements denoted by the letter 'X' at the end of the certificate number. It should also be operated within all of its ratings.

The installation should be designed to protect the equipment from adverse environmental influences which could jeopardize the method of protection; these include temperature extremes, corrosion, mechanical damage and vibration. Electrical equipment, such as pressure transmitters and temperature sensors, connected directly to the process, may become filled with process fluid under failure conditions. The installation of the wiring system should include, for example, impervious barriers to prevent the process material migrating through the wiring system into other areas. The installation should also be designed with regard to ease of inspection and maintenance (see section 15.8). Portable, transportable and temporary equipment should also be suitable for the zone of use.

15.7.1 Protection from sparking

Fault currents cause potential differences between different parts of an electrical installation and plant structure. Although non-hazardous area electrical installation practices minimize these to levels which are not an electrical shock risk, they can be sufficient to create ignition-capable sparking. Consequently equipotential bonding has particular guidance. Only TN-S, TT and IT power distribution systems are recommended for hazardous area use and specific guidance is given for each. SELV and PELV systems may also be used (see section 16.2.2.4).

Other equipotential bonding systems such as those provided for lightning protection, control of static charging, electromagnetic compatibility may also exist in, or associated with, hazardous areas. These should be connected together at one point to minimize potential difference between them. Cathodic protection is not normally suitable for use in hazardous areas.

Table 15.10 IP ratings for combustible dust

Intended zone of use	Dust electrically conductive?	Required IP rating
20	Both	IP 6X
21	Both	IP 6X
22	Yes	IP 6X
22	No	IP 5X

15.7.2   Electrical protection

In most cases the electrical protection provided for the same installation in a non-hazardous area is suitable but there are a number of significant exceptions. Specific guidance is given for rotating electrical machines and transformers which includes overload protection. Multi-phase equipment should be protected against loss of one or more phases. Auto-reclosing protection is not recommended.

15.7.3   Means of isolation

Residual neutral voltages may be ignition-capable so, the means of isolation should also isolate the neutral.

15.7.4   Wiring systems

The following aspects of wiring systems have specific guidance:

- Aluminium conductors
- Non-sheathed single cores
- Unused openings in enclosures
- Circuits traversing hazardous areas
- Openings in walls etc.
- Protection of stranded conductor ends
- Overhead lines
- Mechanical damage
- Electrical connections
- Migration and accumulation of flammable materials
- Fortuitous contact
- Cable jointing
- Unused cores
- Cable surface temperature

Cable types for Zones 1 and 2 are mineral-insulated metal sheathed cables and thermoplastic, thermosetting or elastomeric sheathed cables. Especially robust cable types are recommended for portable and transportable equipment, and where flexibility is required. Conduit systems may be used and specific guidance is given on stopping boxes, fault currents, corrosion, non-sheathed single conductors, condensation drains and IP ratings.

15.7.5   Flameproof equipment

Five additional pieces of guidance are given for flameproof equipment all of which can nullify the method of protection if not carried out.

15.7.5.1   Solid obstacles

The safety of a flameproof enclosure relies on the products of combustion emerging from the enclosure not igniting a surrounding flammable atmosphere. The maximum gap between parts of a flameproof enclosure is specified in the flameproof standard as a fraction of the MESG. Such gaps, having specified maximum dimensions, are known as *flamepaths* and may be in the form of flanges, spigots, shafts or screw threads. The length of the flamepaths necessary in flanges and spigots is indicated in the sectional illustration of an EEx'd' cage induction motor, in Fig. 15.2.

The ignition process is dependent on both the temperature of the emerging gases and the time for which they are in contact with the flammable atmosphere. While it is true that for gases below the AIT of the surrounding atmosphere, will not cause ignition, they can exceed the AIT for a short time and still not cause ignition. The process therefore involves both cooling and rapid mixing and dispersion. A solid obstruction close to the exterior of a flamepath may reduce the speed of mixing and dispersion which, in turn, may cause ignition if the emerging gases are above the AIT. If solid obstructions are sufficiently far from flamepaths, the mixing and dispersion are adequate (see Table 15.11).

It is the Group of the flammable atmosphere which is important, not the Group of the equipment, so that 10 mm would be appropriate for a IIC enclosure in a IIA atmosphere.

Some enclosures are manufactured with obstructions closer than the above. These are acceptable because the testing of the enclosure demonstrates that the emerging gases do not ignite the surrounding flammable atmosphere.

Fig. 15.2 Sectional illustration of an EEx'd' induction motor (courtesy of Invensys Brook Crompton)

Table 15.11 Solid obstructions to flamepaths

Flammable atmosphere group	Minimum distance mm
IIA	10
IIB	30
IIC	40

15.7.5.2 Protection of flamepaths

The standard for flameproof equipment does not require the enclosure to have ingress protection. Consequently, the installer or user of the equipment will often need to take steps to prevent liquid ingress and corrosion of flamepaths. Gaskets, 'O'-rings and other sealing methods are acceptable only if they are provided by the equipment manufacturer and are included in the certification. The installer or user of the equipment may use a light smear of grease to protect a flamepath against corrosion but the grease must be low-viscosity and non-hardening, so as not to obstruct the flamepath, and must not catch fire if an explosion occurs inside the enclosure. No other material, including paint, is acceptable in a flamepath.

Grease-impregnated non-hardening fabric tape may be applied to the exterior of a flamepath provided that no more than one layer with a 25 mm overlap is used. This is acceptable for IIA applications and unacceptable for IIC applications. It is often considered impractical for IIB applications because the guidance in the code of practice requires the installer or user of the equipment to verify that the maximum gap of the flamepath does not exceed 0.1 mm.

15.7.5.3 Cable entries

Irrespective of the method of protection a cable entry must not lower the integrity of the enclosure. It must maintain the strength, IP rating and, where required, the flameproof properties of the enclosure. Cable glands for flameproof enclosures are certified as flameproof and are usually designed for group IIC. A flamepath exists where the gland fits into the enclosure so the maximum radial clearance and length of the mating surfaces are critical. Certified flameproof glands have defined radial tolerances and entry length so that these criteria are met provided the gland is properly fitted. The gland may also have internal flamepaths or explosion seals which will be effective only if the gland is properly installed.

The explosion within the flameproof enclosure will impinge on the end of the cable where it emerges from the gland inside the enclosure. Exposure to repeated aggressive explosions can degrade the cable construction and lead to propagation along the cable through the gland and subsequent ignition of the surrounding flammable atmosphere. Barrier glands are designed to protect the end of the cable from explosion and are recommended where such a problem may arise. In other situations, flameproof compression glands can be used. The contributory factors are taken into account in the gland selection procedure shown in Fig. 15.3.

15.7.5.4 Motors with variable frequency supply

Traditionally it has been the responsibility of the user to ensure that the motor does not overheat as a result of misuse. This has been achieved through the use of devices such as current sensing protection relays. With converter supplies, the situation is

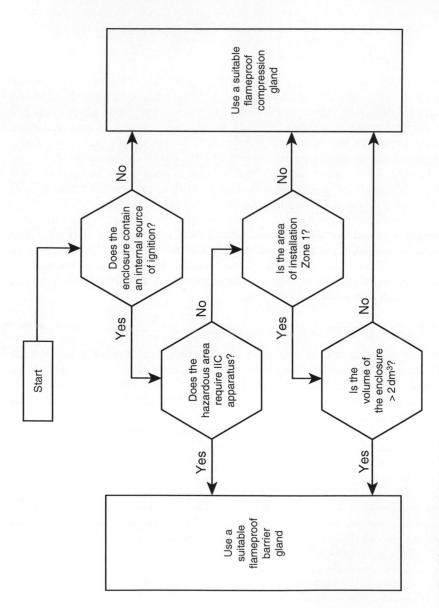

Fig. 15.3 Flameproof gland selection

somewhat more complex and it is necessary for the motor manufacturer to assume the responsibility of rating the motor correctly for variable frequency inverter supplies.

These motors should be either type tested with the intended variable speed drive under the conditions of use or be equipped with internal temperature sensors which are arranged to disconnect the supply before the temperature class is exceeded. Thermistor sensors are usually mounted at critical points in the motor, and used to monitor motor temperature during operation. Trip relays are used to remove the supply from the motor if any one thermistor reaches the tripping temperature.

15.7.5.5 Conduit systems

Conduit should be either solid drawn or seam-welded heavy gauge steel. Specific types of flexible conduit may also be used. If the conduit is to be screwed directly into a flameproof enclosure, the thread should be of 6 g tolerance class and be long enough to allow five threads to be fully engaged in the enclosure wall. An explosion stopper is required on the enclosure wall or within 50 mm to prevent any explosion propagating along the conduit.

15.7.6   Increased safety equipment

Increased safety relies on a high standard of design and construction and the quality of the workmanship during installation is equally important. There is a great deal of detailed guidance in the manufacturers' literature and BS EN 60097-14 summarizes the main features. The three critical areas are cable and conduit entries, wiring terminations and temperature limitation.

15.7.6.1 Cable and conduit entries

The primary objective of cable and conduit entry devices is to maintain the strength and IP rating of the enclosure. One way to achieve this is to use devices certified as increased safety and to use a sealing device, such as a washer, between the entry device and the enclosure. Certain types of standard industrial metal compression glands exceed the minimum requirements for increased safety and may be used instead of a certified equivalent. Where there is any doubt, and with non-metallic glands, a certified device should always be used.

15.7.6.2 Wiring terminations

The majority of increased safety terminals fall into one of three basic design types, each with its own particular guidance for connection. Rail-mounted and so-called 'chocolate block' terminals comprise an insulating body through which a metal bar passes. Each end of the metal bar is recessed within the body and carries a wiring clamp. The wiring clamp may be a thin metal tongue which presses the wire onto the bar when the terminal screw is tightened, or it may be a saddle or yolk arrangement which is pulled up to the bar when the terminal screw is tightened and likewise presses the wire onto the bar. This type of terminal is designed for one wire at each end. A further type comprises a metal bar projecting from, or embedded in, the insulation of a component or part of an item of equipment. They are commonly found moulded into component parts of increased safety equipment, such as ballasts in luminaires. They have a wiring clamp in the shape of an inverted 'U' on top of the metal bar which is pressed onto the bar by a screw passing through both. These are designed to accommodate one or two conductors of the same or different sizes. A third common type of terminal is in the form of a threaded metal pillar with a vertical slot. A metal bar mounted inside a screw-on cap presses the conductors down on to the base of the slot. This design of terminal is intended to accommodate one or more conductors.

Conductors should preferably be fitted with compression ferrules before termination in any type of increased safety terminal, especially if they are stranded conductors or if the terminal is designed for more than one conductor. All strands of a conductor should be clamped securely and the conductor insulation should continue up to the metal part of the terminal.

15.7.6.3   Wiring within enclosures

The principal source of heat in a terminal box is the wiring. The temperature classification is based on the rated current of the terminals and is measured with wiring tails equal in length to the largest internal dimension of the enclosure. It is important that this length, the ratings of the terminals and the current rating of the conductors are not exceeded, or the temperature class may be exceeded.

15.7.6.4   Thermal protection of electrical machines

The thermal protection for increased safety of electrical machines is critical for two main reasons. The temperature in question is the hottest part of the interior, which may in a cage induction motor be the stator windings or the rotor bars. The temperature during starting and stalled operation is taken into account; on a fixed-frequency supply, the time taken to reach the limiting temperature of the temperature class, starting from the normal maximum running temperature, is marked on the rating plate. This is called the t_E time and the protection must disconnect the motor within this time. Current-sensing protection with an inverse time delay characteristic may suffice if the motor run-up time is shorter than the t_E time. Otherwise more sophisticated protection or embedded thermal sensors will be required.

Soft-start circuits and variable frequency/voltage drives need special attention.

15.7.7   Intrinsic safety equipment

Intrinsic safety is system-based and it has to be preserved throughout the entire installation. The two areas requiring particular care are wiring segregation and earthing. Intrinsically safe circuits need to be segregated from all other types of circuits and from each other. In the majority of intrinsically safe circuits only one connection to earth is acceptable. Guidance is given in BS EN 60079-14 in the following areas and should be consulted:

- location and installation of apparatus
- energy storage in cables
- cable types
- earthing of screens
- multiple earthing and electromagnetic compatibility
- cable armour bonding
- installation of cables and wiring
- marking/identification of cables, wiring and terminals
- multi-core cables
- unused cores in cables
- cable and wiring fault considerations

- design and justification of uncertified systems
- the use of uncertified 'Simple Apparatus'
- temperature classification of wiring and cables
- cable entries
- installations in zone 0

15.7.8 Pressurized apparatus

Pressurized apparatus is generally category 2, however it may be restricted to zone 2 applications depending on the action taken by the end-user on loss of pressurization. If the pressurized enclosure does not contain a source of release of flammable material the guidance shown in Table 15.12 should be followed.

If the enclosure contains a source of release of flammable material the documentation for the equipment should be consulted. Consideration of the following is required:

- any 'Special Conditions' of certification denoted by a suffix 'X' to the certificate number
- the nature of the containment of the flammable material
- the nature of the flammable material
- whether the supply of flammable material is shut off on loss of pressurization
- the nature of the equipment within the enclosure
- the external area classification
- whether the pressurizing medium is air or an inert gas
- whether the pressurization technique is 'leakage compensation' or 'continuous dilution'
- the safety implications of sudden disconnection of the supply

The enclosure will normally have to be purged with the pressurizing supply before the power can be turned on. The documentation supplied with the enclosure or the certification should be consulted to determine the purging flow rate and duration as these will be affected by any ducting connected to the enclosure during installation.

15.7.9 Type n equipment

Type n equipment comprises five subtypes which have been listed in Table 15.4.

In *non-sparking apparatus* (nA), the principles of design, construction and installation are similar to increased safety, so the same general guidance applies.

Table 15.12 Action on loss of pressurization

Intended zone of use	Enclosure contains parts not suitable for zone 2 without pressurization[a]	Enclosure contains parts suitable for zone 2 without pressurization
Zone 2	Alarm[b]	None required
Zone 1	Alarm and switch off[c]	Alarm[b]

Notes: (a) parts with a source of ignition in normal operation.
(b) action should be taken to restore the pressurization as soon as practicable and within 24 hours in any case. Precautions should be taken to avoid the entry of a flammable atmosphere into the enclosure while the pressurization is off.
(c) if a more dangerous condition is created by sudden disconnection of the supply, a delayed switch-off may be justifiable or the reliability of the supply should be increased.

In *restricted breathing enclosures* (nR), the quality of the sealing is critical to the protection, so cable types and cable entries need careful specification and installation.

Energy-limited apparatus (nL) may be subgrouped like intrinsic safety and the same selection criteria apply. It may also be system-based rather than apparatus-based. If it is apparatus-based only the normal supply limits are relevant as the cabling is not energy limited. However, if it is system-based, the cabling will also be energy limited. This means that cable types, segregation, supply characteristics, energy storage in cables and earthing requirements are similar to those for intrinsic safety.

Simplified pressurization (nP) is similar to pressurization except there are relaxations on purging and action on loss of pressure. The general principles for pressurized installations apply.

Type nC uses one of several protection techniques, which are enclosed break (similar to flameproof), sealing (using gaskets etc.), hermetic sealing (fusion sealed) and encapsulated. It may be subgrouped.

15.7.10 Personal electrical apparatus

Personnel may inadvertently carry personal electrical equipment such as electronic wrist watches, hearing aids, car remote locking devices, key-ring torches, calculators and so on into hazardous areas. The risk with wristwatches which have no other function such as a calculator is very small and generally acceptable. No other personal electrical equipment should be permitted in hazardous areas unless it is suitably protected.

15.8 Inspection and maintenance

All equipment installed in hazardous areas and all parts of intrinsically safe, or similar, installations require periodic inspection to verify the continued integrity of the methods of protection. It is possible for installations to become unsafe due to deterioration, damage or alteration while the installation remains functional. A programme of initial and periodic inspections and, if necessary, remedial action is recommended in BS EN 60079-17 (reference 15S).

Inspections are generally initial, periodic or sample. An *initial inspection* is recommended before a new installation or re-installed equipment is brought into service. The objective is to verify that the equipment is suitable for its location, that it has been correctly installed and is in good condition. A *periodic inspection* is intended to check that the equipment continues to be suitable and a sample inspection can be used to check the effectiveness of the inspection and maintenance arrangements.

Inspections can also be visual, close or detailed. A *visual inspection* involves no more than an examination of the exterior of the equipment and it does not require the use of tools or the isolation of equipment for electrical safety. A *close inspection* may require access equipment and basic tools but does not involve opening of equipment enclosures or isolation for electrical safety. A *detailed inspection* may be carried out in the field or the equipment may be removed to a workshop; isolation, opening of the enclosure, and a wider range of tools are required, and test equipment may also be used in a hazardous area if suitably protected, or in a workshop.

15.8.1 Inspection strategy

The inspection strategy is designed for optimum use of resources. At first sight, zero defects is a desirable target. If this is achieved there is no way to determine if the grade and frequency of inspection are excessive. Too many defects indicate that the inspections are not sufficiently frequent or thorough. The strategy, therefore, compares the current results with the previous results and adjusts the frequency and/or grade of inspections until a small number of minor defects are found at each inspection. Records need not be kept but may be useful in identifying trends.

15.8.2 Inspection schedules

The BS EN 60079-17 contains sample inspection schedules for the various methods of protection, depending on whether the inspections are initial or periodic. It may be desirable to include work instructions, access requirements and recommended tools in the inspection schedules. A means of prioritizing and recording the findings can also be included. It is important to produce schedules which can be practically used in a plant environment. The schedules may also incorporate electrical testing and any statutory requirements.

15.8.3 General requirements

15.8.3.1 Withdrawal from service

When equipment is removed from the site, the exposed conductors should be terminated in a suitable enclosure, or isolated from all sources of supply, insulated and protected from the environment, or isolated from all sources of supply, earthed and protected from the environment.

15.8.3.2 Isolation

Intrinsically safe installations do not normally need to be isolated to protect personnel from electric shock or to prevent ignition-capable sparking. Consequently, the following maintenance activities are permissible in a hazardous area without isolation:

- disconnection, removal or replacement of any item of equipment or cabling in an intrinsically safe circuit
- adjustment of calibration controls
- removal or replacement of plug-in assemblies or components
- use of test equipment or procedures specified in the system documentation

Safety earths must not be disconnected or tested while the associated intrinsically safe circuits are connected into the hazardous area.

All other installations, except type of protection nL, should be isolated, including the neutral, before work commences. The documentation supplied with the equipment, or the certificate, should be consulted for isolation requirements for type of protection nL.

15.8.3.3 Fault loop impedance or earth resistance

The standard types of test equipment for these measurements uses voltages and currents which are ignition-capable and should not, therefore, be used in a hazardous area without an appropriate permit to work. They may also cause ignition-capable currents to flow in a hazardous area even when used in a non-hazardous area, especially if there are faults in the earthing and bonding system. It is preferable to carry out such checks before flammable materials are brought into a plant or during a major shutdown when all flammable materials have been removed from the plant. If such measurements are duplicated by measurements using intrinsically safe test meters, comparisons can be carried out in the intervening period. If there is then evidence that the earthing and bonding system is deteriorating, it can be confirmed using standard test equipment with suitable precautions.

15.8.3.4   Insulation resistance measurements

Insulation resistance measurements pose similar problems because the standard types of test equipment use voltages which are ignition-capable. A similar procedure to that for fault loop impedance or earth resistance measurements can be used.

Intrinsically safe high voltage insulation resistance meters, working at 500 V, are available but the energy stored in a few metres of cable is ignition-capable at this voltage. Such equipment should only be used strictly in accordance with the manufacturer's instructions.

15.8.3.5   Overloads

The setting of overloads should be checked to verify if they are set to I_N. The characteristics of the devices should be checked to verify that they are such that they will operate in not more than two hours at 1.2 times the set current and will not operate within two hours at 1.05 times the set current.

15.8.4   Specific requirements

15.8.4.1   Flameproof equipment

Although the example inspection schedules in BS EN 60079-17 include checking flamepath dimensions, it is not normally considered necessary to check spigots, spindles, shafts and screw threads and joints which are not normally capable of being dismantled, unless there is evidence of damage or deterioration.

15.8.4.2   Increased safety

It is recommended that the settings of overload and excess temperature protection are checked periodically. If experience shows it to be necessary, the tripping characteristics should be measured periodically to verify that they are within 20 per cent of the set values.

15.8.4.3   Intrinsic safety

There is extensive guidance in BS EN 60079-17 for intrinsic safety covering documentation, labelling, modifications, intrinsically safe interfaces, cables, earth continuity and insulation from earth, connections to earth, and separation and segregation of intrinsically safe circuits from each other and from non-intrinsically safe circuits.

15.8.5   Maintenance

Maintenance is defined as a combination of any of the actions carried out to retain an item in, or restore it to, conditions in which it is able to meet the requirements of the hazardous area installation and perform its required function. Repair, reclamation and overhaul are not included and are best left to specialist organizations complying with

BS EN 60079-19. Maintenance should be integrated with the inspection system so that defects are rectified in a time commensurate with their priority and so that re-installed equipment receives an initial inspection.

References

15A. 1999/92/EC, Minimum requirements for improving the safety & health protection of workers potentially at risk from explosive atmospheres; OJ L23 of 28/01/2000, p. 57.

15B. 94/9/EC, The approximation of the laws of member states concerning equipment and protective systems intended for use in potentially explosive atmospheres; OJ L100 of 19/04/1994, p. 1.

15C. 98/24/EC, The protection of the health and safety of workers from the risks related to chemical agents at work; OJ L131 of 05/05/1998, p. 11.

15D. DSEAR, Statutory Instrument 2002 No. 2776; The Dangerous Substances and Explosive Atmospheres Regulations 2002.

15E. Dangerous Substances and Explosive Atmospheres Regulations 2002, Approved Code of Practice and Guidance L133 – Unloading of petrol from road tankers; HSE Books, ISBN 0 7176 2197 9.

15F. Dangerous Substances and Explosive Atmospheres Regulations 2002, Approved Code of Practice and Guidance L134 – Design of plant, equipment and workplaces; HSE Books, ISBN 0 7176 2199 5.

15G. Dangerous Substances and Explosive Atmospheres Regulations 2002, Approved Code of Practice and Guidance L135 – Storage of dangerous substances; HSE Books, ISBN 0 7176 2200 2.

15H. Dangerous Substances and Explosive Atmospheres Regulations 2002, Approved Code of Practice and Guidance L136 – Control and mitigation measures; HSE Books, ISBN 0 7176 2201 0.

15I. Dangerous Substances and Explosive Atmospheres Regulations 2002, Approved Code of Practice and Guidance L137 – Safe maintenance, repair and cleaning procedures; HSE Books, ISBN 0 7176 2202 9.

15J. Dangerous Substances and Explosive Atmospheres Regulations 2002, Approved Code of Practice and Guidance L138 – Dangerous substances and explosive atmospheres; HSE Books, ISBN 0 7176 2203 7.

15K. EPS Regulations, Statutory Instrument 1996/192; Equipment and Protective Systems Intended for Use in Potentially Explosive Atmospheres Regulations 1996.

15L. BS EN 60079-10:2003, Electrical apparatus for explosive gas atmospheres – Part 10: Classification of hazardous areas.

15M. BS EN 50281-3:2002, Equipment for use in the presence of combustible dust – Part 3: Classification of areas where combustible dusts are or may be present.

15N. IP Code, 2nd edition Aug 2002, Area classification code for installations handling flammable fluids, Institute of Petroleum, Model Code of Safe Practice, Part 15, ISBN 0 85293 223 5.

15O. IGE/SR/25, The Institution of Gas Engineers, Safety Recommendations IGE/SR/25, Communication 1665, Hazardous Area Classification of Natural Gas Installations.

15P. IEC 79-20, Electrical apparatus for explosive gas atmospheres – Part 20: Data for flammable gases and vapours, relating to the use of electrical apparatus.

15Q. BS EN 50281-1-2:1999, Electrical apparatus for use in the presence of combustible dust – Part 1-2: Electrical apparatus protected by enclosures – Selection, installation and maintenance.

15R. BS EN 60079-14:2003, Electrical apparatus for explosive gas atmospheres – Part 14: Electrical installations in hazardous areas (other than mines).

15S. BS EN 60079-17:2003, Electrical apparatus for explosive gas atmospheres – Part 17: Inspection and maintenance of electrical installations in hazardous areas (other than mines).

Chapter 16

Principles of electrical safety

J.M. Madden
Health and Safety Executive

16.1   Injuries from electricity

Every year people are injured or killed as a result of hazardous defects in electrical systems or because they adopt unsafe working practices on electrical systems. The most common types of injury are electric shock and burns, with the burn injuries arising from either current passing through the body or from the effects of arcing and flashovers. In addition to these direct forms of electrical injury, the following secondary types of injury can occur:

- burn injuries and the adverse effects of smoke or fume inhalation from fire of electrical origin
- the effects of an explosion that has an electrical source of ignition
- physical injuries arising from the reaction to electric shock, such as being thrown off a ladder as a result of electric shock and suffering impact injuries from the fall

16.1.1   Electric shock

Electric current passing through the body, particularly alternating current at power frequencies of 50 Hz and 60 Hz, may disrupt the nervous system, causing muscular reaction and the painful sensation of electric shock. The most common reaction is to be thrown off the conductor as a result of the muscular contraction. However, in a small number of instances, the consequence is death from cardiac arrest, or from ventricular fibrillation (where the heart muscle beats in a spasmodic and irregular fashion) or from respiratory arrest.

The physiological effects are largely determined by the magnitude and frequency of the current, the waveform (for example, continuous sine wave, or half wave rectified sine wave, or pulsed waveform), its duration, and the path it takes through the body. An authoritative guide on the topic is published in IEC 60479 (see Table 16.2). The following text concentrates on the most common situation of a shock from a continuous power frequency ac waveform.

The magnitude of the current is the applied voltage divided by the impedance of the body. The overall circuit impedance will comprise the body of the casualty and the other components in the shock circuit, including that of the power source and the interconnecting cables. For this reason, the voltage applied to the body, which is commonly known as the *touch voltage,* will often be lower than the source voltage.

The impedance of the body is determined by the magnitude of the touch voltage (there being an inverse relationship between impedance and voltage) and other factors, such as the wetness of the skin, the cross-sectional area of contact with the conductors, and whether or not the skin is broken or penetrated by the conductors. As a general rule of thumb, at an applied voltage of 230 V at 50 Hz, the total body impedance for a hand-to-feet path will be in the range 1000 Ω to 2500 Ω for most of the population, falling to around 750 Ω at voltages in excess of about 1000 V.

The path that the current takes through the body has a significant effect on the impedance. For example, the impedance for a hand-to-chest path is in the order of 50 per cent of the impedance for a hand-to-foot path. Moreover, the current's path through the body is a significant determinant of the effect on the heart.

Table 16.1 summarizes the physiological effects of current passing through the body. The effects relate to a hand-to-hand shock exceeding 1 s for a person in good health. If the duration were less than 1 s, greater currents could be tolerated without such adverse reactions.

Electric shock accidents are most common on low-voltage systems and are usually subdivided into two categories of direct contact and indirect contact shocks. A direct contact shock occurs when conductors that are meant to be live, such as bare wires or terminals, are touched. An indirect contact shock occurs when an exposed conductive part that has become live under fault conditions is touched, as depicted in Fig. 16.1. Examples of an exposed conductive part are the metal casing of a washing machine and the metal casing of switchgear. This type of accident, which requires two faults to occur (the loss of the earth connection followed by a phase-to-earth fault), is quite common.

When providing first aid to an electric shock casualty, the first action should be to remove the cause by switching-off the supply or otherwise breaking contact between the casualty and the live conductor. Cardiopulmonary resuscitation may be required. If the casualty is suffering from ventricular fibrillation, the only effective way to restore normal heart rhythm is by the use of a defibrillator. Where a defibrillator is not immediately available, the first aider should carry out cardiopulmonary resuscitation until either the casualty recovers or professional assistance arrives.

Table 16.1 The effect of passing alternating current (50 Hz) through the body from hand-to-hand

Current (mA)	Physiological effect
0.5–2	Threshold of perception
2–10	Painful sensation, increasing with current. Muscular contraction may occur, leading to being thrown-off
10–25	Threshold of 'let go', meaning that gripped electrodes cannot be released once the current is flowing. Cramp-like muscular contractions. May be difficulty in breathing leading to danger of asphyxiation from respiratory muscular contraction
25–80	Severe muscular contraction, sometimes severe enough to cause bone dislocation and fracture. Increased likelihood of respiratory failure. Increased blood pressure. Increasing likelihood of ventricular fibrillation (unco-ordinated contractions of the heart muscles so that it ceases to pump effectively). Possible cardiac arrest
Over 80	Burns at point of contact and in internal tissues. Death from ventricular fibrillation, cardiac arrest, or other consequential injuries

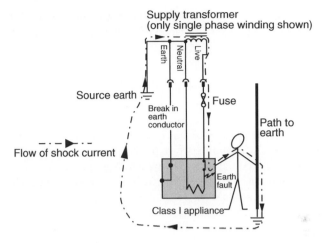

Fig. 16.1 Depiction of a typical indirect contact electric shock

16.1.2   Contact and internal burns

In addition to causing electric shock, current flowing through the body may cause burn injuries at the points of contact and in the muscle and other internal tissues and organs.

The extent of any burn injuries is determined by the current density at the point of contact and in the internal tissues; the higher the current density the more severe will be the injuries. The magnitude of the current will be determined by the same factors as described in section 16.1.1.

It is uncommon to see significant burn injuries when the touch voltage is low, including mains voltage of 230 V ac. The usual physical evidence is small white blister-like marks on the skin on the entry and exit points but there have been accidents where more severe burns have been experienced. At higher voltages, especially in incidents involving contact with an overhead high-voltage power line, the burning is invariably severe and can be the main cause of death.

16.1.3   Arc and flashover burns

Arc burns, also known as flashover burns, are commonly associated with the failure of insulation in electrical equipment, leading to an arc developing in the air between adjacent conductors.

A common cause is metal objects, such as screwdrivers, spanners and other foreign objects forming a short circuit between a phase conductor and earth, or across conductors at different voltages. More often than not the initial fault is between two adjacent conductors, such as a phase conductor and earthed metalwork, but the ionized gases created by the fault allows arcs to develop between other conductors. In three-phase systems, the result is often described as a full three-phase flashover. The typical consequence is the expulsion from the short circuit of a highly energetic arc and hot gases, with temperatures in the plasma typically exceeding 1000°C. A person in the immediate vicinity of the arc will suffer burn injuries which are often severe and life-threatening.

The amount of power that can be supplied into the fault is determined by the voltage and impedance in the fault circuit, and is quoted in megavoltampere or kiloampere (strictly, the second of these parameters is a measure of short-circuit current rather than power). Modern systems, even at low voltage, often have very high fault levels.

A typical fault level in a low-voltage domestic installation is 6 kA or 2.4 MVA and in an 11 kV high voltage installation, it can be as high as 13 kA or 250 MVA.

16.1.4   Fire injuries

Electrical systems that are poorly designed, or that have certain fault conditions, may overheat due to excess current flowing to such an extent that adjacent flammable materials may be ignited. Fires may also be started by arcs and sparks evolved from short-circuit faults, most frequently resulting from a breakdown in insulation. Hot spots in circuits can develop, for example, when poorly made connections have sufficiently high resistance to cause localized heating, which may lead to fire.

Most people caught in fire situations who are killed or seriously harmed are affected by the smoke and toxic fumes emitted from the burning substances, including electrical cables and components, rather than by being burned.

16.1.5   Explosion injuries

If standard electrical equipment is installed in places, called hazardous areas, where a flammable or explosive atmosphere exists, the arcs, sparks, electrostatic discharges or hot surfaces created during normal operation or under fault conditions may have enough energy for them to act as an ignition source, leading to an explosion. Anybody in proximity to the explosion may suffer burns and physical injuries, which may be serious enough to be fatal.

16.2   Precautions against electric shock and contact burn injuries

16.2.1   General principles

Shock and contact burn injuries can be prevented first by ensuring that the electrical system is designed, installed and maintained in accordance with sound engineering principles and in compliance with accepted and published standards, and secondly by ensuring that any work on the system is carried out in a safe manner. This section provides some detail on the hardware design principles, safe systems of work being described at the end. Details can be found in reference 16A.

16.2.2   Prevention of direct contact injuries

Direct contact shock and burn injuries are commonly prevented by ensuring that conductors energized at dangerous voltages cannot be touched. There are, however, some instances where conductors have to be uninsulated and available to touch, such as when live testing work is being done or because they need to be exposed for functional reasons, examples being power pick-up rails on fairground rides and overhead travelling cranes.

The most commonly used preventive techniques are listed. The list is not exhaustive and there are other less commonly used and specialized techniques available.

16.2.2.1   Insulation and enclosures

Direct contact injuries are most commonly prevented by covering the conductors with suitably rated insulating material, as in the case of cables, or by placing them inside an enclosure that is constructed so as to prevent access to the live parts.

Insulation must be suitably rated so that it can withstand the applied voltage and must be selected or protected to withstand external influences, such as impact and abrasion, high and low temperatures, sunlight, water and corrosive liquids. Information on cable insulation is provided in Chapter 9 and general information on the properties of insulating materials and dielectrics is provided in Chapter 3.

Enclosures must be constructed of materials that will withstand the environment in which they are installed. They must also prevent access to the internal live parts and prevent the ingress of liquids and dusts that may be present in the operating environment and which may create a hazard if allowed to get into the enclosure.

IEC 60529 (see Table 16.2) defines an Ingress Protection (IP) code that is used to describe how well an enclosure prevents the ingress of moisture and dust. The code uses two figures; a third can be added to cover mechanical impact protection but it is rarely used. The first figure shows the degree of protection against ingress by solid objects and the second figure shows the degree of protection against liquid ingress. As an example, IP20 describes an enclosure that will prevent finger access but will not prevent the ingress of moisture – a domestic socket outlet would fall into this category. An enclosure with an IP rating of IP55, such as a halogen lamp for general illumination, would be suitable for external use. The highest rated enclosures, at IP68, will prevent the ingress of dust and can be immersed in water.

Enclosures are often also classified according to the criteria set out in IEC 60536-2 (see Table 16.2). There are two main classifications. Class I equipment has a metal casing that needs to be earthed for safety reasons; metal-clad switchgear is an example. Class II equipment has internal electrical conductive parts on which the basic insulation is supplemented by additional insulation to provide an additional insulating barrier; this type of equipment, of which the common plastic-encased power drill is an example, does not require any metal parts of the casing to be earthed on the grounds that it would only become live if both layers of insulation were to fail, an event that is very unlikely to occur.

Another important standard for enclosures is IEC 60439-1 (see Table 16.2). Among other things, this standard describes 'forms' of enclosures according to the protection against contact with live parts belonging to adjacent functional units inside the enclosure, the limitation of the probability of initiating arc faults, and the protection against the passage of solid foreign bodies from one internal unit to an adjacent unit.

There are many other standards covering the design and configuration of enclosures.

16.2.2.2 Safe by position

Uninsulated conductors energized at dangerous voltages can, in principle, be made safe by being placed out of reach. For example, high-voltage overhead lines with uninsulated conductors are raised to heights above ground level at which they are safe from being inadvertently touched or approached so closely by a conducting object (particularly an object with a sharp pointed end) that an arc can develop from the line to the object. Safety is not assured, however, as exemplified by the common occurrence of direct contact with the conductors by the likes of construction and agriculture vehicles, fishing rods and kites.

16.2.2.3 Reduced voltage

The direct contact hazard can be minimized by reducing the shock voltage. In the UK, the most common reduced voltage system operates at 110 V three-phase or single-phase. In the former, the star point of the supply generator or transformer is earthed and, in the latter, the centre point of the output winding is earthed, reducing the shock voltage between a phase conductor and earth to 55 V. Reduced voltage is most frequently used to supply Class I and Class II portable tools in work locations, such as construction sites.

16.2.2.4   Extra-low voltage

The two common types of extra-low voltage systems are *Safety Extra-Low Voltage (SELV)* and *Protective Extra-Low Voltage (PELV)*.

The SELV systems operate at voltages no greater than 25 V ac or 60 V ripple-free dc between conductors or between any conductor and earth. These are considered to be safe voltages for most applications, although lower voltages may be needed for work in wet conditions and confined spaces. In an SELV system, any exposed conductive parts should not be connected to, or be in contact with, the protective conductor of another system, nor with extraneous metal which could be energized by another system. A step-down transformer, to provide the safe low voltage, should be a safety transformer to EN 60742 (see Table 16.2).

PELV differs from SELV only by having its circuits earthed at one point only. The conductors are usually protected either by barriers or enclosures to at least IP2X or with insulation capable of withstanding 500 V dc for 60 s. Where the voltage does not exceed 25 V ac or 60 V ripple-free dc in a dry location within the equipotential zone, these additional enclosure precautions are not required, but otherwise the limits are 6 V ac or 15 V ripple-free dc.

16.2.2.5   Limitation of energy

Limiting the amount of energy that a system can deliver into the body is a method of protection against direct-contact injuries. The electric stock-control fence is one example. In these systems, the fence wires are energized with pulses, with the peak voltage of each pulse being in the order of 5–10 kV, with a pulse duration in the order of 1 ms and pulse repetition frequencies of about 1 Hz. The amount of energy delivered into a 500 Ω load is limited to 5 J per pulse, where 500 Ω represents the typical lowest value of human body resistance at the fence operating voltages. Five joules is estimated to be below the energy needed to cause cardiac fibrillation effects in the large majority of the population, but it is high enough to cause sufficient pain to any animal that may touch the fence wire to deter them from touching it again.

16.2.2.6   Electrical separation

Here the source of supply is usually a safety isolating transformer or its equivalent. The secondary windings are not earthed or otherwise referenced, which explains why such systems are commonly known as *isolated* or *unreferenced* supplies. The principle is illustrated in Fig. 16.2. A person simultaneously touching one pole of the unreferenced supply and earth will not experience an electric shock because there is no complete circuit back to the point of supply, although small amounts of reactive current may flow due to capacitive coupling.

16.2.2.7   Earth leakage protection

Circuit breakers and fuses protect against excess current arising from overload conditions and faults. The most common type of fault is an earth fault, but the current

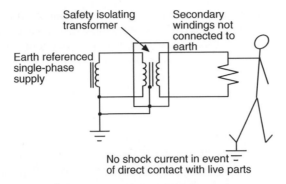

Safety isolating transformer

Secondary windings not connected to earth

Earth referenced single-phase supply

No shock current in event of direct contact with live parts

Fig. 16.2 Principle of electrical separation

flowing due to these faults may be too low to operate the overcurrent protection devices. In addition, the overcurrent protective devices will not operate in the event of somebody making direct contact with a live conductor because the current which flows through the body to earth will be too low to operate the devices, although it will often be high enough to cause fatal electric shocks. These two problems can be obviated by the use of earth leakage protection devices that, in low-voltage systems, will be RCDs as described in Chapter 7. Earth leakage protection on higher voltage systems is provided using protection devices of the type described in Chapter 8, although these do not provide protection against direct conduct with live HV conductors.

RCDs should not be relied upon as the sole means of protecting against injury from direct contact. This is because RCDs have failure modes that can lead to unrevealed dangerous fault conditions and because, for an RCD to operate in the event of direct contact, current of at least 30 mA must flow. This amount of current is large enough to cause muscular contraction whereas in most cases it will almost certainly prevent electrical injury effects, such as ventricular fibrillation, it may not prevent injury arising from the muscular contraction, such as falling off a ladder, or being thrown against a wall.

16.2.3 Prevention of indirect contact injuries

Some of the techniques used to prevent direct contact injuries will also prevent indirect contact injuries; this includes the use of SELV or PELV supplies and Class II equipment. However, by far the most common method is the earthing of exposed conductive parts, coupled with the installation of fuses or circuit breakers that disconnect the supply in the event of an earth fault, a technique known as *Earthed Equipotential Bonding and Automatic Disconnection of Supply (EEBADS)*.

The EEBADS technique requires the exposed conductive parts of the apparatus and equipment to be earthed through a protective conductor that is connected to the main earthing terminal of the installation. Overcurrent protective devices, such as fuses and circuit breakers are placed in the phase conductors. When an earth fault occurs, connecting a live conductor to exposed metalwork and thereby creating the conditions for indirect contact injuries, the fault current flows back to the earthed point(s) of the supply system. If the impedance of the fault circuit, the *earth fault loop impedance*, is low enough to ensure that the fault current is high enough to operate the protective circuit breakers or fuses quickly enough, danger from indirect contact will be prevented. The limiting values of earth loop impedance for different types and ratings of protective devices are listed in

BS 7671 (see Table 16.2); this specifies that supplies to hand-held equipment must be disconnected within 0.4 s and those to fixed equipment must be disconnected within 5 s of the fault occurring. In circumstances where sufficiently low values of earth loop impedance cannot be achieved the disconnection device may need to be an RCD.

Extraneous conductive parts, being metalwork that does not form part of the installation but which can introduce a potential (usually earth) into the installation, must also be bonded together and earthed. This is usually achieved by using suitably sized bonding conductors to connect water, gas and oil pipes (and any other metallic services) to the installation's main earthing terminal. Connecting together the exposed and extraneous conductive parts in this way creates an equipotential zone in which the metalwork is normally held at earth potential; it can raise to higher potentials during fault conditions but, since the touchable conductive parts are all connected together, the risk of electrical injury is reduced to an acceptable level. In areas of higher risk, such as bathrooms and swimming pools, supplementary bonding conductors are used to connect together all the exposed metalwork.

16.3   Precautions against arc and flashover burn injuries

Most arc and flashover burn injuries occur during live work on electrical systems. To avoid the hazard, work should be done whenever possible on apparatus that has been made dead and isolated. If this is not possible then only competent people should undertake the work and precautions must be taken to minimize the risk. Depending upon the particular circumstances, these may include the use of:

- insulating screens or barriers between live parts of different polarity and between live parts and earthed metalwork
- insulated tools
- test equipment probes on which only the contact points are bare, with a maximum length of exposed metal of 4 mm, and the rest of the probe is insulated. The leads should have protection against short-circuit current, such as integral fuses or current-limiting features in the test equipment
- personal protective equipment, such as insulating gloves, and heat-resistant face shields and clothing

16.4   Precautions against fire

The main precautions against fire of electrical origin are as follows:

- each cable must have sufficient current carrying capacity for the load current and must be protected by an excess current protective device (fuse or circuit breaker) with tripping value set to be below the cable's maximum current carrying capacity
- fuses and circuit breakers must be rated for the system fault level
- devices with the potential for creating overload conditions, such as motors, should be protected by excess current protective devices
- system components must be selected and appropriately protected for use in the prevailing environmental conditions to prevent degradation that may lead to insulation failure
- the electrical system must be subject to routine preventive maintenance to detect degradation or failures that may otherwise lead to fire.

Enhanced protection against fire and its effects can also be obtained by using cables that have low smoke and fume emission properties; more information on these cables is provided in section 9.2.5.

16.5   Precautions against explosions

The precautions to be taken against electrical ignition of flammable atmospheres are described in Chapter 15.

16.6   Preventive maintenance and safe systems of work

16.6.1   Preventive maintenance

A common cause of the types of accidents described in this chapter is electrical systems which are not maintained and which are allowed to degrade into a dangerous condition. Typical examples of dangerous features arising from lack of maintenance are:

- enclosures and assemblies that are broken, allowing access to live parts, or in which the seals have degraded, allowing the ingress of dusts and liquids
- cables with damaged sheathing and basic insulation or with mechanical damage that compromises the protective conductors or creates the conditions for arcing as a result of insulation failure
- corrosion of exposed metallic parts
- low electrolyte levels in battery-based tripping supplies or uninterruptible power supplies
- low levels of insulating oils in switchgear, or insulating oils that have degraded insulating properties

Advice on the preventive maintenance of low-voltage distribution systems is published in BS 7671. The IEE publishes guidance on the maintenance of electrical equipment, see references 16B and 16C. Advice on the preventive maintenance of high-voltage systems is available in BS 6626 (see Table 16.2). Manufacturers' literature should also provide information on the maintenance strategies for equipment supplied.

16.6.2   Safe systems of work

Safe working practices on electrical systems comprise precautions on equipment that has been made dead and precautions during live working. These are covered separately in the following sections.

16.6.2.1   Safe isolation procedures

Most electrical work should be done on systems that have been made dead and on which precautions have been taken to prevent them from being re-energized while work is going on. The generic safe isolation procedure is as follows:

A. Identify the circuit or apparatus on which work is to be done and all possible points of supply.

B. Disconnect the supply and secure the point of isolation, typically by using pad-locks to lock devices in the off position, and post caution notices at the point(s) of isolation to warn that work is being done on the disconnected circuit(s).

C. At the point of work, prove the conductors are dead using a suitable voltage indicator, such as test lamps. The voltage indicator must be proven to be serviceable immediately before and after the conductors have been tested for voltage.

D. For high-voltage equipment, apply earths to the conductors on which work is to be done.

E. Ensure that the persons carrying out the work on the isolated systems are aware of the scope and limitations of the work to be done.

Safe isolation procedures such as these should only be carried out by competent people who have been trained to implement them and who are familiar with the equipment involved.

In the cases of high-voltage systems and high fault level low-voltage systems, it is standard practice that the safe isolation procedure is formalized using a safety document, such as a *Permit to Work*. Where systems that have been isolated under a Permit to Work have to be energized to allow testing to be carried out, the testing is often authorized and described in a safety document called a *Sanction for Test*. Details on safety documents are published in BS 6626 and in reference 16D.

16.6.2.2   Live working practices

Live working is most commonly carried out during fault finding and testing, particularly on low-voltage systems. Some more specialized live work is carried out by specialists who have received in-depth training. Live work should normally only be carried out when it can be argued that it is unreasonable to make the conductors dead, and that the risks are acceptable, and only if precautions can be implemented to prevent electrical accidents. Common examples of live working activities are:

- live cable jointing and meter changing on low-voltage distribution systems
- phasing out high-voltage conductors on switchgear, such as circuit breakers and ring main units
- 'rubber glove working' on 11 kV overhead distribution lines, a technique in which linesmen wearing insulating gauntlets and working in cradles on insulated booms on mobile elevating work platforms work on live plant that has been covered with suitable insulating materials
- live line working on the transmission grid overhead lines at voltages from 132 kV upwards, a technique in which the linesman carrying out the work is raised to line potential

The main hazards during live work are electric shock and burns from direct contact with live conductors and flashover/arc burns due to insulation failure.

In order to prevent these types of injuries, live work should only be carried out when it is essential and when it is unreasonable for the conductors to be made dead, and when the risks are acceptable. In the event of live work being undertaken, precautions against injury must be adopted.

The most important precaution is to ensure that the live work is carried out by people who are competent for the task or who are being closely supervised by somebody who is competent. This means that they must have been trained in the task and been assessed as being competent, and must understand the system on which they are working and the means of protecting against injury. They must be provided with the appropriate tools, test equipment and personal protective equipment. Live work should only be undertaken after a risk assessment has been conducted, the risks identified and the appropriate safety precautions have been determined.

The types of precautions that can be taken are as follows:

A. The area of the work activity should be cordoned off or otherwise delineated and protected to prevent the workers being distracted or disturbed. Warning signs should be posted.
B. Only tools and test equipment that are suitable for the job should be used. Test equipment probes should be fused or otherwise protected against short circuit or overload, should have finger guards to stop the fingers slipping down the probe onto the live conductors, and there should only be 2–4 mm of metal exposed at the tips of the probes. Care should be taken to ensure that the test equipment itself, such as Class I oscilloscopes and signal generators, does not introduce additional hazards.
C. Where necessary, flexible insulating materials should be used to shroud off metalwork in the vicinity of the work that may be at other potentials, including earth.
D. Where it will contribute to risk reduction, the live worker should be accompanied.
E. Appropriate and well-maintained personal protective equipment should be worn, such as antiflash clothing and eye protection where there is a risk of flashover or arcing.

More detailed information of safety during live working is published in references 16A and 16D.

16.7 Standards

Table 16.2 National and international standards relating to electrical safety

IEC	EN	BS	Title/subject
		6626	Code of practice for maintenance of electrical switchgear and controlgear for voltages above 1 kV and up to and including 36 kV
364		7671	Requirements for electrical installations. IEE Wiring Regulations, Sixteenth edition
60439-1	60439-1	60439-1	Specification for low-voltage switchgear and controlgear assemblies. Type-tested and partially type-tested assemblies
60479		6519-1	Guide to effects of current on human beings and livestock. General aspects
60529	60529	60529	Specification for degrees of protection provided by enclosures (IP code)
60536-2		2754-2	Classification of electrical and electronic equipment with regard to protection against electric shock. Guide to requirements for protection against electric shock
	60742	3535-1	Isolating transformers and safety isolating transformers

References

16A. Oldham Smith, K. and Madden, J. *Electrical Safety and The Law*. (4th edn), Blackwell
      Science, 2002, 0-632-06001-8.
16B. *Code of practice for in-service inspection and testing of electrical equipment*, IEE,
      2003, 0-852-96764-4.
16C. *Electrical maintenance. Code of practice*, IEE, 2003, 0-852-96769-1.
16D. *Electricity at work. Safe working practices, HSG85,* Health and Safety Executive,
      2003, 0-717-62164-2.

Index

Printed and bound by CPI Group (UK) Ltd, Croydon, CR0 4YY

13/10/2024

01773567-0001